APPLIED MICROPHOTONICS

OPTICAL SCIENCE AND ENGINEERING

Founding Editor
Brian J. Thompson
University of Rochester
Rochester, New York

APPLIED
MICROPHOTONICS

WES R. JAMROZ
ROMAN KRUZELECKY
EMILE I. HADDAD

Taylor & Francis
Taylor & Francis Group
Boca Raton   London   New York

CRC is an imprint of the Taylor & Francis Group,
an informa business

CRC Press
Taylor & Francis Group
6000 Broken Sound Parkway NW, Suite 300
Boca Raton, FL 33487-2742

Library of Congress Cataloging-in-Publication Data

Jamroz, Wes R.
    Applied microphotonics / Wes R. Jamroz, Roman V. Kruzelecky, Emile Haddad.
        p. cm.
    Includes bibliographical references and index.
    ISBN 0-8493-4026-8 (alk. paper)
    1. Photonics--Materials. I. Kruzelecky, Roman V. II. Haddad, Emile. III. Title.

    TA1520.J36 2006
    621.36--dc22   2006044469

Visit the Taylor & Francis Web site at
http://www.taylorandfrancis.com

and the CRC Press Web site at
http://www.crcpress.com

Dedication

This book is dedicated to Dr. Morrel P. Bachynski

Preface

Microphotonics is a new branch of technology that deals with the development and utilization of miniaturized photonic devices and systems. Microphotonics represents the current stage of the technological evolution that was initiated by microelectronics and was followed by the development of optoelectronics and photonics.

Microphotonics encompasses a range of new devices, materials, and processes such as photonic integrated circuits (PICs), microoptical and electromechanical systems (MOEMS), photonic-band-gap structures (PBGs), smart materials, optical computing, and quantum photonic systems. Microphotonic devices are used to enhance the performance of existing electronic systems and to offer greater functionality, miniaturization, and reliability to the future commercial products. The microphotonic systems of the future will be smaller, faster, and more energy efficient. Therefore, it is expected that in this decade microphotonics will begin augmenting and complementing electronic semiconductor technologies and, possibly, replace them completely later.

Microphotonics is widely recognized as a uniquely important technology because of its direct application in a number of markets. It has been recognized for its strategic value as an enabling technology, linking it to many other established and emerging markets and business sectors. Applications of microphotonics may range from telecommunication to space, medicine, security, and military. The marriage of photonic technologies with others, including microelectronics, material sciences, and computing technologies, facilitates a vast array of components, devices, and subsystems that are key enablers of larger systems and future innovations.

There is a vast volume of information on the latest development of microphotonics, both in scientific journals and magazines as well as in daily media. A number of books have been published that deal with various devices and technologies that are part of microphotonics. Nearly every week there are reports on recent advances, such as new nanomaterials, smart materials, and new band-gap structures, that have been made in research laboratories and are directly or indirectly related to microphotonics. It would be impossible to cover the massive advances made in this field in a single book, a field that continues to grow at an enormous rate.

While existing books are written mainly for experienced photonic scientists and they cover very detailed studies of various topics that microphotonics encompasses, the emphasis of this book is on applications. The book intends to provide an introductory review of the newly developed technologies for those who are not experts in the field of photonics and telecommunications. The main purpose is to provide guidance to those who would like to take advantage of this technology but who need more basic information to feel comfortable with these new developments. Therefore, this book offers an introductory approach and tries to avoid overloading the readers with in-depth presentations that are not essential for the readers who are not experts in microphotonics. On the other hand, a bigger picture of the entire field, including industrial trends and some aspects of commercialization, has been included.

The book does not contain an extensive number of references to the original work and scientific literature. It was a deliberate decision that this book would provide basic information without the scholarly detail of bibliographic information. The emphasis has been put on giving the reader a wider perspective on the availability and maturity of the devices, and on their potential applications. In other words, microphotonics is presented from the point of view of what is possible and what is feasible now and in the near future. The reader is provided with insight into the possibilities, advantages, limitations, and trends of the current and future microphotonic systems. Most important, the readers will be able to find out if and how the advancement of microphotonics may affect their own area of interest, whether it is material engineering, military, sensing, biomedicine, or space exploration.

The book begins with a brief overview of the technological and commercial environments in which the current development of microphotonics is taking place. Both of these environments have a great influence on the direction in which the efforts are focused and on what kind of gains are expected (see Chapter 2).

A brief introduction to the physical laws of interaction of electromagnetic radiation with the matter is included in Chapter 3. A limited number of mathematical rules and equations are introduced so that the reader with no background in theoretical physics can fully appreciate the basic formulations on which microphotonics is based. Again, the emphasis is on practical and design aspects. No theoretical presentations are included that are not required for the explanation of practical aspects of the discussed topics.

Chapter 4 introduces the basic element of all microphotonic systems, i.e., the photonic node.

The following chapters are dedicated to the introduction and description of the node components. The review of these various devices is presented with references to traditional designs and then illustrating their gradual evolution into the micro scale. Transmitters are described in Chapter 5, couplers and switches in Chapter 6, multiplexers in Chapter 7, receivers in Chapter 8, and amplifiers and compensators in Chapter 9. These chapters include description of conventional photonic devices or components as well as their corresponding microphotonic counterparts.

Several new technologies that have or are expected to have a great influence on the advances of the microphotonic systems are introduced in Chapter 10. Such new technologies as microoptoelectromechanical systems (MOEMS), PBG structures, and smart materials provide building blocks that can be subsequently interconnected to furnish various microphotonic integrated circuits.

Chapter 11 looks at materials and processes that are used for fabrication of microphotonic devices and components. Integration approaches are also addressed, for they will be the critical factors in the implementation processes, and they will decide the commercial successes of the microphotonic systems.

Chapter 12 is dedicated to advanced microphotonic devices. This review of advanced microphotonic devices covers a wide range of applications, including photonic computers, PBG sensors, cascade lasers, and miniaturized spectrometers.

Chapter 13 is an introduction to quantum photonic systems. Recent developments in quantum photonics have provided a new way to manipulate and encode information that has no classical counterpart. Many designs have been devised to build devices that will be able to harness the quantum world and turn the theoretical advantage of quantum photonics into a practical implementation. These include such systems as quantum communications, quantum computers, and cryptography.

Chapter 14 provides a forward-looking review of the future for microphotonics and its potential applications in such areas as space exploration, and military and satellite communications.

The book will benefit students and engineering and technical staff who are working in the fields of advanced signal and data processing systems, space exploration, and military sensing, as well as those who are involved in the development of smart structures, nanomaterials, and materials in health monitoring systems.

Another group of readers who may appreciate this book are those who are involved in planning, purchasing, and investing in the emerging technologies.

Acknowledgments

We are grateful to many individuals who have either directly or indirectly contributed to this book.

Our special thanks are to Morrel P. Bachynski who for many years provided us with a stimulating environment at MPB Communications, Inc., where we have been able to experience and explore many aspects of industrial research and development.

We would like to thank to Asoke Ghosh who introduced us to the exciting area of photonics.

We express our gratitude to our colleagues from various institutions and organizations who over the years provided us with many ideas and assistance to our work. Especially, we thank Darius Nikanpour, Wanping Zheng, and Linh Ngo Phong from the Canadian Space Agency; Philips Laou from Defence Research and Development Valcartier; Errico Armandillo, Iain Mackenzie, Igor Zayer, Pierrik Vuilleumier, and Giovanni Colangelo from the European Space Agency; Mohamed Chaker and Mohamed Soltani from INRS; Arthur Yelon, Ives-Alain Peter, Richard Beaudry, and Oliver Grenier from Laboratoire de Microfabrication (LMF) at Ecole Polytechnique de Montreal; James Drummond from the University of Toronto; Jim Sloan from the University of Waterloo; Ed Cloutis from University of Winnipeg; Suong V. Hoa and Philippe G. Merle from Concordia University; Serge Samson, François Bussières, Pamela Markland, and Alexandre Emond from EMS Technologies, Ltd.; Hubert Jerominek, Loic Le Noc, and Francis Picard from the National Optics Institute.

We would like to express our appreciation to the staff at CRC Press and Taylor and Francis Company for their helpful cooperation, especially to Theresa Del Forn for her availability and willingness to provide guidance and advice during preparation of the manuscript.

Special thanks are to Najeeb Mohammad for preparing the drawings, to Brian Wong for his painstaking and careful reading of the manuscript, and to Jing Zou for her assistance with experimental data.

Finally, warm words of appreciation are due to Roy Josephs for his constant availability to help and for his morning cups of fresh Blue Mountain coffee.

The Authors

Wes R. Jamroz is director of the Space Photonics Devices Division, MPB Communications, Inc. Wes obtained his Ph.D. from the Technical University of Lodz, specializing in nonlinear optics and laser physics. In 1980 he joined the Laser Physics and Chemistry Group at University of Toronto as a research associate. In 1984 he was awarded an industrial research fellowship from the Natural Sciences and Engineering Research Council of Canada.

In his current position he is responsible for the business development in the area of photonic devices for space applications. His team developed a series of innovative photonic devices for space, such as a smart thermal radiator device for satellites, fiber sensor system for satellite propulsion subsystems, smart sunshield coating for SAR antennas, and a miniaturized NIR spectrometer for space missions.

He has been working closely with universities, governments, and industrial research organizations in Canada, the U.S., Europe, and Japan. He has acquired significant experience with the execution of collaborative projects and coordination of industrial consortia.

He is the author and coauthor of scientific papers, chapters in books, conference talks, invited presentations, workshops, and patents.

Roman V. Kruzelecky is manager of the Hybrid Structures and Mission Instruments Group at MPB Communications, Inc. He obtained his Ph.D. in 1987 from the University of Toronto, specializing in solid-state physics and semiconductor devices. He joined the Department of Engineering Physics at McMaster University in 1988 as a research engineer. At McMaster University he was involved in the design and setting-up of Canada's first gas-source MBE system for InGaAsP materials. He is currently the project manager for the fiber sensor demonstrator payload for the European Space Agency's Proba-2 satellite. He is also leading a team of scientists to develop the Microsatellite Earth Observation Mission (MEOS) to study the global greenhouse gas cycles associated with global warming.

His current interests include smart thin-film materials, waveguide-based MOEMS devices, miniaturized infrared spectrometers, and fiber-optic sensor networks.

Emile I. Haddad is manager of Smart Materials Group at MPB Communications, Inc. He obtained his Ph.D. in 1986 from Ecole Polytechnique de Montreal, specializing in nuclear engineering. In the same year he joined INRS-Energie

Varennes, Quebec, as a post-doctoral fellow. His work was related to VUV and x-ray spectroscopy.

He is currently the project manager for smart thermal radiator implementation into satellite heat management systems and smart sun-shield coating for SAR antennas. He is also leading a team of scientists to develop self-healing structures for spacecrafts.

His current interests include smart thin-film materials and the effects of space environment on fiber-optic devices.

Contents

1 Introduction

1.1 MICROPHOTONICS: A NEW BRANCH OF TECHNOLOGY

Optical beams do not have the time response limitations of electronic currents, do not need insulators, and can contain dozens or hundreds of streams at different frequencies that are immune to electromagnetic interference. They have low-loss transmission and provide large bandwidth, i.e., they are capable of transmitting several channels of data in parallel without interference. They are capable of transmitting signals within the same or adjacent waveguides with essentially no interference or cross talk. This will lead to a new generation of communication systems, data processors, and computers in which wires and electronic circuits will be replaced by waveguides, photonic crystals, smart thin films, and quantum photonic circuits.

Within the next generation, miniaturized devices and photonic systems will carry and store data in the same manner as electrons do on classical computer chips. It is expected that these new devices will invade every piece of electronics, from cell phones to supercomputers, and make them smaller, faster, and more energy efficient. It is said that the second half of the 20th century was transformed by the electronic transistor. It is expected that the 21st century will be transformed by its photonic equivalent. This is the beginning of a new branch of technology — *microphotonics*.

Microphotonics is a branch of technology that deals with wafer-level integrated devices and systems that emit, transmit, detect, and process light and other forms of radiant energy whose quantum unit is the photon. It is part of the broader field of photonics that involves the interaction of light with microscale structures and materials. It is worth mentioning that there is a fundamental qualitative difference between photonic and microphotonic effects and processes. Some of the microphotonic effects can be observed only on the submicron scale. Microphotonic devices typically measure between 0.1 and 100 μm in diameter. In other words, their size allows for their implementation in wafer-level waveguides.

So why is microphotonics generating so much interest? Again, it is useful to draw analogies with electronics. The current explosion in information technology is due to the fact that it is possible to control the flow of electrons in a semiconductor in the most intricate ways. Microphotonics promises to give similar control over photons. Given the impact that semiconductor materials have had on every sector of society, microphotonics could play an even greater role in the 21st century, particularly in the data processing and information transmission industries.

Microphotonics includes the emission, transmission, amplification, detection, modulation, and switching of optical beams. Optical beams are used to transmit, store, display, and scan information. Microphotonic technology is thus closely related to communication and information technology and will continue to be so in the foreseeable future. Microphotonics has advanced from bulk-optics systems, which were organized on an optical bench using many separate optical components, to relatively compact optical waveguide and integrated optical devices that can combine several optical functions on one waveguide substrate. The overall goal is to develop techniques for making photonic microchips, in which streams of electrons are replaced by photons. This includes methods to guide, bend, "switch on," and "switch off" photons, as well as to connect microphotonic chips to a network of optical waveguides and fibers.

The current telecom infrastructure is a mix of optical fiber, copper cables, and satellite links. The core long-haul backbone networks are based mostly on fiber-optic cables. Metro networks consist of both optical fiber and copper wire. The so-called access networks and "last mile" connections to customers are primarily copper wire or satellite links. The ultimate communication network of the future will be a photonic system from start to finish, i.e., an all-optical network. These all-optical networks will gradually become the main infrastructure that will support the future information systems. It will be up to microphotonics to provide the basic technological solutions that will allow for commercial implementation of these all-optical systems.

In current optical communication systems, audio or video signals are encoded as streams of digital data packets. These voltage pulses are applied to a light-emitting diode (LED) or a semiconductor laser, which converts them into short pulses of light that are then sent along an optical fiber network. Many different conversations or video signals can be transmitted using a single wavelength of optical beam by interweaving the data packets from different sources, a technique known as *time-division multiplexing*. These optical data pulses are sorted at the receiving end of the fiber, where they are converted by photodetectors into continuous analog electrical signals that are then transmitted along copper wires.

A simple way of increasing the amount of data that can be transmitted by a single optical fiber is to make the incoming electronic pulses as short as possible. Current optical systems have achieved data rates in excess of 40 Gb/sec. To increase this value significantly, the wavelength range of transmission must be increased. However, an optical fiber is transparent only over a finite wavelength range, so the number of separate conversations that can be transmitted depends on the linewidth of neighboring optical channels (which is currently at the sub-nanometer level). Encoding and sorting thousands of such channels pose quite a challenge.

At the transmitting end of the system, each wavelength channel requires a very stable optical source that only emits in a very narrow range of wavelengths. Although LEDs offer high switching speeds, they emit optical beam over a wide range of wavelengths, which makes them less suitable than lasers for multiplexing systems.

At the receiving end, very-narrow-linewidth filters and optical switches are required to separate individual channels and to route them to the appropriate destinations. Because of the large number of individual components in a multiplexing system, it would be necessary to combine as many of them as possible onto an "integrated" microphotonic chip. A microphotonic chip would reduce both the cost and the number of possible points of failure.

The basic structure of all optical systems is a microphotonic integrated circuit (micro-PIC). Micro-PICs are the monolithic integration of two or more integrated photonic circuits on a single substrate. They are the microphotonics equivalent of microelectronic chips, which integrate two or more transistors on a chip to form an electronic integrated circuit (IC). However, instead of guiding electricity, a micro-PIC routes optical beams. In micro-PICs, waveguides act as the photonic analog of copper circuits, serving as interconnects among various discrete components on a chip. The refractive index of an active or core layer, which is sandwiched between two cladding layers with a lower index of refraction, uses total internal reflection to confine the light. The photons are directed by optical elements, including gratings, lenses, and prisms.

Such integration, however, raises other problems. It would be necessary to develop small-scale optical "interconnects" — tiny planar waveguides that can steer photons around tight corners. Whereas it is straightforward to route electrons around sharp bends in microchips, it is impossible to direct photons in the same way using conventional optical waveguides because of losses at sharp bends.

Data processing is another application that will benefit from the development of microphotonics. Microprocessors are expected to be clocked at about 10 GHz by the end of the decade. Consequently, it is becoming extremely difficult to design enough bandwidth into a conventional printed circuit board using the currently implemented electrical wires because, for example, the frequency-dependent loss on circuit boards rapidly rises above 1 GHz, reducing the signal-to-noise ratio (SNR) and introducing timing errors.

Currently, personal computers operate at frequencies above 1 GHz (10^9 Hz). However, there is a strong market demand for a 100-GHz desktop. Conventional semiconductor technology may have a problem with manufacturing a 10-GHz personal computer. However, by transmitting signals with light rather than electrons, it might be possible to build a computer that operates at hundreds of terahertz (10^{12} Hz). Industrial observers believe that such an awesome processing engine could be built from optical components made from the so-called photonic-band-gap (PBG) structures.

The PBG technology offers an alternative way of guiding light. Most of today's commercial waveguides rely on the effect of total internal reflection. The band-gap structures, however, operate much in the same way as the transistor, i.e., only photons with certain frequencies are allowed to pass through a PBG. It is expected that PBG structures will provide the missing elements required for a full commercial implementation of all-optical data processing and communication networks.

Advances in the field of quantum photonics have provided another thrust for the development of quantum computers. Here, data processing will rely on the manipulation of a single photon. The new field of linear optics quantum computing (LOQC) has been created towards the realization of such a goal. It has already been demonstrated with the performance of simple quantum logic gates.

It should also be mentioned that microphotonic devices are finding applications in a variety of other areas, including sensing, energy, and life sciences.

1.2 HISTORICAL PERSPECTIVE

The development of microphotonics parallels that of microelectronics (see Figure 1.1).

It is widely accepted that the beginning of microelectronics was marked by the invention of the *transistor*, i.e., an electronic device consisting of a semiconductor material, generally germanium or silicon, and used for rectification, amplification, and switching. Its mode of operation utilizes transmission of donor electrons and holes across the junction. The first semiconductor was developed in 1947 by three scientists at Bell Laboratories, namely John Bardeen, William B. Shockley, and Walter H. Brattain.

This was followed by the development of the first IC in the 1950s by Jack Kilby of Texas Instruments and Robert Noyce of Fairchild Semiconductor. ICs are used for a variety of devices, including microprocessors, audio and video equipment, and automobiles.

The microprocessor was invented at the beginning of the 1970s. It was a merger of digital computers and semiconductor ICs. This lead to the rise of the personal computer at the end of that decade.

The introduction of the transistor lead to a progressive miniaturization of electronic devices and components. By late 1980s, it was possible to mass-produce solid-state devices that contained more than 3,000,000 transistors on a single silicon chip. A *chip* is a small piece of semiconducting material (usually

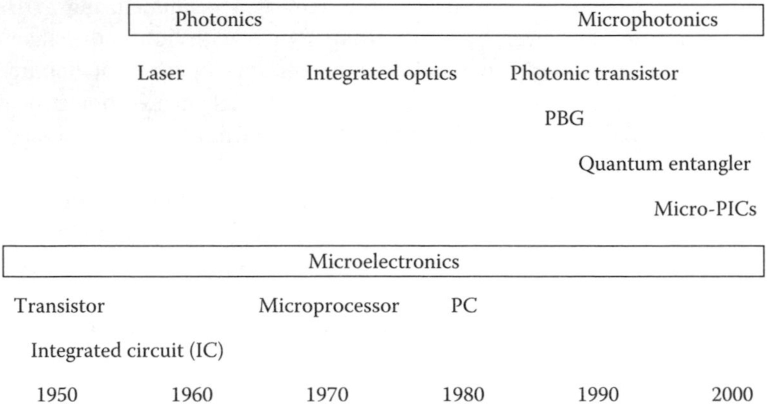

Photonics		Microphotonics
Laser	Integrated optics	Photonic transistor
		PBG
		Quantum entangler
		Micro-PICs

Microelectronics			
Transistor	Microprocessor	PC	
Integrated circuit (IC)			

| 1950 | 1960 | 1970 | 1980 | 1990 | 2000 |

FIGURE 1.1 Technology evolution.

silicon) in which an IC is embedded. A typical chip is less than 0.25 in.^{2} and can contain millions of electronic components (transistors). Computers consist of many chips placed on electronic boards called *printed circuit boards*. Such high-density microcircuits, called *microprocessors*, have led to tremendous advances in computer technology and in computer-based automated systems.

A seemingly unrelated event took place in 1960. It was the demonstration of the first laser. There was a lot of excitement when Theodore H. Maiman demonstrated a rather bulky and not very efficient ruby laser that generated a coherent red beam at 0.694 μm. This was followed in September 1962 by scientific reports on the development of injection diode lasers from three independent laboratories — General Electric, IBM, and Lincoln Lab of the Massachusetts Institute of Technology.

For many years after that, the laser was considered to be "a solution waiting for a problem." One of the most promising and exciting possibilities for a laser beam was its application as a carrier of information in optical communication because of its frequency range.

Around early 1960s, the demand for increased telecommunication volume was growing quite rapidly. It became obvious that conventional coaxial cables and microwave links would reach their saturation point. This new transmission technology arrived at the right moment.

At that time, communication was based on electromagnetic radiation in the radio and microwave range. The newly developed light amplification by stimulated emission of radiation (LASER) offered a light source that was far more monochromatic, directional, powerful, and coherent than any other. It is interesting to note that Albert Einstein had postulated the occurrence of the stimulation emission effect back in 1916.

This newly developed light source was promising to be a very attractive carrier of information. In conventional telecommunication links, electromagnetic waves such as radio waves and microwaves were used as carriers of information. The information to be sent is encoded on a carrier and then sent to the receiver. The receiver decodes the signal to retrieve the original information. Conventional carriers operate at frequencies of 600×10^3 to 20×10^6 Hz (i.e., radio broadcasting) and 50×10^6 to 900×10^6 Hz (TV broadcasting). The higher the carrier frequency, the higher the amount of information that can be sent.

Laser beam frequencies of 5×10^{14} Hz are some 10,000 times higher than those offered by the conventional communication systems (see Figure 1.2).

Therefore, there was a chance for a tremendous increase in the capacity of the existing communication links. A conventional telephone line is capable of sending up to 48 simultaneous conversations. A modern 2.5 Gb/sec optical fiber transmission line is capable of transmitting 35,000 or more simultaneous conversations.

The first attempts to commercialize laser-based communication links turned out to be rather limited. The so-called free-space optical communication links showed a range of problems and limitations. *Free-space link* is an arrangement in which a modulated laser beam travels in an open space between the transmitter

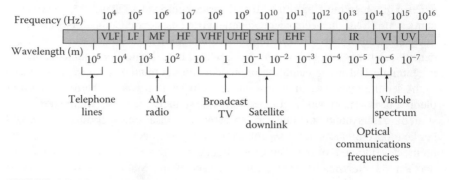

FIGURE 1.2 Electromagnetic spectrum.

and the receiver. Any physical obstacle that happens to intersect the laser beam path would greatly degrade the transmission or even obstruct it completely. Unlike microwave or radio transmission, the laser beam can be blocked by rain, fog, or snow. This is related to the fact that the atmosphere absorbs certain optical frequencies. It was obvious that the laser beam would have to be enclosed in a protective medium before it could be used in commercial communication systems.

It took a decade before the laser became the key tool that bridged together optics and electronics and led to the development of integrated optics. Pioneering work on the concept of integrated optics was performed in the 1970s and 1980s, resulting in the demonstration of various passive devices such as couplers, splitters, and Bragg gratings and active devices such as acousto-optical filters and electro-optical switches.

Commercial communication had to wait for the development of a new type of "cable" that would allow for a more efficient transmission. But this cable was already waiting to be used. The producers of silica glass were looking for a new market for their technology of fiber optics. It did not take long before kilometer-long silicon wires were ready to be tested.

The dramatic reduction of transmission loss in optical fibers coupled with equally important developments in the area of light sources and detectors have brought about the phenomenal growth of a new industry — fiber optics. The birth of optical communications occurred in the 1970s and was triggered by two developments. The first was the development of the semiconductor laser in 1962. The second was the demonstration of a glass fiber with an attenuation less than 20 dB/km, which was the required threshold for a fiber to be used as a viable transmission medium.

With the development of optical fibers with an attenuation less than 20 dB/km, the threshold to make fiber optics a viable technology for telecommunications was crossed. In 1977, AT&T installed the first optical fiber cables in Chicago. The first fiber-optic transmission lines used multimode fibers and lasers operating in the 850-nm wavelength band.

The next generation of lasers operating at 1310 nm allowed transmission in the optical fiber, in which the losses in single-mode fiber (SMF) were in the range of 0.5 dB/km. In the 1980s, the optical window around 1550 nm was developed, in which an SMF had its minimum optical loss in the range of 0.22 dB/km.

Today, the entire telecom infrastructure is based on fiber optics, with the exception of the "last miles to homes," which is still based on coaxial cables.

Commercial implementation of optical fiber communication systems began in the 1980s. Since that time, optical fibers have been deployed with supercomputers, local area networks (LANs), and storage area networks (SANs).

The first systems operated at 90 Mb/sec, i.e., a single optical fiber could handle approximately 1300 simultaneous voice channels. Today, systems commonly operate at 10 Gb/sec and beyond. This translates to over 130,000 simultaneous voice channels. Over the last decade, new technologies such as dense wavelength-division multiplexing (DWDM) and erbium-doped fiber amplifiers (EDFA) have been used to further increase data rates to beyond a terabit per second (>1000 Gb/sec).

Since the first test with optical communications, scientific and technological progress in this field has been so phenomenal that, within a brief span of 30 years the fifth generation of optical fiber communication systems has already been implemented.

It is widely accepted that the beginning of commercial microphotonics was in 1997, when the first micro-PICs were introduced.

1.2.1 PHOTONIC COMPUTING

Since the 1930s, the advantages of photonics had been considered for application in the emerging field of computer science. However, there was a lack of appropriate devices to make such an optical computer. Among these missing devices were: (1) a source of coherent light, (2) optical signal processors, and (3) photonic transistors.

Over time, these missing devices have been developed. Photonic computing again became a hot research area in the 1980s. But that work was slowed down because of material limitation that prevented fabrication of photonic microchips that would be small enough and cheap enough.

Now, photonic computers are back with advances in the manufacturing of all-optical chips. At the same time, advances in the development of optical storage devices hold promise of efficient, compact, and large-scale storage systems.

The last missing element, a photonic transistor, was invented in 1989. Since that time, the entire basis of an optical computer has been growing quite rapidly.

1.2.2 PHOTONIC-BAND-GAP STRUCTURES

In 1987, Eli Yablonovitch and Sajeev John had the idea to create materials that block all incoming light of a particular wavelength. The two scientists independently published their idea of band-gap structures.

In 1991, Eli Yablonovitch made the first PBG structure. His team at Bell Communications in New Jersey mechanically drilled a complex diamond-like three-dimensional array of millimeter-sized air holes into a transparent material; it blocked frequencies in the microwave region. Since then, researchers have been incredibly inventive in devising all kinds of techniques to generate band-gap structures at shorter wavelengths in a wide range of materials.

1.2.3 QUANTUM PHOTONICS

If the last century was the era of electronic and optical data processing technology, the 21st century will be an era of quantum data processing technology, where previously underutilized quantum effects, such as quantum teleportation and entanglement, will be essential resources for information encoding and processing.

The first deep insight into quantum information theory came with Bell's 1964 analysis of the paradoxical thought experiment proposed by Einstein, Podolsky, and Rosen (EPR) in 1935. Bell's inequality drew attention to the importance of correlations between separated quantum systems that have interacted in the past but no longer influence one another. In essence, his argument shows that the degree of correlation that can be present in such systems exceeds that which could be predicted on the basis of any law of physics that describes particles in terms of classical variables rather than quantum states.

In 1993 an international group of six scientists (Bennett, Brassard, Crépeau, Jozsa, Peres, and Wootters) confirmed that it is possible to construct teleportation systems. In subsequent years, other scientists have demonstrated teleportation experimentally in a variety of systems, including single photons, coherent light fields, nuclear spins, and trapped ions.

The first experimental demonstration of entanglement was carried out at the University of Innsbruck in 1997 by the group under Anton Zeilinger. Their setup allowed the teleportation of the polarization quantum state of photons across an optical table. The source of entanglement used was parametric down-conversion. A slightly earlier experiment in Rome, by the group under DeMartini, showed the principle of teleportation that involved two particles instead of three: one carrying the state to be teleported and two entangled ones that provided the "quantum channel."

Researchers at the University of Vienna and the Austrian Academy of Science used an 800-m-long optical fiber to demonstrate teleportation. The optical fiber was fed through a public sewer system tunnel to connect labs on opposite sides of the river Danube.

In 2004, another research group at the University of Innsbruck and a group at NIST (Boulder, CO) demonstrated teleportation of atoms: the electronic quantum state of a calcium (Innsbruck) and beryllium ion (Boulder) was teleported to another one, with all the ions being held in a linear ion trap.

In summary, the development of microphotonics has been led by parallel efforts in several areas, such as micro-PICs, photonic computing, PBG structures, and quantum photonics. The main driving factors for these developments have been information capacity and cost. This has resulted in the development of several key components for microphotonic systems such as:

- Laser diodes
- Electro-optical modulators
- Optical amplifiers
- Wavelength-division multiplexers
- Optical switches
- Optical data storage
- Logical optical circuits

This suite of basic components provides many of the capabilities for processing information optically that have been used to process signals using microelectronic systems, i.e., amplification, switching, multiplexing, and signal modulation and storage. The biggest challenge that microphotonics faces now is related to system integration. Although entire electronic computers can be realized on a fraction of a silicon wafer, current integrated microphotonic components still require an entire wafer for a single functional block, for example, a typical array-waveguide wavelength-division multiplexer.

It is expected that further advances in microphotonics will bridge the gap and take this technology to a new and unprecedented scale.

2 Technological Growth and the Market Push

2.1 LAW OF GROWTH

It has been deduced from biological processes that life and human growth follow a logarithmic scale. This deduction is based on several observations. For example, the gestation period is 10 lunar months, the period of childhood is 100 lunar months, and the full span of life averages 1000 lunar months. A similar logarithmic scale may be used to measure the growth of biological cells as well as civilizations. According to this logarithmic law, the speed of the growth is at its maximum at the start. It follows a pattern similar to that of the exponential growth of "grains of rice on the chessboard," i.e., doubling its growth measure at a given time period. Then, it becomes slower and slower, before it reaches its saturation and ultimate decay. However, before it reaches its decay, a new growth cycle is initiated within its predecessor. And this new cycle provides means that allows overcoming its predecessor's limitations and directing the development into a new direction.

If it is accepted that various forms of development obey this logarithmic law, then an interesting series of analogies becomes available. One of them may be applied to the technological developments within our contemporary civilization. It is interesting to note that a specific form of such a logarithmic law has been identified for the development of the semiconductor industry.

In 1965, Gordon Moore of Intel Corporation made a simple extrapolation from three data points in his company's product history and predicted that the number of transistors per chip would double each year for the next 10 yr. After 10 yr, in 1975, he modified his projection for further progress by predicting that the doubling period would be closer to 18 months [2.1]. The press called this statement "Moore's Law," and the name has been widely accepted by the industry observers.

The trend observed by Moore has been maintained by the semiconductor industry, and it is expected that it will continue at least to the end of this decade. The first microprocessor, Intel 4004, introduced in 1971 had less than 3000 transistors and operated at a clock frequency of 108 kHz. Today, the Intel Pentium IV using the 0.18-μm technology has nearly 42 million transistors.

Table 2.1 shows the growth of the semiconductor industry since the 1970s [2.2]. The number of transistors installed on a single semiconductor chip is used as a measure of growth. From 1972 to 2009, it is expected that the overall doubling

TABLE 2.1
Semiconductor Component Complexity Growth

Component	Year of Introduction	No. of Transistors
i4004 (Intel)	1971	2250
i8008 (Intel)	1972	2500
i8080 (Intel)	1974	5000
i8086 (Intel)	1978	29,000
i80286 (Intel)	1982	120,000
i386™ processor (Intel)	1985	275,000
i486™ DX processor (Intel)	1989	1,180,000
Pentium® processor (Intel)	1993	3,100,000
Pentium II processor (Intel)	1997	7,500,000
Pentium III processor (Intel)	1999	24,000,000
Pentium 4 processor (Intel)	2000	42,000,000
Itanium 2 (Intel)	2002	221,000,000
	Expected 2009	1,000,000,000

period will be 24 months. The graphical presentation of the growth is shown in Figure 2.1.

Moore's Law as applied to the semiconductor component complexity can be expressed analytically by the following equation:

$$\log N = A_1 t + B_1 \quad (1971 \le t \le 2009) \tag{2.1}$$

where N is the number of transistors per integrated circuit (IC) and t is the year in which a chip was introduced. A_1 (= 0.15) and B_1 (= −292) are constants.

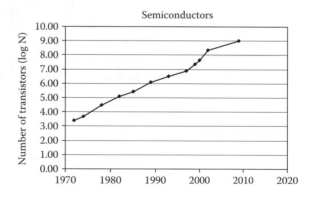

FIGURE 2.1 Semiconductor component complexity growth.

TABLE 2.2
Semiconductor Component Complexity
Growth

Processor	Year of Introduction	Processing Speed (MHz)
i486 (Intel)	1989	25
	1992	50
Power (IBM)	1993	100
Alpha 4 (DEC)	1994	266
Alpha 5 (DEC)	1995	300
	1996	500
Alpha 21264 (DEC)	1997	600
P4 (Intel)	2001	2,200
	Expected 2008	40,000

The practical aspect of the semiconductor chip complexity growth is reflected by the increase of the processing speed of commercially available ICs. As one would expect, the number of transistors per IC is directly related to the processing speed of the processor. The historical trend of this relationship is shown in the Table 2.2 [2.2].

Figure 2.2 shows the graphical presentation of the processor speed growth over the last two decades.

The preceding relationship, i.e., Moore's Law as applied to the growth of the microprocessor speed, can be expressed analytically by the following equation:

$$\log S = A_2 t + B_2 \quad (1989 \leq t \leq 2008) \tag{2.2}$$

where S is the processing speed in MHz and t is the year of a processor introduction. A_2 $(= 0.17)$ and B_2 $(= -336)$ are constants.

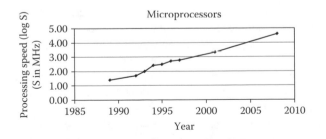

FIGURE 2.2 Semiconductor processor performance growth.

Historical records clearly demonstrate that semiconductor technology was used to build sophisticated microprocessors. However, microprocessors limit the amount of data that they are able to process. Electronic switching limits speeds to about 50 Gb/sec (1 Gigabit per second = 1 billion bits per second). This limitation results from several restrictions related to physical size of microprocessors. For example, there is a restriction of the number of electronic wires that may be used because they have to be separated while at the same time kept very close, otherwise the electrical signals would interfere with one another.

According to Equation [2.2], an operation speed of 50 Gb/sec could be attained by electronic microprocessors around the year 2010. It has been estimated, however, that by that time transistors would have become so small that silicon oxide films would no longer guarantee adequate insulation between conducting regions. This situation would mark the end of the semiconductor growth cycle. It would then be time for implementation of a new technology that would provide solutions for the encountered limitations. One may expect that this will be the time when microphotonics will enter into its maturity stage.

2.2 MOORE'S LAW OF PHOTONICS

The data transmission limitation has led to the idea of a photonic chip, i.e., a chip that uses optical beams rather than electric currents to perform digital processing.

Conventional chips transmit and process pieces of information in serial form, or one piece at a time. However, photonic chips may use parallel streams of data. The result would be much faster networks and computers.

This advantage of using optical beams instead of electrons is also related to the flow speed of light as compared to the speed of electrical current. An electric current flows at about 10% the speed of light.

An additional advantage of a photonic microprocessor is that optical beams, unlike electric currents, may pass through each other without interacting. Therefore, there is no such thing as a short circuit with optical beams, so beams can cross with no problem after being redirected by pinpoint-sized mirrors in a switchboard. Therefore, several laser beams can be used simultaneously with no interference among them, even when they are confined essentially to two dimensions. On the other hand, electric currents must be guided around each other; this makes three-dimensional wiring necessary (see Figure 2.3).

These aforementioned differences translate into much faster data transmission when optical beams are used. This advantage can be implemented both at the device as well as chip levels. Therefore, it is possible to have a photonic microprocessor that will be much faster — and at the same time — much smaller than the electronic device. In the pursuit to probe into cutting-edge research areas, microphotonic technology is one of the most promising, and it may eventually lead to new computing applications as a consequence of faster processor speeds, better connectivity, and higher bandwidth.

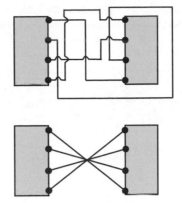

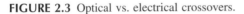

FIGURE 2.3 Optical vs. electrical crossovers.

TABLE 2.3
Capacity Growth for Commercially Deployed DWDM Systems

No. of Channels × Data Rate (Gb/sec)	System Capacity (Single Fiber)	Date of Deployment
8 × 2.5	20 Gb/sec	1996
16 × 2.5	40 Gb/sec	1997
32 × 2.5	80 Gb/sec	1999
80 × 2.5	200 Gb/sec	2000
40 × 10	400 Gb/sec	2000
160 × 2.5	400 Gb/sec	2000
80 × 10	800 Gb/sec	2001
160 × 2.5	1.600 Th/sec	2002
40 × 40	1.600 Tb/sec	2002
80 × 40	3.200 Tb/sec	2004
100 × 40	4.000 Tb/sec	2005
160 × 40	6.400 Tb/sec	2007

The unit of growth in optical data processing is the bit rate capacity. The historical data for data transmission growth are summarized in Table 2.3 [2.3]. Progress in data transmission is illustrated in Figure 2.4.

Over the past decade the bit rate has doubled every 15 months (see Equation 2.3).

$$\log C = A_3 t + B_3 \ (1996 \le t \le 2007) \tag{2.3}$$

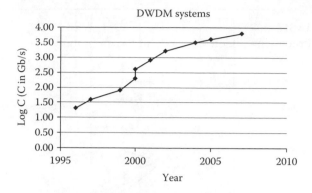

FIGURE 2.4 DWDM system capacity growth.

where C is the system capacity in Gb/sec and t is the year of a system deployment. A_3 (= 0.245) and B_3 (= –487) are constants.

Microphotonic chips can facilitate significantly higher information processing and transmission bandwidths than electronic systems, i.e., in the range above 100 GHz. Over the last few years, photonics has advanced from traditional bulk-optics systems to compact integrated optical devices that can combine several optical functions on a waveguide substrate. Microphotonics offers further significant improvements in terms of manufacturing costs, size, weight, processing speed, and power consumption over distributed bulk-optics systems. Currently, coupling to external optical channels is a major source of optical signal loss and mechanical reliability concerns for microphotonic integrated devices. It is expected that further minimization of external optical interconnects through multifunction microphotonic integration can substantially improve the net system throughput and reliability.

Two main processes take place in a photonic data processor — logic operation and data transmission. These two processes determine the overall system performance.

The increase of the data transmission, both for the communication networks and computers, will affect the architecture of information technology. This increase may be realized by increase in the (1) bandwidth capacity and (2) data rate for each channel.

Bandwidth is growing rapidly, primarily through the introduction of the so-called dense wavelength-division-multiplexing (DWDM) systems. The increase in the data rate or optical channel has been achieved by the introduction of a series of novel photonic devices.

2.3 MOORE'S LAW OF DATA PROCESSING

Processors need memory as well as processing speed to do their work. The ratio of memory to speed has remained constant throughout computing history. The

TABLE 2.4
Processing Speed Capacity of Supercomputers

Supercomputer	Processing Speed (Teraflop/Sec)	Date of Deployment
Human brain	1000	2.2–1.6 million yr ago
IBM's Deep Blue	3	1996
NEC's Earth Simulator	36	2002
IBM's Blue Gene	360	2005
RIKEN's MD GRAPE-3	1400	2006

earliest electronic computers had a few thousand bytes of memory and could do a few thousand calculations per second. The latest and greatest supercomputers can do a trillion calculations per second and can have a trillion bytes of memory. Dividing memory by speed defines a *time constant* [2.4]. The time constant expresses how long it takes for the computer to run once through its memory.

Table 2.4 summarizes the progress made in the development of supercomputers, starting with the famous IBM's Deep Blue computer that successfully challenged world chess champion Garry Kasparov in 1997.

The progress achieved is graphically illustrated in Figure 2.5, in which the computer processing speed growth is shown for the last decade. It is anticipated that supercomputers will reach the power of the human brain (shown as a horizontal line in Figure 2.5) within the coming year. Based on the progress achieved so far, it is projected that such powerful computers — equivalent to the human brain — will arrive in our homes before 2020.

Over the past decade the computer processing speed has doubled every year (see Equation 2.4).

$$\log P = A_4 (t - 1996) + B_4 \quad (1996 \le t \le 2006) \tag{2.4}$$

where P is the computer processing speed in flops and t is the year of a supercomputer deployment. $A_4 (= 0.28)$ and $B_4 (= -553)$ are constants.

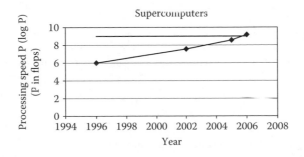

FIGURE 2.5 Progress in the computer processing speed.

As shown in Figure 2.5, the human brainpower is estimated to be in the range of 1 petaflop (i.e., quadrillion operations per second). One of the ways for this estimate is as follows [2.4].

Structurally and functionally, one of the best understood neural assemblies in the human body is the retina of the eye. The retina is a natural microphotonic processor that is built in the form of a transparent thin layer of nerve tissue at the back of the eyeball and on which the eye's lens projects an image. It is connected by the optic nerve, i.e., a million-fiber cable, to regions deep in the brain. The retina is less than a square centimeter in area and a half millimeter thick. The retina seems to process about ten one-million-point images per second. This is an equivalent to a billion operations per second for a computer. It has been estimated [2.4] that the 1500 cm³ human brain is about 100,000 times as large as the retina, suggesting that matching overall human behavior will be equivalent to 100 teraflops or more (teraflops are measure of a computer's processing speed and can be expressed as trillion floating point operations per second; floating point is considered to be a method of encoding real numbers within the limits of finite precision available on computers). Therefore, it is reasonable to assume that the human brainpower is in the range of one petaflop. Some other researchers think that the human brain is a 10-teraflop computer.

In the early 1990s, a new field of science emerged — quantum photonics. This new development has provided the basis for revolutionary advances involving computation, communication, and cryptography. The roots of this field go back about 20 yr, when pioneers such as Charles Bennett, Paul Benioff, Richard Feynman, and others began thinking about the implications of combining quantum mechanics with the classical Turing computing machine [2.5]. Quantum photonics was developed as a consequence of the demonstration of the use of a quantum computer for factorization of very large numbers and the demonstration of quantum photonic entanglers.

The growth of this field may be illustrated by the progress achieved in generation of entangled photonic pairs that form the basis of all quantum photonic devices. This growth is illustrated in Figure 2.6, in which the photon pair

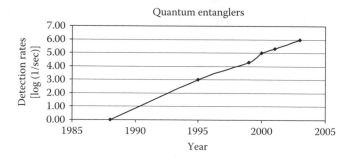

FIGURE 2.6 Progress in entangled-photon-pairs generation rate.

production rate is shown as a function of time. It can be seen from Figure 2.6 that more than a millionfold improvement in the detected rates of entangled photons has been achieved in the past two decades [2.6]. Quantum photonics is an interesting example in which purely fundamental and even philosophical research can lead to a new technology. Quantum photonic applications in communications and computing demonstrate qualitatively new concepts that are much more powerful than their classical counterparts.

In this context, it is interesting to mention that terabit speeds (1 terabit [Tb] is 10^{12}, or 1 trillion bits) will be needed to accommodate the increasing demand for future data transmission systems, both in communication and computation applications. At this processing speed it would be possible to realize the long-standing dream of merging the two worlds of technology — computing and telecommunications [2.7]. The question arises as to whether Moore's type of growth could also govern such a new hybrid technology expansion.

2.4 GENERAL TREND OF TECHNOLOGICAL GROWTH

A practical realization of Moore's law is the result of a complex interaction between technology, economics, and sociology. As mentioned earlier in this chapter, it seems that Gordon Moore in his simple extrapolation identified a more general trend that governs the growth of contemporary industry. It seems that Moore's quantitative observation could be applied to the most dynamically growing industrial sector at any given time.

This general trend can be observed in the data shown in Table 2.5. The growth of five technologies, i.e., semiconductors, electronic processors, telecommunications, supercomputers, and quantum photonics, has been presented in Table 2.5. A measure of growth for each of the technologies has been assigned in the following manner: (1) the number of transistors per chip for the measure of growth

---

TABLE 2.5
General Trend in Technological Growth

Technology	Growth Measure	Time of Growth Doubling (Months)	Applicable Time Period
Semiconductor transistors	Number of transistors per chip	24	1972–2009
Electronic processors	Processing speed (MHz)	22	1989–2008
Telecom systems	System capacity (Gb/sec)	15	1996–2007
Supercomputers	Processing speed (flops)	13	1996–2006
Quantum photonic systems	Entangled-photon generation rates (photon pairs per second)	9	1988–2004

of the semiconductor industry, (2) the processing speed (in MHz) for the electronic processor industry, (3) the system capacity (in Gb/sec) for the telecommunications industry, (4) the processing speed (in flops) for the supercomputer industry, and (5) the generation of entangled photons (in photon pairs/sec) for quantum photonic systems. Time required for doubling of the growth has been used as an indicator of the growth rate for each of the analyzed industries.

The data tabulated in Table 2.5 indicate that indeed a new hybrid communication and computing technology will follow the pattern of exponential growth.

One can therefore assume that such a growth will be realized and sustained through development and implementation of a series of novel microphotonic systems and devices.

2.5 TECHNOLOGICAL CHALLENGES

With today's growing dependence on computing technology, the need for high-performance computers (HPC) has significantly increased. Many performance improvements in conventional computers are achieved by miniaturizing electronic components to very small micron-size scale such that electrons need to travel only short distances within a very short time. This approach relies on the steadily shrinking trace size on microchips (i.e., the size of elements that can be drawn onto each chip). This has resulted in the development of very large scale integration (VLSI) technology with smaller device dimensions and greater complexity. The smallest dimensions of VLSI nowadays are about 0.08 mm. Despite the incredible progress in the development and refinement of the basic technologies over the past decade, there is growing concern that these technologies may not be capable of solving the computing problems of even the current decade. Applications of HPC and visualization technologies led to breakthroughs in engineering and manufacturing in a wide range of industries. With the help of virtual product design and development, costs can be reduced; hence, looking for improved computing capabilities is desirable. Optical computing includes the optical calculation of transforms and optical pattern matching. Emerging technologies also make the optical storage of data a reality.

The speed of computers was achieved by miniaturizing electronic components to a very small micron-size scale, but they are limited not only by the speed of electrons in matter (Einstein's principle states that signals cannot propagate faster than the speed of light) but also by the increasing density of interconnections necessary to link the electronic gates on microchips. The optical computer comes as a solution to miniaturization problems. In an optical computer, electrons are replaced by photons, i.e., bits of electromagnetic radiation that make up light.

The development of microphotonic systems has been led largely by the communications and data processing industries. The current estimates indicate that a conventional copper-based point-to-point serial link will become prohibitively expensive for transmission speeds above 20 Gb/sec. Therefore, it is expected that microphotonics will provide solutions for future high-speed links. In the past, the main driving factors for the development of communication systems have been information capacity and cost. One of the scenarios for the future communication links is one in which the drive for lower costs and higher performance (smaller size, lower power, higher data rate, greater transmit distance, expanded functionality, and expanded flexibility) will be achieved through an increased complexity of the microphotonic devices.

However, the microphotonic solution will not be the end-all solution. Increased complexity in photonic devices will require increasingly sophisticated electronic control solutions, and at high data rates, there will be pressure to more closely integrate photonic and electronic components. Most industry observers agree that, in the foreseeable future, the vast majority of improvements for cost-sensitive applications can be expected in both microelectronics and microphotonics. Therefore, the commercial implementation of the future high-speed links will be governed by a parallel increase in the complexity of both domains, i.e., photonic and electronic.

There are a number of challenges that are to be overcome. These are related to the link architecture (point-to-point or shared media, encoding scheme, etc.), the transmission medium (polymer waveguide or silica fiber), the integration platform (silicon, GaAs, InP, polymer, and silica), and the level of integration (monolithic photoelectronic, monolithic photonic integration with separate electronics, or photonic platform with hybrid electronic and photonic components).

Today's major drawback of microphotonics is the fact that these devices are still very expensive to manufacture. Some commercial microphotonic devices are made from materials such as indium phosphide and gallium arsenide. This makes them difficult to manufacture and assemble, and consequently they are expensive. This high cost limits the use of microphotonic devices to special applications such as long-haul telecommunications and wide-area networks.

2.5.1 SILICON MICROPHOTONICS

There is a new branch of microphotonics called *silicon microphotonics*, which has been championed by the researchers at Intel Corporation [2.8]. Silicon microphotonics focuses its effort on the development of devices using silicon as a base material and, therefore, on using standard, high-volume silicon manufacturing techniques. The main aim is to provide the industry with a total affordable integrated microphotonic package. Microprocessor manufacturers are focusing on ways to siliconize photonics and bring the benefits of volume manufacturing expertise to microphotonics. The goal is to make integrated, inexpensive microphotonic devices out of silicon instead of the exotic materials used today.

The overall approach is to transfer as much microelectronics communication technology to microphotonic silicon. In this way it will be possible to drive down costs and to realize mass market deployment. For example, by demonstrating that lasers, detectors, and modulators can be made out of silicon using standard manufacturing processes in an existing fabrication plant, it has been possible to remove a significant cost barrier. The next step will be to integrate entire photonic devices on a chip with digital intelligence. This should pave the way to manufacture commercial microphotonic ICs.

It is expected that greater bandwidth in silicon microphotonic devices can be achieved by multiplexing data streams onto multiple wavelengths of light on one optical fiber or waveguide. This approach could bring silicon microphotonics into an age in which enormous amounts of data can be exchanged at high speeds on a single waveguiding channel.

2.5.2 MICROPHOTONIC INTEGRATED CIRCUITS

As telecom manufacturers are feeling the market push to provide more bandwidth at economical price, microphotonic integrated circuits (micro-PICs) are emerging as the most promising solution providers. Micro-PICs are also referred to as planar lightwave circuits (PLCs). Microphotonic chips are the monolithic integration of two or more optical ICs on a single substrate. They are the photonic equivalent of microelectronic chips that integrate two or more transistors on a chip to form an electronic IC. In microphotonic chips, waveguides act as the photonic analog of copper circuits, serving as interconnects among various discrete components on a chip.

The trend toward all-optical component integration is often compared to the evolution from discrete transistors to ICs in the semiconductor industry. Integrating multiple optical functions into a single device could bring down the cost of a circuit, as it would enable the fabrication of a large number of components at the same time. By using manufacturing techniques closely related to those employed for electronic ICs, a variety of microphotonic elements can be placed and interconnected on the surface of a silicon wafer or similar substrate. As an example, PLCs using silica-on-silicon or indium phosphide can greatly reduce the number of discrete elements and hence the associated labor, packaging, and testing costs of individual components. If automation is applied to these wafer-based processes, it can lower costs even further. PLC technology has only recently emerged and is advancing rapidly with leverage from the more mature tools of the semiconductor-processing industry [2.9].

The fabrication of micro-PICs involves building devices in or on a substrate using high-yield, batch semiconductor-manufacturing processes such as deposition, photolithography, and etching. Multiple components or functions can be combined on a single substrate to build multifunctional chips. Both active and passive microphotonic components can be integrated onto one substrate. Passive components are involved in the transmission, splitting, and combining of optical rays. Active components have both optical and electrical properties, and require electrical power to emit, receive, and detect optical signals [2.9].

The scaling of ICs to smaller physical dimensions became a primary activity of advanced device development. This trend is based on two of the strongest drives governing today's electronic circuits business: (1) switching speed and (2) cost reduction. On the one hand, it is the market push for greater device performance measured in terms of switching speed, and on the other hand, it is the desire for greater manufacturing profitability, depending on reduced cost per device built.

The switching speed increases as the inverse square of the transistor physical channel length. This dependence encourages the shrinking of the transistor dimensions. The shrinking of the channel length causes a decrease in the current required by the load devices to achieve switching as their gate area and, hence, their physical channel length decreases [2.9]. Thus, there is a twofold benefit in device dimension reduction. As the drive-current requirement to switch the load devices depends on the total load capacitance and area, there is a strong motivation to reduce the size, and not only its physical channel length, of the complete device.

The reduction in device cost is also directly related to the size reduction. As a general rule, semiconductor devices are manufactured by the batch or by single-wafer processes, whereas they are sold as individual devices. Hence, the more devices one can pack onto each wafer, the higher the potential economic return. This is more conventionally described in terms of functional cost, expressed as dollars per transistor. Maintaining a continuously falling functional cost with time is seen as the key factor in driving an ongoing demand for electronics and semiconductor products [2.9]. To meet this demand in the face of achievable increases in revenue and capital investment, the industry is forced to maintain an essentially flat manufacturing cost per square centimeter of Si wafer area while expanding the total number of transistors built or the number of transistors per chip. This is directly related to the previously discussed Moore's Law.

To really take advantage of smaller sizes, wafers need to be larger. Over the last half century, Si wafer diameters increased from about 20 mm to 300 mm and have provided more than a factor of 200 increase in available device area per wafer. At the same time, the critical dimensions of devices have shrank from 20 μm to around 0.1 μm, and even less.

Currently, industry leaders have started to discuss 450-mm wafer sizes [2.9]. It is believed that 300-mm wafer technology has already paved the way to the anticipated 450-mm wafers.

2.6 NEW GROWTH CYCLE

The exponential growth in computer speed and power in microprocessor chips for the past three decades that has been described in this chapter has resulted from a simple scaling down of the gate length of active transistors. Transistor physics will continue to scale down for many more generations. Recently, however, a fundamental problem in this process has been recognized, i.e., the so-called interconnect problem [2.9]. It has been recognized that the time electrons require to move between transistors in metal interconnections does not scale down at all. Consequently, as the transistor is getting smaller, electrons spend more and more time moving idly around the chip rather than performing the expected functions. This is shown in Figure 2.7, in which the switching time as a percentage of total gate delay time has been plotted against the year in which such an achievement was realized [2.10]. In other words, Figure 2.7 illustrates the

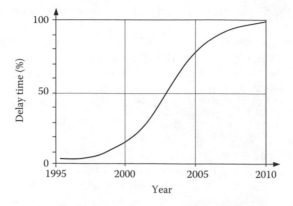

FIGURE 2.7 Time delay caused by electron transport in the metal interconnects between transistors.

percentage of the overall time that electrons perform a useful function and the fraction of the overall time spent by electrons on idle movements in the metal interconnects. As shown in Figure 2.7, by the year 2010 the interconnect delay time will totally dominate electron activity. This problem provides an ultimate and imminent limit to the further increase in electronic computer speed [2.10].

The pioneering work of researchers at IBM [2.11] pointed the way to a theory indicating the interrelationship of device parameters for optimum results. This work showed that the physical thickness of films, as well as lateral dimensions, must be scaled together. Later on, the National Technology Roadmap for Semi-conductors, (ITRS) [2.12], provided a considerably more detailed guide to device-scaling requirements to keep the industry on the Moore's Law curve, which has guided the IC industry's growth.

Until quite recently, the burden of producing ever-faster switching devices in higher quantities has fallen most heavily upon photolithographic technology, which has been driven to develop new techniques in order to pattern smaller structures. In the last decade, it has been realized that, after this long period of geometric scaling, the industry has arrived at a point at which the concomitant scaling of the thickness has left it only a few atomic layers thick, beyond which the material no longer possesses its inherent physical characteristics. This has brought about the search for new approaches to the problem.

The process of scaling of ICs has now moved well beyond geometrical shrinking, leading to the need for significant new materials and devices to replace the components and structures that have been used in the technology from its outset. It is quite apparent that it is microphotonics that has provided a seed for the next technological growth cycle. The development of photonic-band-gap structures and quantum photonic devices has been an indication that a new developmental cycle has been initiated.

The ability of the industry to maintain its growth along this new growth cycle will depend upon the degree of success in utilization of microphotonics.

REFERENCES

1. Moore, G.E., Cramming more components onto integrated circuits, *Electronics,* 38, http://www.intel.com/silicon/moorespaper.pdf (August 23, 2004).
2. http://www.intel.com/research/silicon/moorelaw.htm (August 26, 2004).
3. http://www.dtc.umn.edu--odlyzko-doc-internet.moore.pdf.url (August 26, 2004).
4. Moravec, H., When will computer hardware match the human brain?, *J. Transhumanism,* 1, 1–12, 1998.
5. Bennett, C.H. et al., Quantum Information Science: An Emerging Field of Interdisciplinary Research and Education in Science and Engineering, report of the NSF Workshop, Arlington, VA, October 28–29, 1999.
6. Kumar, P., Kwiat, P., Migdall, A., Nam, S.W., Vuckovic, J., and Wong, F.N.C., Photonic technologies for quantum information processing, *Quantum Inf. Process.,* 3, 215–231, 2004.
7. http://abcnews.go.com/sections/Business/SiliconInsider/SiliconInsider040226-1.html (August 27, 2004).
8. Salib, M., Liao, L., Jones, R., Morse, M., Liu, A., Samara-Rubio, D., Alduino, D., and Paniccia, M., Silicon photonics, *Intel Technol. J.,* 8,144–160, 2004.
9. Brown, G.A., Zeitzoff, P.M., Bersuker, G., and Huff, H.R., Scaling CMOS: Materials and devices, *Materials Today,* 4, 20–25, January 2004.
10. Homewood, K.P. and Lourenço, M.A., Light from Si via dislocation loops, *Materials Today,* 34–39, January 2005.
11. Dennard, R.H., Gaensslen, F.H., Yu, H.N., Rideout, V.L., Bassous, E., and Le Blanc, A.R., Design of ion-implanted MOSFETs with very small physical dimensions, *IEEE J. Solid-State Circuits,* SC-9, 256–268, 1974.
12. http://public.itrs.net (November 11, 2004).

3 Fundamentals of Interaction of Light with Matter

3.1 WAVE EQUATION

Light is a form of electromagnetic radiation. It is characterized by the electric field intensity E and the electric displacement vector D. These two vectors are related to the magnetic field intensity vector H and the magnetic flux density vector B by Maxwell's equations:

$$\nabla \times H = \sigma E + \varepsilon \partial E / \partial t \tag{3.1a}$$

$$\nabla \times E = -\mu \partial H / \partial t \tag{3.1b}$$

where ∇ is the vector operator $[\partial/\partial x, \partial/\partial y, \partial/\partial z]$ and the constant σ is electrical conductivity.

The two electric vectors and two magnetic vectors are related to each other in the following way:

$$D = \varepsilon\, E \tag{3.2a}$$

$$B = \mu\, H \tag{3.2b}$$

where the constant ε is the dielectric permittivity, the constant μ is the magnetic permeability. The dielectric permittivity could be real or complex. In general, this parameter is a function of frequency. The aforementioned relationships are more complex when the constants ε and μ are second-order tensors. In practice, the tensor ε is expressed in a scalar form, and it is usually known as the dielectric constant $\varepsilon_r = \varepsilon/\varepsilon_0$ (where ε_0 is the vacuum value of the permittivity). The Equation 3.2a and Equation 3.2b completely describe the electromagnetic field in linear, isotropic media and in the absence of currents.

Maxwell's equations allow deriving a form of the wave equation:

$$\nabla^2 E = 1/c^2 \left(\partial^2 E / \partial t^2 \right) \tag{3.3}$$

where $c = (\varepsilon \, \mu)^{-1/2}$ is the velocity of light in the medium with dielectric constant $\varepsilon/\varepsilon_0$. Accordingly, the velocity of wave propagation in free space is a constant given by

$$c_0 = (\varepsilon_0 \, \mu_0)^{-1/2} = 300,000,000 \text{ m/s} \qquad (3.4)$$

Equation 3.3 is known as the *wave equation*. A general solution of the wave equation indicates that the electric field is a function of time t and space coordinates. Assuming that the electric field oscillations are sinusoidal, one possible solution of Equation 3.3 for a plane wave polarized in the x direction and propagating in free space in the z direction may be presented in the following form:

$$E_x = E \exp i(\omega t \pm k_0 z)$$

$$\text{with } \omega = 2\pi\upsilon = \frac{2\pi}{\lambda} \qquad (3.5)$$

where E is amplitude of the electric field intensity, υ is its temporal frequency (expressed in cycles per second), λ is wavelength, $i = \sqrt{-1}$, k_0 is the wave vector, and $k = \omega\sqrt{\mu\varepsilon}$.

The relationship 3.5 is graphically illustrated in Figure 3.1. Figure 3.1 shows the electric and magnetic fields that oscillate sinusoidally as a function of time and space coordinates.

3.2 BAND GAP IN SOLIDS

Interaction of radiation with matter depends both on the properties of the electromagnetic radiation, such as its frequency and intensity, and on the material properties. In most cases interaction phenomenon can be described with simple classical models. However, there are certain aspects of interaction of light with solids that are best explained on the basis of a band theory and quantization of the electronic energy levels.

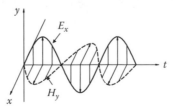

FIGURE 3.1 Electromagnetic plane wave as a function of time and space coordinates.

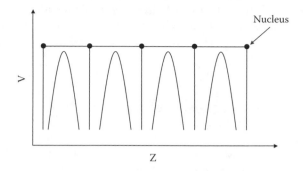

FIGURE 3.2 Periodic potential in one-dimensional crystals.

A band theory states that the energy levels that electrons can occupy in a solid are distributed over several energy bands and that a region of forbidden energy that electrons cannot occupy separates any two allowed energy bands.

Figure 3.2 illustrates a periodic potential in one-dimensional crystal. This figure shows that the electrostatic potential fluctuates from one atom to the next. The value of the potential is maximum at the nucleus and it is minimum in between.

The well-known Schrödinger equation is used to describe bound electrons in an atom. This equation may be written in the following form:

$$\frac{h^2}{2m}(\frac{\partial^2}{\partial x^2} + \frac{\partial^2}{\partial y^2} + \frac{\partial^2}{\partial z^2})\Phi + (E_{tot} - V)\Phi = 0 \qquad (3.6)$$

where m is mass of the electron, h is Planck's constant (normalized by 2π), V is the potential energy, and E_{tot} is total energy of the electron. The solution of this equation is the wave function Φ, which depends on space coordinates. This equation describes a special case of the potential V being independent of time. The wave function contains all the relevant information about the behavior of the electron, such as its velocity, energy, position, etc.

For a one-dimensional (z) crystal, the following solution of the preceding equation can be derived:

$$\Phi = A \exp\pm[i\frac{\sqrt{2m(E-V)}}{h}]z \qquad (3.7)$$

where A is a constant and the plus minus sign in the exponent corresponds to electron motion in the positive or negative direction. This solution may be compared with the time independent part of the electromagnetic plane wave solution (Equation 3.5), i.e.,

$$E = E_0 \exp (\pm i \, k_0 z) \tag{3.8}$$

where k is the wave vector. Therefore, one can obtain the following expression
for an electron wave vector:

$$k = \frac{1}{h} \sqrt{2m(E_{tot} - V)} \iff E_{\text{tot}} = \frac{h^2 k^2}{2m} + V \tag{3.9}$$

where E_{tot} is the total energy of the electron and V is its potential energy. The
difference $(E - V)$ corresponds to kinetic energy of an electron in the one-
dimensional crystal. The expression in Equation 3.9 indicates that E_{tot} is a parabolic function of the electron wave vector (see Figure 3.3). Figure 3.3 illustrates
the two top energy bands in a crystal, i.e., the unoccupied conduction band and
the valence band. In the case of a semiconductor, at absolute zero temperature,
the valence band is completely filled and the conduction band is empty. These
two bands are separated by the gap E_g. If thermal energy is provided, then some
of the valence band electrons may acquire sufficient energy to jump the band
gap, E_g, and occupy the empty levels in the conduction band. Therefore, as the
temperature increases, additional electrons are excited to the conduction band,
and the conductivity of such a material increases.

According to the Pauli exclusion principle, each quantum state can be occupied by no more than one electron of each spin. Therefore, at absolute zero, the
electrons occupy the lowest energy states in such a manner that the occupied
states are just sufficient to account for the number of electrons in the system. At
any temperature other than zero ($T \neq 0$) some electrons will occupy higher energy
levels. The probability $f_e(E)$ of finding an electron of any one spin in a given
energy level E is given by the following expression:

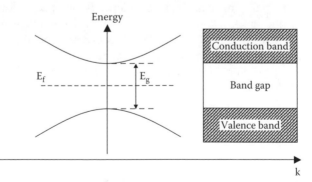

FIGURE 3.3 Energy band diagram for semiconductors.

$$f_e(E) = \frac{1}{[\exp(E - E_f)/kT]+1} \qquad (3.10)$$

where E_f is the Fermi energy level, which corresponds to the energy value that divides the filled and vacant levels at absolute zero (see Figure 3.3). In the case of semiconductors it is possible to introduce dopants, which provide additional electrons, i.e., n type (donor). These additional electrons may create new energy levels in the band gap. Therefore, in such a case, the Fermi level may lie closer to the conduction band. If the dopants have less valence electrons than the atoms of the parent material, as in the case of p type, then some of the energy levels in the valence band remain unoccupied. Here, the Fermi level may shift towards the valence band. There may be a special case of the semiconductor being very heavily doped so that even at $T = 0$ the allowed energy level falls within the conduction (n-type) band. Such semiconductors are called *degenerate semiconductors*, and they are used in semiconductor lasers.

3.3 INDEX OF REFRACTION

Equation 3.4 indicates that in free space the velocity of light c_0 is independent of the frequency ω. In general, for the case of a material medium the parameters $\varepsilon > \varepsilon_0$ and $\mu > \mu_0$, and therefore the velocity of light is always slower than in free space. The ratio of the velocity in free space to that in a material is known as the index of refraction n of the material:

$$n = \sqrt{\frac{\varepsilon\mu}{\varepsilon_0\mu_0}} \qquad (3.11)$$

If either ε or μ or both are complex, then the index of refraction will be complex, i.e.,

$$n^* = n - i\,\kappa \qquad (3.12)$$

where n describes the phase velocity of the wave and κ is the extinction coefficient. A complex index of refraction implies absorption of energy from the propagating wave. The electric field intensities for a plane wave (Equation 3.5) expressed in terms of complex index of refraction, i.e., propagating in a medium characterized by the index of refraction (n^*), can be presented as

$$E_x = E \exp i[\omega t - k_0(n - i\kappa)z] = E \exp (-k_0\kappa z) \exp i(\omega t - nz) \quad (3.13)$$

The preceding equation describes the absorption of electric field amplitude as the wave propagates through a material medium. The absorption coefficient

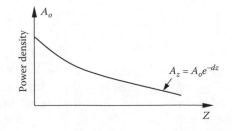

FIGURE 3.4 Optical attenuation in an absorbing material.

α is defined as a measure of the power density attenuation per unit length. Its value is determined as

$$\alpha = 2\,k_0 \kappa = \frac{4\pi\kappa}{\lambda_0} \tag{3.14}$$

In practice, the absorption of a plane wave propagating along the z axis in an absorbing medium is calculated in the following way:

$$A = A_o\,[1 - \exp(-\alpha z)] \tag{3.15}$$

The absorption effect is graphically illustrated in Figure 3.4. A_o is the power density at $z = 0$. The depth of penetration is defined as the length within which the power density drops to $1/e$ of its initial value.

3.4 POLARIZATION

Another quantity that is of great importance in the description of the flow of electromagnetic power in space is the so-called Poynting vector S:

$$S = E \times H \tag{3.16}$$

This vector is a useful parameter that allows determining the behavior of the electromagnetic waves when they encounter a boundary separating two media whose optical properties are different. Because the cross product is a vector, it gives the direction in which electromagnetic power is flowing (see Figure 3.5). The Poynting vector defined in Equation 3.16 indicates the magnitude of the energy flux density (energy flow rate per unit area per unit time) as well as the direction of energy flow.

The polarization direction is defined as the vibration direction of the electric field vector E. The plane containing the electric field and the propagation direction is called the *plane of polarization* (Figure 3.6). This type of polarization is called

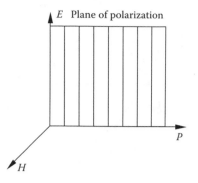

FIGURE 3.5 Poynting vector and polarization plane.

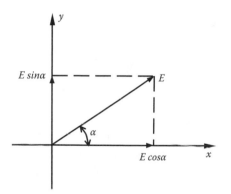

FIGURE 3.6 Polarization components.

linear polarization because the electric field vector is described by only one component.

Light emitted by separate atoms and molecules is always polarized. In practice, however, any macroscopic source of the light consists of large number of separate emitters, and the direction of the electric field at any moment of the time is not well defined. Such electromagnetic field is called *randomly polarized* or *unpolarized*.

In practice, it is convenient to designate the components of polarized beam as parallel and perpendicular to the plane of incidence. For example, Figure 3.6 represents a linearly polarized beam propagating in the direction perpendicular to the paper. The electric field vector is oriented at the angle α to the plane of incidence ($y = 0$). The linearly polarized beam can be represented as the sum of two linearly polarized components. One is polarized in the direction parallel to the plane of incidence (i.e., the x axis) and the other one in the direction perpendicular to the plane of incidence (i.e., the y axis). Traditionally, the parallel component is called the p component, and the perpendicular component is called

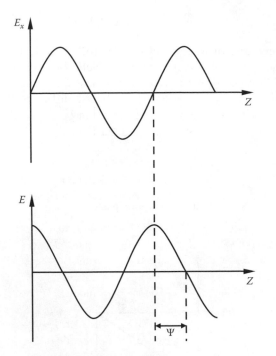

FIGURE 3.7 Phase difference between polarization components.

s component (from the German word *senkrecht* [3.1]). Accordingly, the *s* component and the *p* component can be given by

$$E_p = E \cos \alpha \tag{3.17a}$$

$$E_s = E \sin \alpha \tag{3.17b}$$

Depending on the amplitude of these two waves and their relative phase, the combined electric vector traces out an ellipse in one cycle, and therefore it may be said that it is *elliptically polarized*.

We may use Equation 3.17 to illustrate the concept of circular polarization. Let us assume that two electromagnetic waves of the same frequency polarized perpendicular to each other and with a phase difference of Ψ are propagating along the *z* axis (see Figure 3.7). Their electric field amplitudes may be presented in the following way:

$$E_x = E_{x1} \cos(\omega t - k_0 z)$$

$$E_y = E_{y2} \cos(\omega t - k_0 z + \Psi)$$

The resulting electric field intensity at any point on the *z* axis is given by

$$E = E_{x1} \cos(\omega t - k_0 z) + E_{y2} \cos(\omega t - k_0 z + \Psi)$$

For a special case of when $\Psi = +\pi/2$ and $E_{x1}/E_{y2} \neq 1$, the resultant electric vector will trace out an ellipse and rotate in the clockwise direction at an angular frequency of ω. A beam of this type of polarization is called *right elliptically polarized beam*. In a similar case, when the phase difference Ψ is equal $(-\pi/2)$, the vector will rotate in the counterclockwise direction, and such beam is called *left elliptically polarized beam*.

Another case may be considered when $E_{x1}/E_{y2} = 1$. This is called *circularly polarized*. And again, depending on the sign of the phase shift, there will be either a right or left circularly polarized beam. Elliptical and plane polarization can be converted into each other by means of natural or artificially produced birefringent materials.

It is possible to convert an unpolarized beam into a linearly polarized beam by using polarizers. A polarizer acts as a polarization filter. It transmits only the component polarized along its polarization axis and suppresses the other component. As the result the transmitted beam will be linearly polarized (see Figure 3.8).

If the linearly polarized beam is incident onto a polarizer, then the intensity of the transmitted light (I) will depend upon the angle α between the direction of the polarization and the orientation of the polarizer. The intensity of the optical beam is proportional to E^2; therefore, it can be described by the following equation:

$$I = I_0 \cos^2 \alpha \tag{3.18}$$

The polarization of the optical beams is widely explored in many fiber-optic, waveguide-based, and microphotonic devices.

3.5 REFLECTION AND TRANSMISSION

When an electromagnetic beam is incident upon a boundary separating two media of different optical properties, then a part of the beam will be reflected and a part

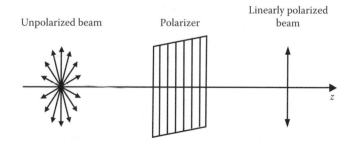

FIGURE 3.8 Optical polarizer.

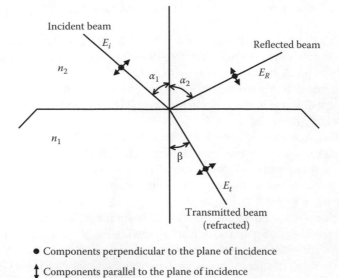

● Components perpendicular to the plane of incidence

↕ Components parallel to the plane of incidence

FIGURE 3.9 Reflection and transmission at an optical boundary.

of the beam will be refracted (transmitted). There are very well-defined relationships between directions, intensities, polarization, and phases of the incident, reflected, and refracted beams. There are three basic laws that describe these relationships: (1) the law of reflection, (2) Snell's law, and (3) the Fresnel formulas [3.1].

1. The law of reflection states that the angle of incidence is equal to the angle of reflection, i.e.,

$$\alpha_1 = \alpha_2 \qquad (3.19)$$

where α_1 is the angle of incidence and α_2 is the angle of reflection. The angles are defined with the respect to the normal to the surface (see Figure 3.9). Furthermore, this law states that the normal to the reflecting surface and the directions of propagation of the incident, reflected, and refracted beams lie in the same plane.

2. The law of refraction (the so-called Snell's law) states that the angle of incidence and the angle of refraction are governed by the following formula:

$$n_1 \sin \alpha_1 = n_2 \sin \beta \qquad (3.20)$$

where n_1 and n_2 are the indices of refraction of medium 1 and medium 2, respectively, and β is the angle of refraction. As has been indicated in Equation 3.12, the indices of refraction may be complex numbers.

3. The relationships between the amplitudes of the electric vector components for the incident, reflected, and refracted beams, as well as for their direction of polarization, are described by the Fresnel formulas. These formulas can be presented in the following way:

$$T_p = \frac{2n_1 \cos \alpha_1}{n_2 \cos \alpha_1 + n_1 \cos \beta} I_p \qquad (3.21a)$$

$$T_s = \frac{2n_1 \cos \alpha_1}{n_1 \cos \alpha_1 + n_2 \cos \beta} I_s \qquad (3.21b)$$

$$R_p = \frac{n_2 \cos \alpha_1 - n_1 \cos \beta}{n_2 \cos \alpha_1 + n_1 \cos \beta} I_p \qquad (3.21c)$$

$$R_s = \frac{n_1 \cos \alpha_1 - n_2 \cos \beta}{n_1 \cos \alpha_1 + n_2 \cos \beta} I_s \qquad (3.21d)$$

where I_p, T_p, and R_p are the incident, transmitted, and reflected amplitudes of the electric vector, respectively, of the parallel component of the polarization. I_s, T_s, and R_s are the incident, transmitted, and reflected amplitudes of the electric vector, respectively, of the perpendicular component of the polarization.

3.6 TOTAL INTERNAL REFLECTION

The fundamental phenomenon that is widely applied in photonic devices is the effect of total internal reflection (TIR). This effect forms a basis for the so-called waveguide structures. These are structures that allow for a very efficient transmission of the light beam through various channels that are made in light-guiding materials.

TIR is related to a particular case of Snell's law (see Equation 3.20). As has been discussed in the preceding section, Snell's law describes the relationship between the incident, reflected, and refracted rays at the interface of two media that are characterized by the index of refraction n_1 and n_2. When a ray of light is incident at the interface of two media, the ray is partially reflected and partially refracted, as is shown in Figure 3.10a. According to Snell's law an incident beam is inclined toward the normal to the interface when it enters from one medium into another with higher index of refraction. If a light wave is incident at the interface of a medium whose $n_2 < n_1$, then the ray will bend away from the normal, as is shown in Figure 3.10b. The angle of incidence for which the angle of refraction is 90° is known as the *critical angle*. When the angle of incidence exceeds the critical angle (see Figure 3.10c), TIR will occur. In such a situation

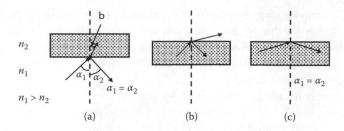

FIGURE 3.10 Snell's law: (a) light wave incident on a medium with higher index of refraction n_2 ($n_2 > n_1$), (b) light wave incident on a rarer medium ($n_2 < n_1$), (c) total internal reflection.

the beam is totally reflected at the boundary, with no light transmitted across the boundary.

3.7 OPTICAL WAVEGUIDES

TIR is the basis for the so-called optical waveguiding structures. These structures are the fundamental components of all microphotonic devices. The transmission of optical signals through the optical fibers is based on the waveguiding effect, and hence optical fiber may be used to illustrate the waveguiding principle. For example, Figure 3.11 shows a typical optical fiber. A concentric dielectric cylinder with the index of refraction n_2 surrounds a dielectric cylinder of index of refraction n_1. The two indices obey the relation $n_2 > n_1$.

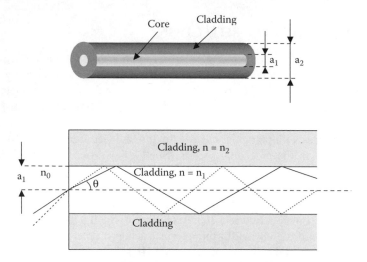

FIGURE 3.11 Waveguiding optical fiber: (a) a typical glass fiber consists of a cylindrical core that is clad by a material of slightly lower refractive index, (b) light entering at the core-cladding interface at an angle greater than the critical angle is guided inside the core of the fiber.

It should be pointed out that the outer cylinder is not really essential for the waveguiding effect. There are a couple of reasons why it is desirable to use this so-called clad fiber rather than a bare fiber [3.2]. The fields of a dielectric waveguide are not fully contained inside the dielectric region of index n_1, but they extend to the outside region, where they decay exponentially (see Section 3.17 for a more detailed description of this phenomenon). Because the fiber must be supported in space, it is advantageous to surround the inner core with an outer cladding to avoid scattering and field distortion by the supporting mechanism. Because the field decays exponentially inside the region characterized by index n_2, practically no field exists outside the cladding.

The second reason for using cladding is related to the mechanism of field guidance in the fiber. At a given frequency, an optical fiber is capable of supporting a finite number of modes. If the diameter a_1 (see Figure 3.11a) of the core is much larger the wavelength λ of the guided light beam, a very large number of guided modes is possible. For the practical purposes of optical transmission, it is desirable to limit the number of guided modes. For example, single-mode operation is possible by properly selecting the fiber dimensions. The dimensions of the inner core that induce single-mode operation depend on the ratio of n_1/n_2. The larger this ratio, the smaller must be a_1 to ensure that only one guided mode can be transmitted. In practice, the cladding is usually made of pure silica, whereas the core is usually silica doped with germanium. Doping with germanium results in a typical increase of refractive index from n_2 to n_1.

One additional important variety of single-mode fiber is the polarization-maintaining fiber (PM fiber). All other types of single-mode fibers are capable of carrying randomly polarized light. PM fiber is designed in such a way that it preserves propagation of a linearly polarized beam. This characteristic of the fiber is of great importance to some photonic devices, for example, external modulators.

Figure 3.12 shows a cross section of a typical PM fiber. The polarization maintenance is achieved by adding two stress rods. The purpose of the rods is to create a directional stress in the core of the fiber. The stress induces changes in the core index of refraction and therefore induces conditions that are favorable for maintaining of the direction of polarization.

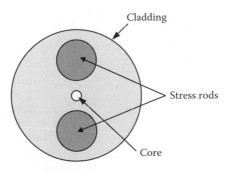

FIGURE 3.12 Polarization-maintaining (PM) fiber.

3.8 DISPERSION IN DIELECTRICS

The simplest model that can be used to describe the electromagnetic wave frequency dependence on the dielectric constant is based on the harmonic oscillation of the electronic cloud with respect to the nucleus. The following equation represents the motion for a bound electronic charge:

$$eE \exp(i\omega t) = m \frac{d^2 x}{dt^2} + \gamma \frac{dx}{dt} + fx \tag{3.22}$$

where E is the electric field intensity of the incident radiation; e, the charge of an electron; m, the mass of the electron; γ, the damping constant; ω, the frequency of the incident radiation; and f, the spring force constant.

The steady-state solution of this equation is given by

$$x = \left(\frac{eE}{m} \right) \left(\frac{1}{\omega_0^2 - \omega^2 + i\Gamma\omega} \right) \exp(i\omega t) \tag{3.23}$$

where $\omega_0^2 = f/m$ is the natural frequency of oscillation and $\Gamma = \gamma/m$ is the damping or scattering frequency.

The aforementioned equation illustrates the fact that application of electric field to a medium may cause either the displacement of the electronic cloud bound to an atom or ion, the displacement of ions, or both. The displacement of ions and bound electrons from their equilibrium position gives rise to polarization P. At the same time, the electronic oscillation experience damping because of their interaction with adjacent ions or because of lattice vibrations. The damping causes the polarization and the dielectric constant to assume complex values. The electric polarization P and the electric filed are related in the following way:

$$P = E (\varepsilon_0 - i \varepsilon) \tag{3.24}$$

For a lossy dielectric, P will be complex; therefore, the electrical permittivity as well as the index of refraction will be also complex. Taking into account the solution (Equation 3.23), it is possible to derive the equation that describes the index of refraction as a function of frequency:

$$\varepsilon_r = \frac{\varepsilon}{\varepsilon_0} = (n - i\kappa)^2 = 1 + \frac{Ne^2}{\varepsilon_0 m} \left(\frac{1}{\omega_0^2 - \omega^2 + i\Gamma\omega} \right) \tag{3.25}$$

where N is the number of charged particles per unit volume. For bound electrons, the resonant frequency ω_0 generally lies in the ultraviolet region. Therefore, the index of refraction for most gases at optical and lower frequencies is close to unity. Equation 3.25 assumes that there is only one kind of oscillator. However, atomic systems are complex and have more than one resonant frequency, depending upon the configuration of their electronic cloud. Therefore, Equation 3.25 can be modified to take into account these characteristics. The modification requires replacement of ω_0 by ω_j and summing over all the different resonant frequencies, corresponding to both electronic and ionic oscillations:

$$(n - i\kappa)^2 = 1 + \sum_j \frac{N_j e^2 f_j}{\varepsilon_0 m_j} \left(\frac{1}{\omega_j^2 - \omega^2 + i\Gamma_j \omega} \right)$$

(3.26)

where j stands for the jth oscillator (electron or ion) and f_j is the oscillator strength.

In the case of a solid, the resonant frequency ω_0 is modified by the local field correction (known as Lorentz-Lorentz correction) in the following way:

$$\omega_j^2 = \omega_{0j}^2 - \frac{e^2}{3m} \frac{N_j}{\varepsilon_0}$$

(3.27)

This correction arises because of the fact that the electric field at the ionic site is, in general, different from the applied field due to polarization fields produced by other charges [3.1]. It is apparent that this change shifts the resonant frequency toward lower values. The effect of resonant frequency on the index of refraction may be illustrated by assuming the case of a simple oscillator with no damping (i.e., $\Gamma = 0$ in Equation 3.25). In such a case the index of refraction is given by

$$n^2 = 1 + \frac{Ne^2}{\varepsilon_0 m} \left(\frac{1}{\omega_0^2 - \omega^2} \right)$$

(3.28)

Figure 3.13 shows an example of the resonant characteristic of light transmission through a silica fiber. There is a transmission window around 1550 nm where the material absorption is negligible. This window is used in most fiber communication systems.

Attenuation represents the most important characteristic of an optical fiber and determines the information-carrying capacity of a fiber-optic communication system. The attenuation A of an optical beam is defined in the following way:

$$A = 10 \log \frac{P_1}{P_2}$$

(3.29)

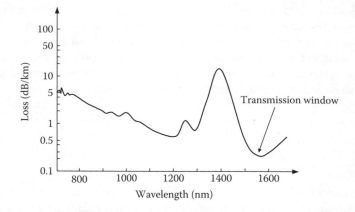

FIGURE 3.13 Wavelength dispersion in silica fiber.

where P_1 and P_2 represent input and output power, respectively. The attenuation is usually measured in decibels (dB). For example, if the output power is only half of the input power, then the loss is 3 dB. On the other hand, in an optical amplifier a power amplification of about 1000 represents a power gain of 30 dB.

There is an important difference between the loss term introduced through finite conductivity of the medium (σ in the Equation 3.1a) and the complex permittivity (see the Equation 3.2a). The conductivity term introduces losses due to absorption of radiation by conduction electrons and their subsequent scattering by phonons, ionized impurities, grain boundaries, etc. [3.1]. The absorption is, therefore, spread out over a very broad frequency range. Losses due to the complex permittivity arise due to excitation of lattice vibrations and their subsequent damping by crystal imperfections, etc. The absorption in dielectrics, therefore, may be very high in selected narrow frequency ranges, as is seen in Figure 3.13.

3.9 DISPERSION IN SEMICONDUCTORS

The plasma dispersion effect in semiconductors is related to the density of free carriers, which changes both the real and imaginary parts of the refractive index (see Equation 3.12). This is described by the Drude–Lorenz equations that relate the concentration of electrons (N_e) and holes (N_h) to the absorption coefficient and refractive index n (α and n are defined in Equation 3.14). From the application point of view, the most convenient way of expressing the dispersion in semiconductors is to express the changes of $\Delta \alpha$ and Δn as the functions of electron (ΔN_e) and hole densities (ΔN_h):

$$\Delta n = -C \left(\frac{\Delta N_e}{m_e} + \frac{\Delta N_h}{m_h} \right) \tag{3.30}$$

$$\Delta \alpha = 2 \frac{Ce}{c} \left(\frac{\Delta N_e}{m_e^2 \mu_e} + \frac{\Delta N_h}{m_h^2 \mu_h} \right) \qquad (3.31)$$

where Δn and $\Delta \alpha$ are the real refractive index and the absorption coefficient variations, e is the electron charge, m is the effective mass, μ is the free-carrier mobility, ΔN is the free-carriers concentration variation, and C is a constant. The subscripts e and h refer to electron and hole contributions, respectively.

The band gap in semiconductors is relatively small, i.e., on the order of 1 eV. This means that the absorption due to band-to-band transitions occurs mainly in the visible and infrared regions. Doping of semiconductors can modify this small gap. In addition to direct band-to-band transitions, phonon-assisted transitions can be observed in semiconductors. These two types of transitions, i.e., direct and indirect, are illustrated in Figure 3.14.

As is shown in Figure 3.14, phonons, i.e., particles corresponding to lattice modes of vibrations, are generated during the band-to-band indirect transition.

3.10 WAVE PROPAGATION IN NONLINEAR MEDIA

The application of an electric field to a material can result in a change of the real and imaginary refractive indices. For example, the electric flux density D (see Equation 3.2b) can be written as

$$D = \varepsilon E = \varepsilon_0 E + P \qquad (3.32)$$

where P is the polarization. Polarization arises because of charge separation induced by the electric field of the incident radiation. Figure 3.15 shows, as an example, an atom with spherical electron charge distribution. In a neutral state, i.e., in unpolarized state, the center of gravity of electron cloud overlaps the

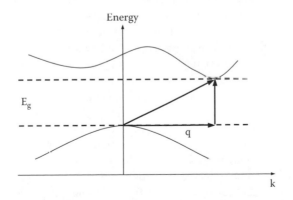

FIGURE 3.14 Indirect transition in semiconductors (q is the phonon wave vector).

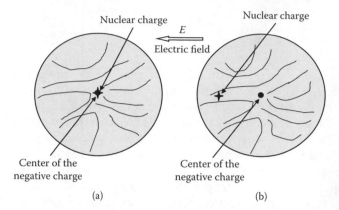

FIGURE 3.15 Induced polarization: (a) unpolarized atom, (b) polarization of an atom induced by electric field E.

positive nuclear charge (Figure 3.15a). If the atom is in an electric field E, then the electron cloud will be displaced with respect to the nucleus (Figure 3.15b). This will generate an electric polarization P. This polarization plays an important role in the interaction of electromagnetic waves with matter.

In practice, waveguide materials are not quite linear. Their nonlinear or perturbed properties play an important role in current optical systems. In these cases it is convenient to introduce the electric susceptibility tensor χ that describes the relationship between the electric field intensity E and polarization P:

$$P_i(\omega_i) = \varepsilon_0 \left[\chi_{ij} + \Delta\varepsilon(r) + \chi_{ijk} E_k(\omega_k) + \chi_{ijkl} E_k(\omega_k) E_1(\omega_1) + \ldots \right] E_j(\omega_j)$$
(3.33)

The term $\Delta\varepsilon(r)$ in Equation 3.33 represents the part of the dielectric constant that may vary periodically from point to point in space. This term applies to a medium in which the dielectric constant ε varies along the direction of wave propagation [3.4]. The perturbation $\Delta\varepsilon(r)$ can be artificially induced and can be periodical in its character. The perturbation can be induced temporarily (e.g., due to thermo-optical or elasto-optical effects) or it can be permanent. This relationship is of great importance for today's optical communication systems. The so-called fiber Bragg gratings are based on such artificially induced defects in waveguiding media.

The linear susceptibility tensor χ_{ij} is a second-rank tensor, whereas χ_{ijk} and χ_{ijkl} are the higher-rank tensors that describe higher-order nonlinear effects (the Einstein's tensor summation is used in the Equation 3.33). The higher-order nonlinear effects can be observed in certain materials when the incident electric field is very strong and causes deviation from the linear dependence between the polarization and the electric field [3.3]. This is schematically illustrated in Figure 3.16. Figure 3.16a shows a P–E relationship for a linearly polarized material,

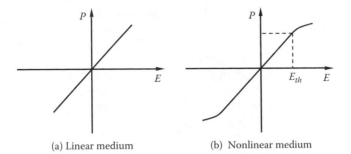

(a) Linear medium   (b) Nonlinear medium

FIGURE 3.16 Linearly and nonlinearly polarized materials.

whereas Figure 3.16b illustrates nonlinearity under a strong electric field E ($E > E_{th}$, where E_{th} is the nonlinearity threshold).

The tensor χ_{ij} only gives rise to polarization components at the incident frequencies. This tensor describes the normal refractive properties of a material, including any kind of birefringence. The lowest-order term producing nonlinear effects is described by χ_{ijk}. According to the Equation 3.33 the value of the nonlinear term $\Delta\, \varepsilon(\omega)$ may be expressed as

$$\Delta\, \varepsilon(\omega) = \Delta\, n^2 = \chi^{(3)} E + \chi^{(4)} E^2 + \dots \tag{3.34}$$

The first term in the preceding equation, i.e.,

$$\Delta\, n^2 = \chi^{(3)} E \tag{3.35}$$

describes the so-called linear electro-optical Pockels effect. The tensor $\chi^{(3)}$ has nonzero components only in noncentrosymmetric materials, such as piezoelectric crystals. Therefore, isotropic materials, such as cubic crystals, liquids and gases, by virtue of their centrosymmetry do not display the Pockels effect.

The Pockels effect is the basic mechanism in electro-optical cells. A schematic of an electro-optical Pockels cell is shown in Figure 3.17.

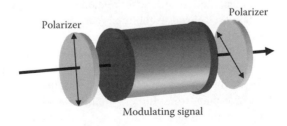

Polarizer

Polarizer

Modulating signal

FIGURE 3.17 Electro-optical Pockels cell.

The Pockels cell is placed between crossed polarizers. An external electric field applied to the Pockels cell induces difference in the index of refraction of the optical beams traveling through the cell. This difference results in a voltage-dependent transmission at the cell output.

Another nonlinear effect that is governed by the tensor $\chi^{(3)}$ is the so-called spontaneous parametric down-conversion (PDC). This effect is a special case of the three-wave mixing. In this nonlinear effect a pump photon decays into two photons. This effect may be considered to be the spontaneous reverse of second-harmonic generation. The PDC effect is subject to the laws of conservation of energy and momentum. These laws determine the properties of the created photons, i.e., the polarization, wavelength, and direction of propagation. In down-conversion there is an interaction between a high-frequency pump field and two lower-frequency down-converted fields. Therefore, the energy conservation requires that $h\nu_p = h\nu_1 + h\nu_2$. The momentum conservation dictates that the wave vectors should comply with $hk_p = hk_1 + hk_2$. The second criterion can be met by choosing appropriate birefringent crystal geometries such that optical beams are projected in specific directions. These various geometries of the nonlinear crystals are required to optimize the so-called phase matching (i.e., to comply with momentum conservation) between pumping and down-converting photons. Spontaneous PDC is possible using a continuous as well as pulsed laser beams.

The electrically induced changes in the refractive index may be on the order of $\sim 10^{-4}$. However, the third-rank tensor χ_{ijk} is nonexistent in glasses. Therefore, the nonlinear effects in glasses, e.g., in silica fiber, can only result from the fourth-rank tensor χ_{ijkl}:

$$\Delta n^2 = \chi^{(3)} E^2 \tag{3.36}$$

Equation 3.36 describes the changes in the refractive index that are proportional to the square of the applied field (E^2). This effect is known as the *quadratic electro-optical effect* or *Kerr effect*, and the corresponding devices are known as *Kerr cells*. The changes in the refractive index that can be caused by this effect could be in the range of $\sim 10^{-7}$.

The Pockels effect and the Kerr effect are related to a change in real refractive index. The changes that are related to the imaginary part of the refractive index are known as electroabsorption (see the following section).

The fourth-rank tensor χ_{ijkl} also describes such effects as the third-harmonic generation, Raman effect, and four-wave mixing. The essence of higher-order optical nonlinearities is that a medium produces electromagnetic oscillations at the frequencies that correspond to multiple combination of the frequency of the applied field. Figure 3.18 illustrates the mechanism of four-wave mixing.

A particular case of the nonlinear optical effect is illustrated in Figure 3.19. This effect is known as the *Raman effect*. Two electromagnetic waves of two frequencies are incident on the medium. These waves are termed as the pumping beam of frequency ω_p and as the Raman (or Stokes) beam of frequency ω_s. The

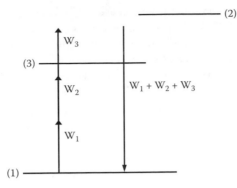

FIGURE 3.18 Four-wave mixing.

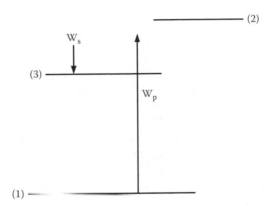

FIGURE 3.19 Raman effect.

Stokes frequency ω_s either experiences gain or loss depending whether the population of state 1 is greater than that or less than that of state 3. In Raman effect, photons are absorbed from the pumping beam ω_p, producing photons at frequency ω_s (see Figure 3.19). Simultaneously the population of state 3 will increase. Alternately, if the system is initially inverted, i.e., the population of state 3 exceeds that of state 1, then photons at the frequency ω_s will be absorbed, leading to the production of photons at frequency ω_p and an increase in the ground state population. This effect is used in the so-called Raman lasers.

The optical properties of materials may also be affected by the application of a magnetic field.

The magneto-optical effect, which is known as the *Faraday effect*, depends on both the optical electric field and the static magnetic field. Therefore, this effect is described by a third-rank magneto-optical susceptibility tensor. The Faraday effect may be used to induce a phase shift between the two circularly polarized optical beams. If a magnetic field is applied perpendicular to the direction of propagation, then it may cause a birefringence proportional to the square of the applied field. Depending on whether the effect occurs in vapors or in liquids, the effect is known as the *Voigt* or the *Cotton–Mouton effect.*

The magneto-optical effect can also cause a shift of the resonant frequencies (i.e., term ω_j in Equation 3.26). This effect can be described by the following equation:

$$\varepsilon_r = \frac{\varepsilon}{\varepsilon_0} = (n - i\kappa)^2 = 1 + \frac{Ne^2}{\varepsilon_0 m}\left(\frac{1}{\omega_0^2 - \omega^2 + \dfrac{eB}{m}\omega} \right) \qquad (3.37)$$

where B is the magnetic flux density. Therefore, in this case the magnetic field affects the dispersion curves for two circularly polarized optical beams. This effect is known as the *Zeeman effect.*

The optical nonlinearities in wide-band-gap crystals arise mainly from the bound electrons and the strong coupling between the electronic states and lattice vibrations. Thus, the nonlinear susceptibility coefficients consist of a term representing purely electronic process as well as a term corresponding to the process in which the electronic polarizability of the ions is influenced by lattice displacement. Wide-band-gap crystals usually exhibit a relatively small nonlinearity. However, doping them with impurities could enhance the nonlinearity of these materials. For example, doping could enhance the electro-optical Kerr constant by three to five orders of magnitude.

Nonlinear optical materials find applications in optical communications, signal processing, and quantum photonic devices. These applications require materials that exhibit large nonlinear susceptibilities. Apart from the importance of high optical nonlinearity, a number of other factors such as reliable crystal growth technique, mechanical strength, and optical homogeneity are also vital.

3.11 ELECTROABSORPTION

A high electric field can affect photon absorption in a semiconductor. The electric field may push electrons and holes to opposite directions, making the energy separations smaller. This is schematically illustrated in Figure 3.20.

This effect is known as the *Franz–Keldysh effect.* The Franz–Keldysh effect is observed in conventional bulk semiconductors. As the result of this effect, a photon with an energy that is lower than the band gap can be absorbed by the semiconductor. The corresponding absorption curve is shown in Figure 3.21.

A similar effect is also observed in quantum-well structures. It is known as *quantum-confined Stark effect* (QCSE; see Figure 3.22). In QCSE the electric field reduces overlap and results in a corresponding reduction in absorption and luminescence. The QCSE-induced absorption change is much larger than that of

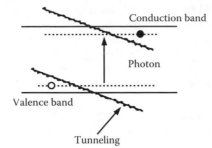

FIGURE 3.20 Franz–Keldysh effect.

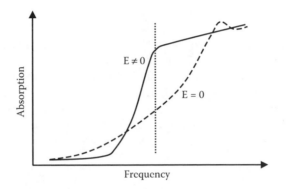

FIGURE 3.21 Electroabsorption.

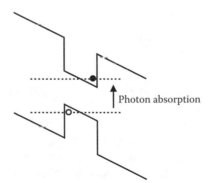

FIGURE 3.22 Quantum-confined Stark effect.

Franz–Keldysh effect. Both of these electroabsorption effects are prominent near the band gap of semiconductors. The electroabsorptive effects have been used in optical beam modulation.

3.12 BRAGG REFLECTION

When optical beam propagates through a material with periodical regions of higher and lower refractive indices, it is partially reflected at each interface. If the spacing between those regions is such that all the partial reflections add up in phase, then it is possible that the resulting reflection can be nearly 100% of the incident beam.

This concept is similar to that of Bragg diffraction, which was discovered for x-ray diffraction in crystals. By virtue of coherent scattering, each set of crystallographic planes may give rise to a diffraction peak at a certain frequency related to the interplanar distance. The condition for high reflection, known as the *Bragg condition*, is given by

$$m \; \lambda_{Bragg} = 2 \; n_{lattice} \; D_{(hkl)} \; \sin \Theta \; (m = 1, 2, 3, \ldots) \tag{3.38}$$

where λ_{Bragg} is the Bragg wavelength, $D_{(hkl)}$ is the distance between atomic crystalline planes labeled by the Miller indices *(hkl)*, $n_{lattice}$ is the index of refraction, Θ is the angle of the incident radiation, and m is the diffraction order.

As a consequence of the interference, photons in the Bragg diffraction peak are not allowed to propagate through the crystal, and they are reflected. This condition will occur only for specific wavelengths. For all other wavelengths, the out-of-phase reflections end up canceling each other, resulting in high transmission.

An important difference between x-ray and optical diffraction is the bandwidth of the Bragg peaks. X-ray diffraction peaks are extremely narrow $(\frac{\Delta\lambda_{Bragg}}{\lambda} \approx 10^{-6})$ and mainly instrumentally broadened. In the case of optical diffraction, the condition for a given reflection is met for a larger range of frequencies $(\frac{\Delta\lambda_{Bragg}}{\lambda} \approx 10^{-2})$. This difference is related mainly to the vastly different dielectric constants occurring for these two electromagnetic wave regions. Indices of refraction for x-rays scarcely differ from unity, whereas at optical wavelengths they are rather larger.

A fiber Bragg grating (often referred to as FBG) is an optical fiber for which the refractive index in the core is perturbed forming a periodic or quasi-periodic index modulation profile. Successive coherent scattering from the index variations reflects a narrow band of the incident optical field within the fiber.

Equation 3.38 can be modified in such a way that it can express the relative change in the Bragg center wavelength $\Delta\lambda_{Bragg}$ as a function of the changes in the index of refraction Δn and the grating period ΔD:

$$\frac{\Delta\lambda_{Bragg}}{\lambda} = \frac{\Delta n}{n} + \frac{\Delta D}{D} \tag{3.39}$$

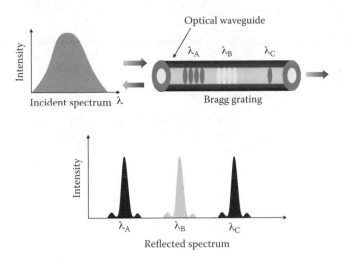

FIGURE 3.23 Bragg reflection.

This equation indicates that the required shift in the Bragg center wavelength can be achieved either by inducing a gradient in the refractive index Δn of the grating, by a change in the grating period ΔD, or both. By modulating the quasi-periodic index perturbation in amplitude and (or) phase, one may obtain different optical filter characteristics (see Figure 3.23).

One can estimate from Equation 3.38 that for a silica fiber and for the Bragg wavelength in the communication band of 1550 nm the grating period D is about 535 nm. FBGs are typically between 1 mm and 25 mm long. It is also possible to make FBGs with nonuniform periods, for example, linearly varying along the length of the grating. Such "chirped" gratings can have much broader bandwidths, i.e., up to tens of nanometers.

An FBG can be used to make an optical add-drop multiplexer (OADM). The principle of operation of an OADM is illustrated in Figure 3.24. An optical signal with different wavelengths $\lambda_1 \ldots \lambda_n$ is sent through the circulator and then via a fiber into an FBG. The FBG passes all but one wavelength. Other wavelengths, i.e., λ_2, λ_3, etc., continue down the fiber, whereas, the rebounded signal λ_i goes back into an optical spectrum analyzer.

OADMs are considered key components in wavelength-division-multiplexing (WDM) systems. OADMs enable great connectivity and flexibility, and they are expected to play an important role for the development of new wavelength-routing techniques. Several types of fiber-optic OADMs based on different configurations have been designed and used in commercial communication systems. Most of them include Bragg gratings because of their optimum spectral characteristics for the WDM.

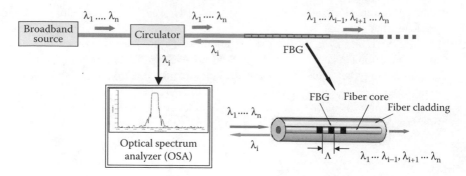

FIGURE 3.24 Principle of operation for an optical add-drop multiplexer (OADM).

3.13 PHOTONIC-BAND-GAP STRUCTURES

Photonic-band-gap (PBG) structures, also known as photonic crystals, are optical materials with periodic variations in refractive index that manipulate light through multiple diffraction. Multiple interference between light waves diffracted from each unit cell of the structure may create a "photonic band gap" — a forbidden frequency range, analogous to the electronic band gap of a semiconductor, within which no propagating electromagnetic modes exist.

The PBG structures were predicted in 1987 by Yablonovitch [3.5]. Simultaneously, John [3.6] predicted that disorder in a PBG structure would cause localization of light, where photons become trapped and can no longer propagate. These pioneering works triggered a great interest in further investigation of these structures [3.7]. The first PBG structure was demonstrated experimentally by Yablonovitch. It was made by drilling holes in dielectric materials. These types of structures are known as *yablonovites*.

A conceptual design of the one-, two-, and three-dimensional PBG structures are illustrated in Figure 3.25. There are examples of such structures existing in nature. One of the most beautiful, naturally occurring effects is that of butterfly wings. Other example is the distinctive color of opal, which is due to the reflection caused by the PBG of the crystal.

These structures are analogous to crystalline atomic lattice in solid-state electronics, where the lattice structures act on the electron wave function to produce the electronic band gaps. In the current electronic systems, electrons are

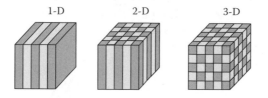

FIGURE 3.25 Photonic-band-gap structures.

routed through electronic gates or transistors. However, electrons are charged particles that interact with each other when brought into close contact, they produce heat, and their movement is limited. PBG structures allow overcoming these limitations because uncharged photons are used as the carriers of data.

An analogy between Maxwell's equations for the electromagnetic wave and Schrödinger's equation for electrons propagating in a periodic potential has been used in the process of investigation of the light localization in the PBG structures. From Maxwell's equations (Equation 3.1), the following wave equation for the propagation of light in a periodically modulated dielectric medium can be obtained:

$$\bar{\nabla} \times (\bar{\nabla} \times \bar{E}) = [\varepsilon_0 - \varepsilon_{var}(r)] \frac{\omega^2}{c^2} \bar{E} \qquad (3.40)$$

where the dielectric constant $\varepsilon(r)$ is given by the following equation:

$$\varepsilon(r) = \varepsilon_0 + \varepsilon_{var}(r) \qquad (3.41)$$

In this equation ε_{var} represents the part of the dielectric constant that varies periodically from point to point in space. By introducing the Hermitian linear differential operator θ

$$\Theta \equiv \nabla \times \{\nabla \times\}$$

the wave equation (Equation 3.40) can be written in the following form:

$$\Theta \bar{E} = [\varepsilon_0 - \varepsilon_{var}(r)] \frac{\omega^2}{c^2} E \qquad (3.42)$$

This wave equation resembles closely the quantum mechanical Schrödinger's equation for an electron in a potential $V(r)$ that varies randomly in space:

$$\bar{\Theta} \Psi = V(r) \Psi \qquad (3.43)$$

where Ψ is the wave function. The wave function contains all the information about the behavior of the electron, such as its velocity, energy, position, etc. The previously described equation [3.6] is a simplified version of Equation 3.43.

Knowing that the periodicity of the crystalline potential is at the origin of the energy bands in crystals, it is possible to use this analogy to photons. By the comparison of the Schrödinger equation (Equation 3.43) and that of Maxwell (Equation 3.42), it is possible to draw a parallel between the periodicity of crystalline potential to the periodicity of the dielectric permittivity. These

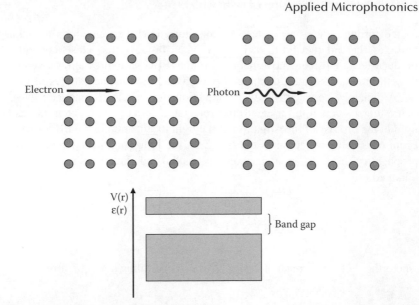

FIGURE 3.26 Band-gap concept.

equations are two linear systems of eigenvalues whose solutions are determined by the properties of the potential $V(r)$ or of the dielectric function $\varepsilon(r)$. The only difference is the nature of the wave equation, which is vectorial in the case of the photons and is scalar in the case of the electrons. A graphical illustration of the band-gap concept for electrons and photons is illustrated in Figure 3.26.

The nature of the PBG structures is the result of a synergetic interaction between two resonance scattering mechanisms [3.8]. The first is the Bragg scattering from a macroscopic periodic structure of the dielectric material. This macroscopic resonance occurs whenever the period D of the dielectric structure is an integral multiple of half of the optical wavelength. The periodicity of the PBG structure has the length on the order of optical wavelengths. This relationship is described by the Bragg equation (Equation 3.38). When Θ is 90°, then:

$$m\lambda_{macro} = 2\, n_{lattice}\, D \quad (m = 1, 2, 3, \ldots) \tag{3.44}$$

The second mechanism is a "microscopic" resonance from a single unit cell of the material. This microscopic resonance occurs when one quarter of the wavelength coincides with the diameter $(2r)$ of a single dielectric well of the index of refraction n_{well}. This effect is conceptually illustrated in Figure 3.27.

The Figure 3.27 shows the PBG formation mechanism for a square well potential. When one half of the optical wavelength fits into the width of the square well, then the transmission of the light is maximum. On the other hand, when one quarter of a wavelength fits into the width of the well, then the least amount of light is transmitted and the maximum amount of light is reflected. This resonance condition can be expressed in the following way:

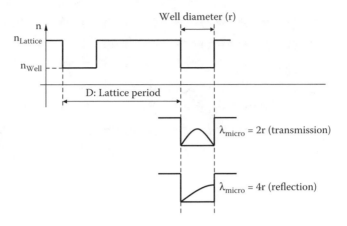

FIGURE 3.27 Formation of PBG.

$$m\lambda_{micro} = 4\, n_{well}\, r \quad (m = 1, 2, 3, \ldots) \tag{3.45}$$

The formation of the PBG is facilitated when the geometrical parameters of the structure are such that both the macroscopic and microscopic resonances occur at the same wavelength, i.e., when $\lambda_{micro} = \lambda_{macro}$. Therefore, the condition for the resonance wavelength for the first order ($m = 1$) can be given by

$$\frac{r}{D} = \frac{1}{2}\frac{n_{lattice}}{n_{well}} \tag{3.46}$$

This expression allows defining the ratio of the potential well diameter ($2r$) to the structure period D. This ratio is called the *filling ratio*.

Under special circumstances, Bragg diffraction prevents a certain range of wavelengths from propagating in any direction inside the crystal. The biggest advantage of the PBG structures lies in the possibility to customize their design to meet the application needs. It is possible to form a narrow-channel waveguide within a PBG structure by removing a row of holes from an otherwise regular pattern (see Figure 3.28). Light will be confined to within the line of defects for wavelengths that lie within the band gap of the surrounding photonic crystal.

For example, in the periodic lattice of air holes, some of the air holes can be made a bit larger or a little smaller than the rest. These features break the periodicity in exactly the same way as incorporating dopant atoms or defects modify the perfect periodicity of a semiconductor crystal. In this way it is possible to create allowed energy levels in the otherwise forbidden energy gap. Therefore, it is possible to introduce a pattern of sharp bends that will either cause the light to be reflected backwards or directed round the bend.

The conventional waveguides that could be used to transmit light signals between different devices are key components in integrated optical circuits.

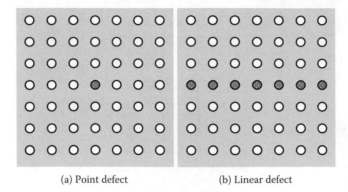

(a) Point defect   (b) Linear defect

FIGURE 3.28 Two-dimensional PBG structure with defects.

However, the development of such small-scale optical interconnects has so far been inhibited by the problem of guiding light efficiently round very tight bends. Conventional waveguides work by the process of TIR. The contrast between the refractive index of the core of the waveguide and the surrounding cladding material determines the maximum radius through which light can be bent without any loss. For example, for conventional waveguides this bend radius is a few millimeters. However, interconnects between components on a dense integrated optical circuit require bend radii of 10 μm or less.

Structural defects in a photonic crystal allow for spatially localizing electromagnetic modes at energies within the gap. In this way, coupling defects together can form waveguides. A waveguide operating at a frequency within a PBG cannot leak — there are no propagating electromagnetic modes in the surrounding photonic crystal capable of carrying energy away. In principle, this allows the fabrication of waveguides that are capable of turning light around corners in a distance on the order of the optical wavelength, or have a minimum bend radius on the order of 100 nm.

Figure 3.29 shows a 90° PBG bend. The incoming light will follow the path of the defects that are implanted into the PBG structure. One may apply Equation 3.43 to estimate the value of the PBG geometrical parameters. For a lattice period of 0.6 μm, with the index of refraction $n_{lattice}$ = 3.4 (e.g., Si in the wavelength communication range of 1.5 μm), and with air holes that are drilled in the lattice (i.e., n_{well} = 1), the required size of the hole diameters will be in the range of 0.15 μm. This example indicates the challenge associated with the fabrication of such structures. High-resolution electron beam lithography and highly anisotropic dry etching are required for the fabrication of photonic crystals.

The fabrication of photonic crystals with three-dimensional periodicity is still a challenge. Two-dimensional crystals consisting of parallel cylinders are easier to realize, and the feasibility of such structures is now well demonstrated at the submicron lengths. For example, a two-dimensional PBG can be used as an OADM. A concept of such a device is illustrated schematically in Figure 3.30.

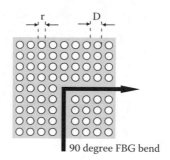

FIGURE 3.29 90° bender.

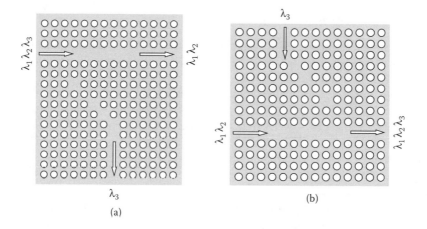

FIGURE 3.30 PBG multiplexers: (a) channel drop filter, (b) channel add filter.

As an important quantum optical consequence, spontaneous emission of excited atoms or molecules inside the PBG structure may be completely inhibited. Moreover, controlled defects in a photonic crystal may result in localized states in the band-gap structure that can act as photon traps (see Figure 3.31).

If there is no available mode for the photon emitted by the decay of an excited atom, then the spontaneous emission is suppressed for photons in the PBG. This characteristic offers novel approaches for manipulation of the electromagnetic field and creation of high-efficiency light-emitting structures. A new class of lasers is being constructed with a highly reduced threshold current and a very low noise level. These are the so-called zero-threshold lasers.

3.14 PHOTONIC CRYSTAL FIBERS

One completely new microstructured material that was recently made is the PBG fiber [3.9]. The PBG fiber (also called *photonic crystal fiber* [PCF]) can be made

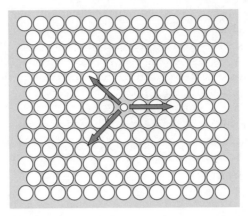

FIGURE 3.31 Photons trapped in a photonic crystal.

with parameters impossible to achieve in standard fibers, which has led some researchers to suggest that the PCF could become the ultimate transmission waveguide for communication links.

One drawback with conventional optical fibers is that different wavelengths of light can travel through the material at different speeds. Over long distances, time delays can occur between signals that are encoded at different wavelengths. This phenomena — known as dispersion — is worse if the core is very large because light can follow different paths or "modes" through the fiber. A pulse of light traveling through such a fiber broadens out, thereby limiting the amount of data that can be sent.

PBG fibers are completely different from conventional fibers because they make use of the unusual properties of two-dimensional periodic microstructured materials in which the solid pure-silica core region is surrounded by a cladding region that contains air holes (see Figure 3.32). These holes form the PBG

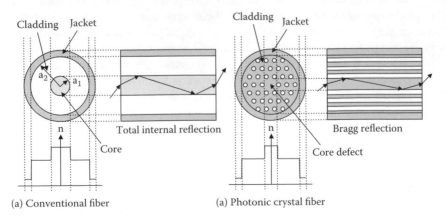

FIGURE 3.32 PBG fiber.

structure. The fiber behaves in many ways like standard step-index fibers, but it has a number of advantages. PCFs are made of undoped silica, which provides very low losses, sustains high powers and temperature levels, and withstands nuclear radiation. Air in the cladding yields a fiber with a huge index step because of the large difference in the refractive index n between air ($n_{air} = 1$) and silica ($n_{Si} = 1.45$). This index difference translates to fibers with numerical apertures (NA) as high as 0.9.

A single PCF fiber may support single-mode operation over wavelength ranges from around 300 nm to better than 2000 nm. Also, because the mode-field area in PCFs can be larger than 300 μm^2, which is several times larger than the 80 μm^2 provided by standard fibers, PCFs can transmit higher powers without running into nonlinear or damaged barriers.

Designers can manipulate the dispersion characteristics of PCFs to create fibers having zero, low, or anomalous dispersion at visible wavelengths. The dispersion can also be flattened. Combining these features with small mode-field areas results in outstanding nonlinear fibers. By altering the pattern of air holes or the materials used, it is possible to manipulate other characteristics of PCFs, such as the single-mode cutoff wavelength, the NA, and the nonlinear coefficient. The design flexibility is very large, and designers can use many different and unusual air hole patterns to achieve specific PCF parameters.

3.15 STIMULATED EMISSION IN SEMICONDUCTORS

Stimulated emission forms the basis for the amplification of electromagnetic radiation. The laser (acronym for light amplification by stimulated emission of radiation) is the fundamental active device of microphotonics. A laser is based on three main components: an active material that is able to generate and to amplify light by stimulated emission of photons, an optical cavity that provides the optical feedback to sustain the laser action, and a pumping mechanism that is able to excite the active material such that population inversion can be achieved. For example, in an injection diode laser the pumping mechanism is provided by carrier injection via a p–n junction and the optical feedback is provided by a Fabry–Perot (FP) cavity. The use of electrical injection makes the device particularly interesting for integration with microelectronics. A brief introduction to the phenomenon of the stimulated emission is presented in this section.

The atoms, ions, and molecules involved in the interaction with electromagnetic waves must be viewed as a quantum mechanical system that possesses numerous discrete energy levels. When the system interacts with an electromagnetic field, transitions between these various energies take place and energy will be emitted or absorbed at a frequency characteristic to the material.

A simple two-energy-level atom is illustrated in Figure 3.33. The energy levels are designated as E_u (upper) and E_l (lower). Electromagnetic radiation may be absorbed or emitted at a characteristic frequency v_{u-l} according to the relation:

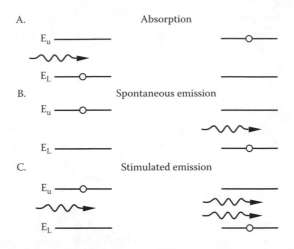

FIGURE 3.33 Interaction of photons with two-level system.

$$\nu_{u-l} = \frac{E_u - E_l}{h} \qquad (3.47)$$

An atom can be raised from a lower to a higher energy level by absorbing a photon (Figure 3.33a). When in the excited state, the atom may decay spontaneously by emitting an electromagnetic wave in a random manner, as shown in Figure 3.33b. There is a possibility that an excited atom may be stimulated to emit radiation by an external photon (see Figure 3.33c). In this later case, the emitted wave is in the same direction as the stimulating photon. This form of stimulated emission is utilized in light-amplifying devices such as lasers. It results when an electron is in a higher level, such as level 2 in Figure 3.33c, and a photon collides with the atom. During the collision the photon stimulates the atom to radiate a second photon having exactly the same energy. Hence, one photon leads to two identical photons, which, in effect, leads to an amplification process. A photon has been gained at the expense of the energy stored within the atom.

The lasing action may be generated if population inversion inside a material is established. For population inversion to occur in the semiconductor, the fraction of occupied states to unoccupied states near the bottom of the conductor band must exceed a similar fraction for states near the top of the valence band. Such energy-level characteristics correspond to a degenerate semiconductor with a Fermi level in the conduction band for an n-type material and in the valence band for a p-type material [3.1].

Inversions in semiconductors can be produced by joining a p-doped semiconductor with an n-doped semiconductor in a way similar to that of producing a p–n junction in a transistor. When a voltage is applied across the junction with the positive voltage on the p side, the electrons are pulled through the junction

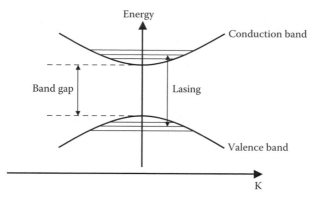

FIGURE 3.34 Lasing effect in a semiconductor.

toward the positive electrode and the holes are attracted to the negative side, producing an electrical current flow across the junction.

Electrons and holes are attracted to each other because of opposite charges and meet within the junction. When they meet, they recombine and emit radiation and can also produce a population inversion. This inversion occurs between energy levels located above and below the semiconductor band gap (see Figure 3.34), i.e., the gap in energy below which the material is transparent. This energy typically corresponds to a wavelength in the infrared, and hence most semi-conductors radiate in the infrared. A very high current density is applied within the junction region to produce a population inversion. The energy band gap determines the laser wavelength.

At very low currents, population inversion does not occur even though recombination radiation is emitted. In fact, such non-laser-like emission is used for production of light-emitting diodes (LEDs).

3.16 THE SAGNAC EFFECT

The *Sagnac effect* states that light beams propagating in opposite directions in a rotating frame experience an optical path length difference. If two pulses of light are sent in opposite directions around a stationary circular loop of radius R, they will travel the same distance at the same speed, so they will arrive at the end point simultaneously. This is illustrated in Figure 3.35a.

Figure 3.35b indicates what happens if the loop itself is rotating during this procedure. The symbol α denotes the angular displacement of the loop during the time required for the pulses to travel once around the loop. For any positive value of α, the pulse traveling in the same direction as the rotation of the loop must travel a slightly greater distance than the pulse traveling in the opposite direction. As a result, the counterrotating pulse arrives at the "end" point slightly earlier than the corotating pulse. Quantitatively, if we let ω denote the angular speed of the loop, then the circumferential tangent speed of the end point is $v = \omega R$, and

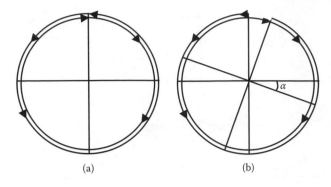

(a)   (b)

FIGURE 3.35 Sagnac effect.

the sum of the speeds of the wave front and the receiver at the "end" point is
$c - v$ in the corotating direction and $c + v$ in the counterrotating direction. Both
pulses begin with an initial separation of $2\pi R$ from the end point, so the difference
between the travel times is

$$\Delta t = 2\pi R \left(\frac{1}{c - v} - \frac{1}{c + v} \right) = \frac{4\pi R v}{c^2 - v^2} = \frac{4A\omega}{c^2 - v^2} \qquad (3.48)$$

where $A = \pi R^2$ is the area enclosed by the loop. If the path can be folded over
N times, such as using N coils of fiber, the Sagnac effect can be cumulatively
enhanced.

This phenomenon applies to any closed loop, not necessarily circular. For
example, suppose a beam of light is split by a semitransparent mirror into two
beams, and those beams are directed in a square path around a set of mirrors in
opposite directions as shown in Figure 3.36. Just as in the case of the circular
loop, if the apparatus is unaccelerated, the two beams will travel equal distances
around the loop and arrive at the detector simultaneously and in phase. However,
if the entire device (including the source and detector) is rotating, the beam
traveling around the loop in the direction of rotation will have farther to go than
the beam traveling counter to the direction of rotation, because during the period
of travel the mirrors and detector will all move (slightly) toward the counter-
rotating beam and away from the corotating beam. Consequently, the beams will
reach the detector at slightly different times and slightly out of phase, producing
optical interference "fringes" that can be observed and measured.

Sagnac first demonstrated this effect in 1911, and it is now called the Sagnac
effect. Because of the incredible precision of interferometric techniques, devices
such as this are capable of detecting and measuring extremely small amounts of
absolute rotation.

The invention of lasers has led to practical small-scale devices for measuring
rotation. These devices, often called *laser gyroscopes*, were first introduced in

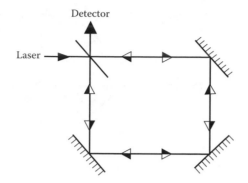

FIGURE 3.36 Sagnac effect in a square path.

1963, and have been steadily improved ever since. Today they are routinely used in guidance and navigation systems for commercial airliners, nautical ships, spacecraft, and in many other applications. The best such devices currently available are capable of detecting rotation rates as slight as 0.00001°/h.

3.17 EVANESCENT WAVES

TIR occurs when light traveling through a medium is reflected from the interface with a lower density medium at an angle exceeding the critical angle (see Section 3.6).

As has been mention in Section 3.7, a more detailed theoretical analysis of the interface shows that an electromagnetic wave is present beyond the TIR interface. This field travels along the boundary, but it does not travel away from the boundary. Its amplitude decreases as the exponential function of the distance from the interface. This effect is schematically illustrated in Figure 3.37. The

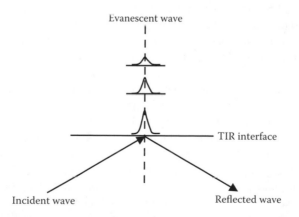

FIGURE 3.37 Evanescent wave at a TIR interface.

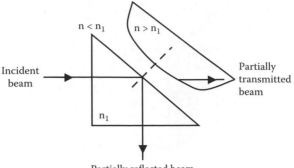

FIGURE 3.38 Partial internal reflection.

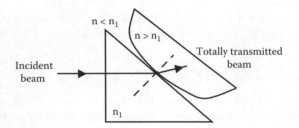

FIGURE 3.39 Refraction at a TIR interface.

distance over which the wave amplitude falls off is often less than the wavelength of the light itself. If nothing interferes with the wave, then its intensity tends to go to zero. Therefore, this wave does not actually carry any energy out of the medium. This wave is known as an *evanescent wave*.

If some other material is placed very close to the TIR interface, then the evanescent wave can itself become a propagating electromagnetic wave inside this material. This effect can be demonstrated experimentally by bringing another material (with a higher index of refraction) very close to the TIR surface of a prism. This is illustrated in Figure 3.38 and Figure 3.39.

As the surface of the other material is brought into close proximity to the TIR surface, there occurs some partial transmission across the interface (Figure 3.38). This demonstrates that there is an electromagnetic energy field outside the prism itself that allows for interaction of the two surfaces across the small air gap.

If the surfaces are in contact, as is shown in Figure 3.39, then the entire optical beam is transmitted as a refracted beam.

The light-guiding effect occurs when light propagates along or is constrained by the physical boundaries of a waveguide. As has been described in Section 3.7, this may be observed, for example, in a single-mode fiber in which the denser core has a refractive index n_1 and the cladding is less dense with a refractive index n_2 (see Figure 3.40).

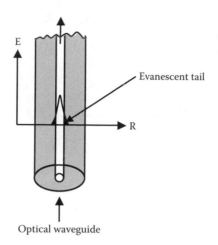

FIGURE 3.40 Evanescent tail.

In such a case, the electromagnetic energy is distributed between the core and the cladding. That part of the energy that is in the cladding is called the *evanescent tail*.

If one considers two parallel single-mode waveguides in close proximity, then it is possible that the evanescent tails of each waveguide may have considerable overlap. In such a case the electromagnetic wave energy is shared between the two waveguides. This interaction is dependent on waveguide separation and their interaction length. This effect is known as *evanescent wave coupling*, and it is used in a number of photonic devices.

For example, removing part of the cladding and bringing the cores of two fibers close together so their polished faces are in contact makes an evanescent wave coupler. By this process, the two cores behave as if they were contained within the same cladding.

3.18 SMART THIN-FILM COATINGS

Several so-called smart materials have been identified that can exhibit change from metallic to insulator behavior (i.e., metal–insulator transition) as a result of composition (doping), electric field, temperature, or the application of pressure. These mechanisms can cause a change in the electron occupancy of allowed band levels to shift the Fermi energy from forbidden levels (nonmetallic behavior) to conduction states (metallic behavior).

There are two main mechanisms to facilitate a metal–insulator transition: (1) localization of conduction electrons and (2) a relative energy shift in the overlap of occupied and empty electron bands. Localization of the electron wavefunction in a semiconductor or metal can be induced by introducing disorder and/or by

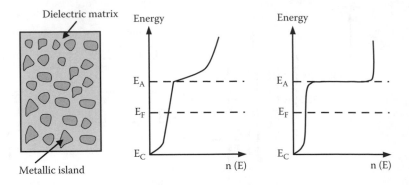

FIGURE 3.41 Formation of localized states.

spatially isolating metallic clusters in a dielectric matrix (see Figure 3.41). Such localization results in the formation of a band of localized states at the bottom of the conduction band, below energy E_A, termed the *mobility edge*. Electron states below E_A are not spatially extended. Within one of these states, the electron wavefunction decays exponentially with distance from the localized site. The resulting electrons require additional thermal energy, W_{hop}, on the order 0.1 to 0.2 eV to facilitate hopping from site to site. This results in a temperature-activated electron mobility of the form:

$$n_n(T) = n_{hop} \exp\left(\frac{-W_{hop}}{k_B T}\right) \tag{3.49}$$

where n_{hop} depends on the spatial overlap of the electron wavefunctions at neighboring localized sites and is typically several orders of magnitude smaller than the electron mobility in extended states. The electron occupancy of the states follows Fermi–Dirac statistics and is determined by the position of E_F. If $E_F < E_A$, the material will exhibit nonmetallic behavior with a reduced optical reflectivity. If $E_F > E_A$, the material will exhibit metallic behavior with a relatively higher optical reflectivity. By shifting the position of E_F above E_A, for example, by increasing the temperature, one can induce a transition from nonmetallic to metallic behavior.

Many of the transition metals such as W, Mn, La, and V are characterized by partially-filled d orbitals, which contribute to the metallic bonding and electrical conduction. The transition metals can readily form a variety of complexes involving the d orbitals (two e_g and three t_{2g} orbitals) [3.10]. Chemical bonding of the transition metal can produce an energy splitting, Δ_o, of the d orbitals into a higher energy e_g-orbital pair and a lower energy t_{2g}-orbital triplet because of electrostatic electron–electron interactions (see Figure 3.42).

Depending on the resulting electron configuration and chemical structure, the resulting complex can occupy t_{2g} orbitals and relatively empty e_g orbitals at lower

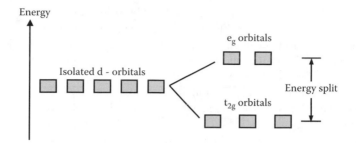

FIGURE 3.42 Energy splitting of the *d* orbitals.

temperatures, resulting in nonmetallic characteristics. The energy splitting of the *d* orbitals because of the formation of the chemical complex results in an effective band gap, Δ_o, for optical absorption and conduction. As the temperature increases, electrons from filled lower *d* orbitals are promoted to the empty e_g orbitals, resulting in conduction electrons and holes.

As has been shown in Equation 3.28, the index of refraction *n* of a metal depends on the free-electron density *N*, the effective mass m_e of the free electrons, and electron plasma frequency ω_0.

The metal is transparent for $\omega > \omega_0$ and is reflective below the plasma frequency. By introducing localization and a resulting effective band gap for free carriers, both n_e and m_e can be made to have exponential dependences on temperature and electric field.

Vanadium oxide (VO_n) exhibits one of the largest observed variations in electrical and optical characteristics because of metal–insulator transition. The transition temperature increases with the oxygen content, varying from 126 K for VO, to 140 K for V_2O_3, and to 341 K for VO_2. The metal–insulator transition in V_xO_n is associated with a change in structure from an orthogonal rutile structure above the transition temperature to a monoclinic structure below the transition temperature (see Figure 3.43).

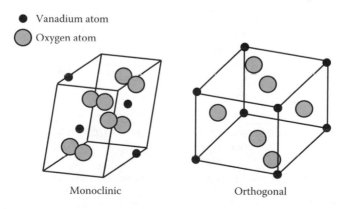

FIGURE 3.43 Structural transition in vanadium oxide.

Thin VO_2 films exhibit a three-to-four order-of-magnitude change in resistance at the metal–insulator transition in films as thin as 670 Å. The transition temperature varies between 58°C and 68°C, depending on the O_2 partial pressure. These results indicate that the metal–insulator transition characteristics of thin VO_2 films can be comparable to those of bulk samples.

The broadband optical transmittance characteristics of VO_2 on Si in the insulating and metallic states as measured by FT-IR are shown in Figure 3.44. The VO_2 coating exhibits in the insulating state a very high transmittance that extends to beyond 10 μm in the infrared [3.11]. The optical absorption by the VO_2 is very low in this insulating state, indicating a relatively clean band gap, free of defect states. In the metallic state, the optical transmittance of VO_2 on Si is very low, typically below 0.5%. This indicates that a high-quality smart coating can provide optical switching over a very broad spectral range.

Doping the vanadium oxide coating with tungsten or molybdenum, ($V_{(1-x)}M_xO_2$, where M is Mo or W) can lower the switching temperature; for example, 3.5 % Mo decreases the phase transition temperature from 67°C to 40 °C, whereas 1.5 % W decreases the phase transition down to 35°C.

It has been demonstrated that it is technically feasible to induce a very fast metal–insulator transition (~1 nsec) by using an electric field [3.12]. This effect can be used as the basis for voltage-controlled thin-film switching. This result indicates that the metal–insulator transition in VO_2 is related to a critical electron density, similar to the Mott transition. Because the driving mechanism is an electric field effect, the power requirement is very moderate.

The transition capability of thin films makes it very attractive for switching and modulating of optical signals in photonic devices.

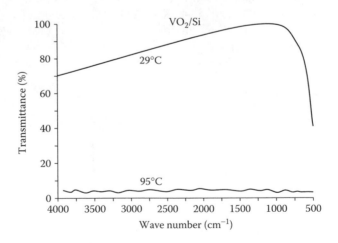

FIGURE 3.44 Optical transmittance of VO_2 on Si in the insulating (29°C) and metallic (95°C) states.

3.19 QUANTUM PHOTONIC EFFECTS

Ever since its invention in the 1920s, quantum mechanics has given rise to countless discussions about its meaning and about how to interpret the theory correctly. These discussions have been focused on issues such as the Einstein–Podolsky–Rosen (EPR) paradox, quantum nonlocality, and the role of measurement in quantum mechanics. In recent years, however, research into the very foundations of quantum mechanics has also led to a new field — quantum photonics. In the last few years there has been an explosion of theoretical and experimental research, creating a fundamental new discipline of quantum photonics. Quantum photonics allows the construction of qualitatively new types of logic gates, absolutely secure cryptosystems, high-capacity communication systems, and fast quantum computers.

The Heisenberg uncertainty principle dictates that it is not possible to measure accurately a particle's exact state without effecting a change in that state. According to this principle, the more accurately a particle's state is measured, the more the measuring process itself disturbs it. Therefore, the uncertainty principle of quantum mechanics forbids any measuring process from extracting all the information in an atom or a particle. However, there is an alternative to finding out the full state of a particle by causing it to interact with another particle in a known state. This subtle and indirect measurement is based on the *EPR correlation*.

EPR correlation is more commonly known as the entanglement effect. The entanglement is a quantum effect in which the states of two or more particles are interconnected with each other even though the individual particles may be spatially at different locations. For example, it is possible to have two particles in a single quantum state in such a way that when one is observed to be spin-up, the other one will always be observed to be spin down, and vice versa. This observation can be made despite the fact that it is impossible to predict which set of spin positions will be observed. As a result, measurements performed on one particle seem to be instantaneously influencing other particle entangled with it. It has to be pointed out, however, that information about the states cannot be transmitted through the entanglement faster than the speed of light.

This effect was discussed in a famous paper published in 1930 by Albert Einstein, Boris Podolsky, and Nathan Rosen. Albert Einstein was quoted as describing this effect as "spooky action at a distance" because of its very nature, i.e., altering the quantum state of one particle of an entangled pair causes the other particle to be affected without any communication between them. The quantum entanglement essentially allows two particles to behave as one, regardless of how far apart they are. Experiments on photons and other particles have repeatedly demonstrated existence of such correlations. Photons can be entangled so that if one is vertically polarized, for instance, then the other photon in the pair is always horizontally polarized.

In 1960 John Bell showed that a pair of entangled particles, which were once in contact but then separated, could exhibit behavior that indicated strong correlation. The simplest possible examples of entanglement are represented by the so-called Bell states. The Bell state is defined as a maximally entangled quantum state of two states. A term "qubit" has been introduced to describe this effect in a more precise manner. A qubit is a unit used in quantum computation.

3.19.1 Qubits

An elementary quantity of information is measured by a bit. The bit can take on one of two values, either "0" or "1." Therefore, a bit can be represented by, for example, a certain voltage level, as is implemented in logical circuits. In practice, it is desirable to have the two states separated by a relatively large barrier so that the values of the bit are clearly distinguishable.

The novel feature of quantum photonics is that a quantum system can be in a superposition of different states. This means that the quantum bit can be in both the 0 state and in the 1 state at the same time. In 1995 Ben Schuhmacher of Kenyon College in the U.S. coined the word "qubit" to describe a quantum bit.

A *qubit* (quantum bit) is a unit defined with respect to a two-level quantum system, whose two basic states — described in Dirac notation — are $|0\rangle$ and $|1\rangle$. A qubit state is a linear quantum superposition of those two states:

$$|\Psi\rangle = a|0\rangle + b|1\rangle \qquad (3.50)$$

where a and b are called the amplitudes of the state. If both a and b are not zero, then the above equation represents a qubit in a superposition of 0 and 1 state.

The qubit continuous nature can be replicated by any analog quantity. It is interesting to note that this state is different from the state of a classical bit, which can only take the value 0 or 1 (see Figure 3.45).

Therefore, a qubit is a unit of information that can be expressed as 0, 1, or superposition of 0 and 1. Qubits can be shown graphically by using a sphere (Figure 3.46). The qubit states $|0\rangle$ and $|1\rangle$ are on the poles of the sphere. Any

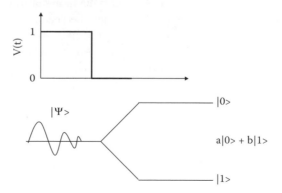

FIGURE 3.45 Classical and quantum bits.

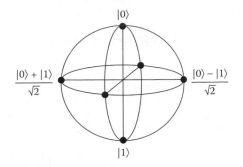

FIGURE 3.46 Generalized Poincaré sphere representing qubits.

superposition of $|0\rangle$ and $|1\rangle$ is located on a shell of generalized Poincaré sphere [3.13].

A tensor product can represent the state of n qubits:

$$|0\rangle \otimes |1\rangle = |01\rangle$$

For example, in a case of two qubits $|\Psi\rangle$ and $|\Phi\rangle$ that are described by

$$|\Psi\rangle = a|0\rangle + b|1\rangle$$

$$|\Phi\rangle = c|0\rangle + d|1\rangle$$

their state may be expressed by using the tensor product:

$$|\Psi\rangle \otimes |\Phi\rangle = ac|00\rangle + ad|01\rangle + bc|10\rangle + bd|11\rangle$$

It has to be pointed out that the state that is the result of the tensor product operation is separable and a measurement of one of the initial state will not affect the other state. In other words, the tensor product operation does not lead to particle entanglement.

In general, an entangled state of two-level system may be expressed as

$$|\Psi\rangle = \frac{1}{\sqrt{2}}\left(|0\rangle_1 |1\rangle_2 + e^{i\theta}|1\rangle_1 |0\rangle_2\right)$$

where θ is the phase shift.

3.19.2 BELL STATES

A particular state of two particles A and B may be described in the following way:

$$\left|\Phi_{AB}^{+}\right\rangle = \frac{1}{\sqrt{2}}\left(\left|0\right\rangle_{A}\left|0\right\rangle_{B} + \left|1\right\rangle_{A}\left|1\right\rangle_{B}\right) \tag{3.51}$$

where Φ is the probability of occurrence for the given states of the particles A and B.

This expression means that the qubit A can be in the state 0 as well as in the state 1. If measured, then one would find out that its value would be perfectly random, i.e., either possibility having 50% probability. But if the qubit B is measured, then the result would be the same as for the qubit A, i.e., its value would be also perfectly random with either possibility having 50% probability. However, if both measurement results are compared, then it will be seen that they are correlated. Therefore there is something missing in the description of the qubit pair given by the preceding equation. It has been assumed that there is an additional parameter that may be called a *hidden variable*. This hidden variable is related to the fact that quantum theory allows qubits to be in quantum superposition, i.e., in $\left|\Phi^{+}\right\rangle$ and $\left|\Phi^{-}\right\rangle$ simultaneously in either of the following example states:

$$\left|\Phi^{+}\right\rangle = \frac{1}{\sqrt{2}}\left(\left|0\right\rangle + \left|1\right\rangle\right) \quad \text{or} \quad \left|\Phi^{-}\right\rangle = \frac{1}{\sqrt{2}}\left(\left|0\right\rangle - \left|1\right\rangle\right) \tag{3.52}$$

Therefore, a full set of states for two qubits can be presented by the following 4 expressions:

$$\left|\Phi_{AB}^{+}\right\rangle = \frac{1}{\sqrt{2}}\left(\left|0\right\rangle_{A}\left|0\right\rangle_{B} + \left|1\right\rangle_{A}\left|1\right\rangle_{B}\right)$$

$$\left|\Phi_{AB}^{-}\right\rangle = \frac{1}{\sqrt{2}}\left(\left|0\right\rangle_{A}\left|0\right\rangle_{B} - \left|1\right\rangle_{A}\left|1\right\rangle_{B}\right) \tag{3.53}$$

$$\left|\Psi_{AB}^{+}\right\rangle = \frac{1}{\sqrt{2}}\left(\left|0\right\rangle_{A}\left|1\right\rangle_{B} + \left|1\right\rangle_{A}\left|0\right\rangle_{B}\right)$$

$$\left|\Psi_{AB}^{-}\right\rangle = \frac{1}{\sqrt{2}}\left(\left|0\right\rangle_{A}\left|1\right\rangle_{B} - \left|1\right\rangle_{A}\left|0\right\rangle_{B}\right)$$

These expressions describe the so-called Bell states that are known as the maximally entangled two-qubit states. It can be seen from the preceding equation that an entangled state is one whose vector $|\Psi\rangle$ cannot be written as a product of the single states of its subsystems.

3.19.3 EPR CORRELATION

The working of the EPR correlation in the case of two photons may be summarized in the following way. The quantum state of photon may be described by

$$|\Psi\rangle = a|\leftrightarrow\rangle + b|\updownarrow\rangle = a|H\rangle + b|V\rangle \tag{3.54}$$

where H and V indicate the two possible orthogonal quantum states, e.g., the horizontal state and vertical state. If two photons A and B are brought together and entangled, then according to Equation 3.52 the quantum state of these two entangled photons that are in mutually orthogonal states can be described by the following equation:

$$|\Psi_{AB}^-\rangle = \frac{1}{\sqrt{2}}\left(|H_A\rangle|V_B\rangle - |H_B\rangle|V_A\rangle\right) \tag{3.55}$$

This equation shows that any changes in the quantum state of one photon will affect the quantum state of the other photon, regardless of their locations. This is schematically illustrated in Figure 3.47, where A and B indicate the photons before the entanglement and $A^{(b)}$ and $B^{(a)}$ are the entangled photons.

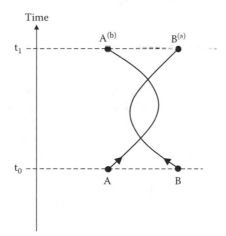

FIGURE 3.47 Single-stage entanglement.

Introducing the third photon C may show another aspect of the EPR correlation. This is known as *entanglement extension*. It describes a process in which an entangled state can be transmitted to another state.

The quantum state of the photon C may be described by

$$|\Psi_C\rangle = a|H_C\rangle + b|V_C\rangle$$

Let assume that the previously entangled photons A and B have been separated and placed at different locations. Each photon must be kept isolated from its surroundings to ensure that the entanglement is maintained. Now it is possible to entangle the photon C with the photon $A^{(b)}$ or $B^{(a)}$. For example, the entanglement of the photons A and C will lead to the following multiple quantum states of equal probability of occurrence:

$$|\Psi_{AC}'^{\pm}\rangle = \frac{1}{\sqrt{2}}\left(|H_A\rangle|V_C\rangle \pm |H_C\rangle|V_A\rangle\right) \tag{3.56a}$$

$$|\Phi_{AC}'^{\pm}\rangle = \frac{1}{\sqrt{2}}\left(|H_A\rangle|H_C\rangle \pm |V_C\rangle|V_A\rangle\right) \tag{3.56b}$$

Therefore, the three photons A, B, and C form a quantum system that can be described by

$$|\Psi_{ABC}\rangle = \frac{a}{\sqrt{2}}\left(|H_C\rangle|H_A\rangle|V_B\rangle - |H_C\rangle|H_B\rangle|V_A\rangle\right) + \tag{3.57}$$

$$\frac{b}{\sqrt{2}}\left(|V_C\rangle|H_A\rangle|V_B\rangle - |V_C\rangle|H_B\rangle|V_A\rangle\right)$$

The entangled system of three photons is shown in Figure 3.48. This two-stage entanglement has led to a situation in which the photon B is correlated to the photon C despite the fact that they have never been in contact.

Another example of an extended (multistage) entanglement is illustrated in Figure 3.49. At the starting point t_0, there are two pairs of previously entangled photons, i.e., the first pair $A^{(b)}$ and $B^{(a)}$ and the second pair $C^{(d)}$ and $D^{(c)}$. These four photons can be mutually correlated by entangling photon $B^{(a)}$ with $C^{(d)}$. This means that now the photons $A^{(b)}$ and $D^{(c)}$ are entangled as well although the two were never in contact.

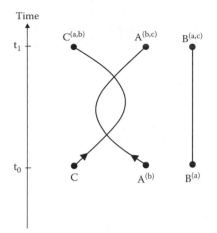

FIGURE 3.48 Two-stage entanglement.

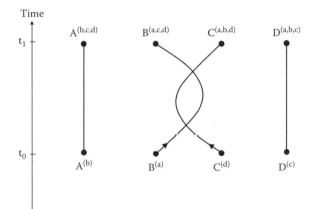

FIGURE 3.49 Extended entanglement.

3.19.4 QUANTUM GATES

A number of so-called quantum gates have been devised that are analogues to classical logic gates. These gates allow for operations and manipulations of qubits.

The basic quantum gates operate on one or two qubits. The gates can be described by using matrices. Among the most common quantum gates are: the quantum identity operator I, the quantum "NOT gate" (also known as Pauli-X gate), the Pauli-Z gate, the Pauli-Y gate, and the Hadamard gate. The working of the quantum gates may be described in the following way.

A quantum state can be represented by a vector:

$$|0\rangle = \begin{bmatrix} 1 \\ 0 \end{bmatrix} \text{ and } |1\rangle = \begin{bmatrix} 0 \\ 1 \end{bmatrix} \tag{3.58}$$

Therefore, a general form of the Equation 3.53 may be presented as

$$|\Psi\rangle = a|0\rangle + b|1\rangle = \begin{bmatrix} a \\ b \end{bmatrix} \tag{3.59}$$

The quantum identity operator I is defined as

$$I = \begin{pmatrix} 1 & 0 \\ 0 & 1 \end{pmatrix}$$

which is equivalent to

$$I|\Psi\rangle = |\Psi\rangle$$

The quantum NOT gate is defined as follows:

$$\text{NOT}|0\rangle = |1\rangle, \text{ and NOT}|1\rangle = |0\rangle$$

And the NOT gate matrix (i.e., the Pauli-X matrix) may be presented as

$$X \equiv \begin{bmatrix} 0 & 1 \\ 1 & 0 \end{bmatrix}$$

If the quantum NOT gate is applied to a superposition state expressed by Equation 3.58, then that operation will lead to the following result:

$$\text{NOT}\left[a|0\rangle + b|1\rangle\right] = a|1\rangle + b|0\rangle \tag{3.60}$$

or, when applied to a qubit, its operation may also be presented in a matrix form:

$$X\left(a|0\rangle + b|1\rangle\right) = \begin{bmatrix} 0 & 1 \\ 1 & 0 \end{bmatrix} \times \begin{bmatrix} a \\ b \end{bmatrix} = \begin{bmatrix} b \\ a \end{bmatrix} \tag{3.61}$$

It is important to note that the Pauli-X gate works on one qubit only. The same applies to the Pauli gates Z and Y. The quantum gates Z and Y are defined as follows:

$$Z \equiv \begin{bmatrix} 1 & 0 \\ 0 & -1 \end{bmatrix}$$

$$Y \equiv \begin{bmatrix} 0 & -i \\ i & 0 \end{bmatrix} \equiv i\,X\,Y$$

The Hadamard gate (H) also operates on a single qubit. The following matrix represents the H gate:

$$H \equiv \frac{1}{\sqrt{2}} \begin{bmatrix} 1 & 1 \\ 1 & -1 \end{bmatrix}$$

For example, if photon polarization is chosen as the qubit basis, then it is possible to have the horizontally polarized state of a photon $|0\rangle$ and the vertically polarized photon state $|1\rangle$. If the plane in which the polarization is measured is rotated by 45°, then the states in this new basis $|0'\rangle$ and $|1'\rangle$ are related to the original states in the following way:

$$|0'\rangle = \left(\frac{1}{\sqrt{2}}\right)(|0\rangle + |1\rangle)$$

$$|1'\rangle = \left(\frac{1}{\sqrt{2}}\right)(|0\rangle - |1\rangle)$$

The preceding expressions describe a particular case of a Hadamard transformation.

There is another type of gates that also operate on a single qubit. They affect a phase shift in a qubit state. The following 2×2 matrix represents the phase gate:

$$R(\theta) \equiv \begin{bmatrix} 1 & 0 \\ 0 & e^{2\pi i \theta} \end{bmatrix}$$

where θ is the phase shift.

The following expressions summarize the operations of the basic quantum gates:

$$\text{Not gate:} \quad a|0\rangle + b|1\rangle \quad \xrightarrow{\quad X \quad} \quad b|0\rangle + a|1\rangle$$

This operation is equivalent to $X|\Psi\rangle = |\Psi \oplus 1\rangle$ (i.e., bit flip).

$$\text{Z gate:} \quad a|0\rangle + b|1\rangle \quad \xrightarrow{\quad Z \quad} \quad a|0\rangle - b|1\rangle$$

This operation is equivalent to $Z|\Psi\rangle = (-1)^{\Psi}|\Psi\rangle$ (i.e., phase flip).

$$\text{Hadamard gate:} \quad a|0\rangle + b|1\rangle \quad \xrightarrow{\quad H \quad} \quad a\frac{|0\rangle + |1\rangle}{\sqrt{2}} + b\frac{|0\rangle - |1\rangle}{\sqrt{2}}$$

$$\text{Phase gate:} \quad a|0\rangle + b|1\rangle \quad \xrightarrow{\quad P \quad} \quad a|0\rangle + e^{2\pi i \Theta}b|1\rangle$$

The so-called controlled gate is a gate that can act upon a two-qubit quantum state. This operation is such that the first qubit is used as a control qubit, whereas the other one is a data qubit.

A graphical representation of the controlled-U gate is shown in Figure 3.50. Let assume that the U gate operation on single qubit and is represented by

$$U \equiv \begin{bmatrix} x_{00} & x_{01} \\ x_{10} & x_{11} \end{bmatrix}$$

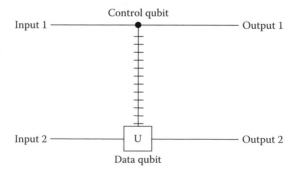

FIGURE 3.50 Controlled-U gate.

Therefore, the controlled gate operations may be described as

$$|00\rangle \rightarrow |00\rangle$$

$$|01\rangle \rightarrow |01\rangle$$

$$|10\rangle \rightarrow |1\rangle U |0\rangle = |1\rangle \left(x_{00} |0\rangle + x_{01} |1\rangle \right)$$

$$|11\rangle \rightarrow |1\rangle U |1\rangle = |1\rangle \left(x_{10} |0\rangle + x_{11} |1\rangle \right)$$

The preceding set of operations may be represented in a matrix form:

$$C(U) \equiv \begin{bmatrix} 1 & 0 & 0 & 0 \\ 0 & 1 & 0 & 0 \\ 0 & 0 & x_{00} & x_{01} \\ 0 & 0 & x_{10} & x_{11} \end{bmatrix}$$

Any operation of a quantum computer can be reduced to a set of basic quantum gates. The gates that are contained in such a set are called *universal quantum gates*. A common set of universal gates that can be applied to a two-qubit system includes: the Hadamard gate (H), the phase shift gate (R), and a special case of the controlled-U gate that is known as the controlled-NOT gate (CNOT). The matrix for the CNOT gate has the following form:

$$CNOT \equiv \begin{bmatrix} 1 & 0 & 0 & 0 \\ 0 & 1 & 0 & 0 \\ 0 & 0 & 0 & 1 \\ 0 & 0 & 1 & 0 \end{bmatrix}$$

i.e., a controlled-U gate with

$$x_{00} = 0$$

$$x_{01} = 1$$

$$x_{10} = 1$$

$$x_{11} = 0$$

Therefore, the operation of the CNOT gate can be described in the following way:

$$CNOT\left|00\right\rangle \rightarrow \left|00\right\rangle$$

$$CNOT\left|01\right\rangle \rightarrow \left|01\right\rangle$$

$$CNOT\left|10\right\rangle \rightarrow \left|1\right\rangle U\left|0\right\rangle = \left|11\right\rangle$$

$$CNOT\left|11\right\rangle \rightarrow \left|1\right\rangle U\left|1\right\rangle = \left|10\right\rangle$$

The operation of the CNOT gate is generalization of the classical XOR logical operator, which results in true if one of the operands (not both) is true. It is usually indicated by the symbol $\oplus$ (Exclusive OR). The following expression means A or B but not both:

$$CNOT\left|A, B\right\rangle = \left|A, B \oplus A\right\rangle$$

A graphical representation of the CNOT gate is illustrated in Figure 3.51. The CNOT gate is invertible gate, i.e., there is no information loss related to its action.

Let us consider a particular case in which the inputs are limited to the states $\left|1\right\rangle$ and $\left|0\right\rangle$. In such a case, the gate operation on the lower input line is only operative when there is $\left|1\right\rangle$ on the upper input. If there is $\left|0\right\rangle$ on the upper input, then the lower qubit passes through unchanged. The CNOT gate performance, for this particular case, is summarized in the truth table (see Table 3.1).

The performance of the CNOT gate may be summarized in the following way: if the input states are in basis states, the output 1 state is the same as the

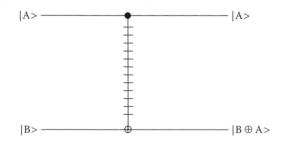

FIGURE 3.51 CNOT gate.

TABLE 3.1
Truth Table for CNOT Gate

Input 1 (Control)	Input 2 (Data)	Output 1 (Control)	Output 2 (Data)
$\lvert 0 \rangle$	$\lvert 0 \rangle$	$\lvert 0 \rangle$	$\lvert 0 \rangle$
$\lvert 0 \rangle$	$\lvert 1 \rangle$	$\lvert 0 \rangle$	$\lvert 1 \rangle$
$\lvert 1 \rangle$	$\lvert 0 \rangle$	$\lvert 1 \rangle$	$\lvert 1 \rangle$
$\lvert 1 \rangle$	$\lvert 1 \rangle$	$\lvert 1 \rangle$	$\lvert 0 \rangle$

input 1 and the output state 2 is the XOR of the two input states 1 and 2. This is called the *controlled-NOT* gate because the state carried by the control qubit is not disturbed, whereas the state carried by the data qubit is changed only if the state of the control qubit is $\lvert 1 \rangle$.

3.19.5 QUANTUM CIRCUITS

Quantum circuits are analogous to classical logic circuits. A quantum circuit is an array that consists of a set of quantum gates; an input is fed into it, resulting in the output. Using the quantum gates allows for building a number of functional circuits. Any multiple qubit logic circuit may be composed from CNOT and a single qubit gate. For example, let us analyze the performance of the following circuit that consists of three CNOT gates (see Figure 3.52).

We can analyze the performance of this circuit on qubits whose states are described by $\lvert A \rangle$ and $\lvert B \rangle$. The first CNOT gate will result in the states $\lvert A \rangle$ (upper link) and $\lvert B \oplus A \rangle$ (lower link). The second CNOT gate converts the upper qubit into the state $\lvert A \oplus (B \oplus A) \rangle = \lvert B \rangle$, while the lower qubit passes unchanged as $\lvert B \oplus A \rangle$. After passing through the third and the last CNOT gate, the qubit is changed to $\lvert (B \oplus A) \oplus B \rangle = \lvert A \rangle$. The final result is illustrated in Figure 3.53. This

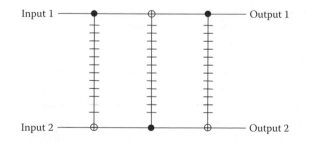

FIGURE 3.52 CNOT-based quantum circuit.

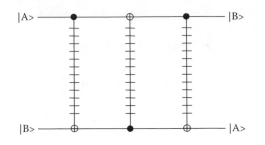

FIGURE 3.53 Qubit swap circuit.

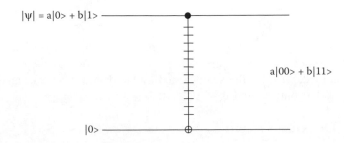

FIGURE 3.54 Entanglement circuit.

result shows that the states of the qubits have been swapped. This type of quantum circuit is called *swap circuit*.

Another quantum circuit that consists of the CNOT gates is illustrated in Figure 3.54. The initial qubit states are: $(1) |\Psi\rangle = a|0\rangle + b|1\rangle$ and $(2) |\Psi\rangle = |0\rangle$.

After passing through the CNOT gate, the resulting qubits state is described by $|\Psi'\rangle = a|00\rangle + b|11\rangle$. This indicates that the two qubits have been entangled. This type of particle manipulation by the CNOT gate corresponds to the process has been shown in Figure 3.47.

3.19.6 BELL MEASUREMENTS

Some of the applications require a set of classical measurements. It is the combination of manipulation of the particles, classical measurements that allows for construction of various devices that may be used in the quantum computing, and quantum communications.

The Bell measurement is a joint quantum mechanical measurement of two qubits that determines which of the four Bell states (see Equation 3.52) the two qubits are in. It has to be mentioned that if the qubits were not in a Bell state before, then the measurement itself would project them into one of the Bell states. The two measured particles would then be in a known entangled state, specified by the measurement result. Therefore, the Bell measurement itself is an entangling operation. In experiments that use photons as qubits, the Bell measurements can only be partly realized as only two of the Bell states can be distinguished with optical techniques.

Let assume that the state of a qubit is described by

$$|\Psi\rangle = a|0\rangle + b|1\rangle$$

The measurement may provide the one of following set of data:

$M = 0$ with the probability $|a|^2$

$M = 1$ with the probability $|b|^2$

where $|a|^2 + |b|^2 = 1$ and M is the classical bit representing the measurement result (see Figure 3.55)

The measurement can be used to distinguish between orthogonal states (e.g., polarization states). If one of the qubits is in one of the orthogonal states, then the measurement can determine which state it is. However, the measurement M changes the state of the system and cannot provide a snapshot of the entire system. After the measurement, the original superposition state cannot be recovered.

Some of quantum circuits may be used to perform the measurements. For example, the quantum circuit shown in Figure 3.56 illustrates a schematic arrangement of the Hadamard gate and the CNOT gate that may be used for the Bell measurements.

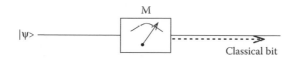

FIGURE 3.55 Quantum measurement.

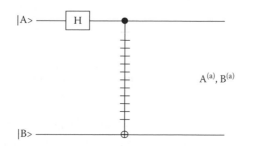

FIGURE 3.56 Bell measurement circuit.

TABLE 3.2
Bell Measurements Matrix

A	B	Result
$|0\rangle$	$|0\rangle$	$\frac{1}{\sqrt{2}}(|00\rangle + |11\rangle)$
$|0\rangle$	$|1\rangle$	$\frac{1}{\sqrt{2}}(|01\rangle + |10\rangle)$
$|1\rangle$	$|0\rangle$	$\frac{1}{\sqrt{2}}(|00\rangle - |11\rangle)$
$|1\rangle$	$|1\rangle$	$\frac{1}{\sqrt{2}}(|01\rangle - |10\rangle)$

Table 3.2 summarizes the Bell measurements matrix for the circuit presented in Figure 3.56.

Figure 3.57 shows a particular Bell measurement for the first case, i.e., for the states $|0\rangle$, $|0\rangle$. After passing through the Hadamard gate the state is described by $\frac{1}{\sqrt{2}}(|0\rangle + |1\rangle)$. After passing through the CNOT gate, the state of the qubits may be described as

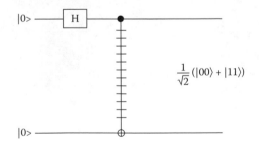

FIGURE 3.57 Bell circuit operation.

$$\text{CNOT}\left(\frac{|0\rangle+|1\rangle}{\sqrt{2}}|0\rangle\right) = \left(\frac{|00\rangle+|10\rangle}{\sqrt{2}}\right) = \frac{1}{\sqrt{2}}\left(CNOT|00\rangle + CNOT|10\rangle\right) =$$

$$= \frac{1}{\sqrt{2}}\left(|00\rangle+|11\rangle\right)$$

which corresponds to the first of Bell measurement. A summary of Bell measurements is given in Table 3.2.

In the last few years the quantum operations on qubits have become the subject of extensive studies because they offer the possibility for a number of practical applications. It is expected that the EPR correlation will form a basis for advanced quantum photonic communication, computation, and cryptographic systems. A more detailed description of quantum photonic systems in given in Chapter 13.

3.20 FABRY–PEROT CAVITIES

There are a number of applications that require FP cavities [3.14]. The applications span from large 4–in.-O.D. FP filters for astronomical hyperspectral imaging to miniature devices for tunable laser diodes, accelerometers, and pressure and temperature sensors.

A typical FP cavity is shown in Figure 3.58. The cavity consists of a medium of refractive index n_g sandwiched between two partially reflecting mirrors $R1$ and $R2$. Incident light entering the cavity undergoes multiple reflections at the two partially reflecting mirrors $R1$ and $R2$.

For normally incident light, the phase change at $R2$ for a round trip within the cavity ($R2$ to $R1$ and $R1$ to $R2$) is given by

$$\delta = 2\pi/\lambda\ 2\ n_g\, d_g$$

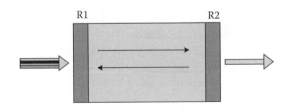

FIGURE 3.58 Schematic of a basic Fabry–Perot cavity.

where n_g is the refractive index of the cavity and d_g is the gap spacing between $R1$ and $R2$. Assuming $R1 = R2 = R$, the relative transmitted intensity due to multiple reflections is given by

$$T_{FP} = I_t/I_i = 1/[1 + F\sin^2(\delta/2)]$$

where

$$F = 4R/(1 - R)^2$$

is termed the *coefficient of finesse* for the cavity. The peak in transmission occurs when there is constructive interference such that the phase change per round trip is

$$\delta = 2\pi m$$

where m is an integer. The transmission characteristics of a FP cavity are periodic with $\delta\lambda$, resulting in a maximum free spectral range (FSR) for a given order at which the peak in transmittance for λ_i will overlap with the peak in transmittance of a wavelength λ_j. This can be calculated as

$$\text{FSR (FP)} = \lambda^2/(2n_g d_g)$$

If used in conjunction with a spectrometer using a detector array, the FSR should be equal to a multiple of the detector pixel channel spacing $\Delta\lambda_{ch}$, giving the condition

$$\text{FSR (FP)} = \lambda^2/(2n_g d_g) = m\Delta\lambda_{ch} = m\ w_s\ (\Delta\lambda/\Delta x)$$

where w_s is the detector pixel spacing and $(\Delta\lambda/\Delta x)$ is the spectrometer output dispersion.

The potential methods of actuation for tuning the FP cavity include:

- Electro-optical variation of the refractive index of a solid-state gap
- Thermal expansion of a spacer between the mirrors
- Electrostatic actuation (bending a cantilever)
- Piezoelectric voltage actuation using smart materials
- Thermorefractive effect

Some basic design parameters for FP cavities are summarized in Table 3.3. FP devices are very often used as the basis of the operation of laser diodes. They are also widely employed for various microphotonic pressure and temperature sensors.

TABLE 3.3
Summary of Fabry–Perot Filter Design Parameters

Parameter	Formula	Typical Experimental Values
Free spectral range	$FSR_R = \dfrac{\lambda^2}{2dn}$	2 to 8 nm, (depending on a spectrometer operating diffraction orders (i.e., m = 1 or m = 2)
FWHM	$FWHM_R = \dfrac{\lambda(1 - R)}{n\pi\sqrt{R}}$	0.5 to 1 nm at 1550 nm
Finesse	$F_R = \dfrac{\pi\sqrt{R}}{(1 - R)}$	About 20
Gap tuning	Depending on application	$\Delta d = 2$ μm

REFERENCES

1. Charschan, S.S. (Ed.), *Lasers in Industry*, Laser Institute of America, Toledo, OH, 1972.
2. Marcuse, D., *Light Transmission Optics*, 2nd ed., Van Nostrand Reinhold, New York, 1982.
3. Jamroz, W. and Stoicheff, B., *Progress in Optics*, Emil Wolf (Ed.), Vol. 20, North-Holland, Amsterdam, 1983.
4. Kashyap, R., *Fiber Bragg Gratings*, Academic Press, New York, 1999.
5. Yablonovitch, E., Inhibited spontaneous emission in solid-state physics and electronics, *Phys. Rev. Lett.*, 58, 2059–2062, 1987.
6. John, S., Strong localization of photons in certain disordered dielectric superlattices, *Phys. Rev. Lett.*, 58, 2486–2489, 1987.
7. Joannopoulos, J.D., Meade, R.D., and Winn, J.N., *Photonic Crystals: Molding the Flow of Light*, Princeton University Press, Princeton, NJ, 1995.
8. John, S., Toader O., and Bush, K., www.physics.utoronto.ca/~john/john/Encyclopedia.pdf (May 6, 2004).
9. Birks, T.A., Roberts, P.J., Russell, P. St. J., Atkin, D.M., and Shepherd, T.J., Full 2-D photonic bandgaps in silica/air structures, *Electron. Lett.*, 31, 1941–1942, 1995.
10. Mahan B.A., *University Chemistry*, 3rd ed., Addison-Wesley, Reading, MA, 685–743, 1975.
11. Kruzelecky, R.V. et al., *Integrated Thin-Film Smart Coatings with Dynamically-Tunable Thermo-Optical Characteristics*, presented at the *32nd International Conference on Environmental System*, Paper 02ICES-170, San Antonio, TX, 2002.
12. Stefanovich, G., Pergament, A., and Stefanovich, D., Electrical switching and Mott transition in VO_2, *J. Phys. Condens. Matter*, 12, 8837–8845, 2000.

13. Tittel W. and Weihs G., Photonic entanglement for fundamental tests and quantum communication, *Quantum Inf. Computation*, 1, 1–54, 2001.

14. Patterson, J.D., Micro-Mechanical Voltage Tunable Fabry–Perot Filters Formed in (111) Silicon, NASA Technical Paper # 3702, 1997.

4 Photonic Node

4.1 MICROPROCESSOR

A microprocessor is the main part of all communication networks, computers, and most workstations. It is the brains of the system. Sometimes referred to simply as the *central processing unit* (CPU), the microprocessor is where most operations take place. In terms of computing power, the CPU is the most important element of a system. On large machines, microprocessors require one or more printed circuit boards. On personal computers and small workstations, the microprocessor is housed on a single chip. Microprocessors also control the logic of almost all digital devices, from clock radios to fuel-injection systems for automobiles.

Since the 1970s there has been intense research aimed at making a photonic microprocessor. Such a photonic microprocessor would contain a variety of integrated optical components to carry out light generation, modulation, manipulation, detection, and amplification (Figure 4.1).

A photonic microprocessor is a system in which data or signal processing is done entirely by using optical signals. In a true photonic microprocessor, every device, be it a switch or an amplifier, works with infrared or visible signals. The conversion to and from electrical impulses is not done except at the source and detector.

Depending on a specific application, a photonic microprocessor may be an autonomous system such as a sensing system or a part of a more complex architecture such as a communication network. There are many possible combinations of functions that can be integrated in a photonic microprocessor.

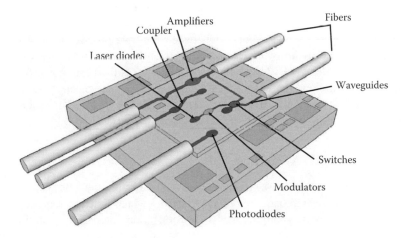

FIGURE 4.1 Photonic microprocessor.

Significant advances in photonic technology, particularly the development of optical amplifiers and dense wavelength-division multiplexing (DWDM), have transformed the design of telecommunications networks such as metro, regional, long-haul, and undersea networks. Single-channel systems that once required optical–electrical–optical (OEO) regeneration at 40-km intervals have been displaced by optically amplified DWDM systems with regeneration intervals of up to 3000 km. These advances tend to drive networks away from electronically switched "opaque" architectures, i.e., an architecture of optical links connected through a set of electrical switches, in which the optical signals are detected, converted to electrical, and retransmitted as optical impulses.

Over the course of the past decade, the number of channels that can be carried on a single optical link has risen from 1 to more than 100, while data rates per channel have increased from 622 Mb/sec to 10 Gb/sec.

A subtle, but potentially far-reaching, advantage of photonic systems over electronic media results from the fact that the optical signals actually travel faster than electricity. It is known that electric currents propagate at about 10% of the speed of light (30,000 km/sec), whereas the energy in optical systems travel at the speed of light in glass, i.e., 300,000 km/sec. This allows for much shorter data transmission delay times between the end points of a network. This advantage is significant within photonic microprocessors. Optical data processing and transmission have several additional advantages over electrical transmission. For example, the increased bandwidth provided by optical signals allows for much larger data capacity as compared to the traditional electronic systems. In addition, a single strand of optical channel (e.g., an optical waveguide) can carry multiple channels at different wavelengths, each channel having its own set of modulating signals.

Industry observers agree that the most likely applications for photonic microprocessors will be in devices in which an extreme amount of data is required in a very small space. Two such applications are usually mentioned, i.e., (1) data buses (from processor to memory or between multiple processors in a server) and (2) in the backplane of server racks [4.1]. These applications are somehow contradictory to the widely held belief that photonics is the best choice for long-distance transmission whereas conventional copper cables are the better choice for short links. However, it is expected that a copper-based point-to-point serial link will become prohibitively expensive above 20 Gb/sec.

4.2 COMMUNICATION NODE

In a communication network, a microprocessor may play the role of a node. A node is a point at which data are received or from which they are sent. Although the term often is used synonymously with workstation, interconnection points in a network also are called nodes. A node can be a computer or other signal processing device. The nodes can be interconnected in various ways. For example, the most common arrangements used in computer networking topologies are bus, star, and ring (see Figure 4.2).

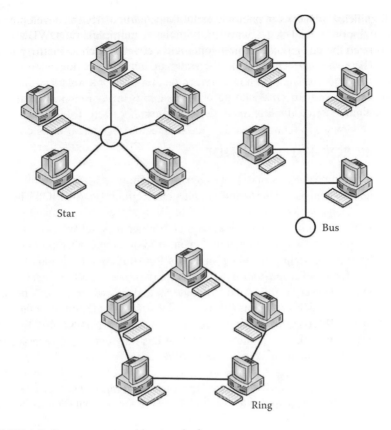

FIGURE 4.2 Computer networking topologies.

A bus network topology has each node directly connected on a main communication line. One end has a controller, and the other end has a terminator. Any node that wants to communicate with the main computer must wait its turn for access to the transmission line. When a node uses the network, the information is sent to the controller, which then sends the information down the line of nodes until it reaches the terminating node. Each node in the line receives the same information.

Star networks incorporate multiport star couplers. A main controlling node interconnects with all the other nodes in the network. As with the bus topology, the failure of one node does not cause a failure in the network. Both the bus and the star network topologies use a central node that controls the system inputs and outputs.

Ring networks operate in a manner similar to bus networks with the exception of a terminating node. In this configuration, the nodes in the ring link to a main communication cable.

These various topologies are configured according to the network applications, i.e., (1) long haul, (2) metro core, (3) metro access, (4) enterprise, and (5) residential.

Long-haul networks are point-to-point links connecting points that are separated by several thousand kilometers without signal regenerators. Metro core networks are based on ring and mesh topologies, and they are used in metropolitan areas where the link length is on the order of hundreds of kilometers. Metro access networks are links that are no longer than few tens of kilometers. Enterprise networks are used for intrabuilding or intracampus links. Residential networks are required to bring the link to residential homes.

4.3 MICROPHOTONIC NODE

Photonic microprocessors may be arranged on a single platform that is called a microphotonic chip or a microphotonic integrated circuit (micro-PIC). There can be many arrangements of nodes in a single microprocessor and many different ways of dividing the functions between multiple network subsystems.

Most applications that can greatly benefit from photonic microprocessors are related to sensing, data processing, and data transmission. They may be based on various forms of a local area network. Source-to-receiver distances may be very short; however, there can be considerable data switching and rerouting between the various subsystems. This is especially true for the systems that require a high degree of redundancy and rerouting to counter catastrophic failure of components. For example, a photonic intranet bus may consist of several redundant optical lines interconnecting various nodes.

Figure 4.3 illustrates a schematic diagram of a photonic intranet bus. A photonic system may contain most of the generic components, i.e., nodes and data processing subsystems, as well as the central processor, data storage, receiver and transmitter modules, and a set of application-specific subsystems.

Figure 4.4 depicts a typical schematic of a photonic microprocessor that may be used to transmit a number of different channels of high-bandwidth information over a single channel. Although the shown schematic closely resembles a transceiver of a typical communication network, its design can be tailored to other types of applications.

Figure 4.4 depicts a typical scenario in which a number of different channels of high-bandwidth information are transmitted and processed. As shown, a number of individual wavelengths are multiplexed together using an N × M switch array. A basic node operation, as schematically depicted in Figure 4.4, entails the following functions:

- Adding (sourcing) an optical signal to the ring network
- Dropping (receiving) selected data from the ring network
- Amplifying and regenerating the data being transmitted to subsequent nodes

At the heart the microprocessor is a functional module. All microprocessor components are designed around this module. If the microprocessor's main role is to function as a communication node, then the functional module will be used

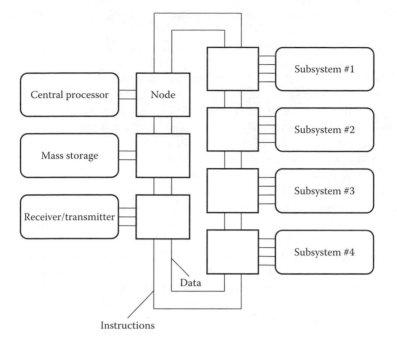

FIGURE 4.3 Photonic intranet bus.

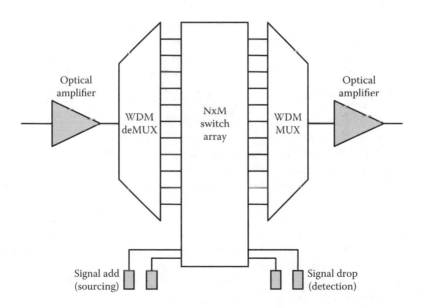

FIGURE 4.4 Integrated photonic processor.

to transmit signals over a distance. In the case of an optical network, this module will consist of a transmission fiber. If the microprocessor's main function is sensing, then the functional module will include a set of sensing devices.

As shown in Figure 4.4, a number of individual optical wavelengths are multiplexed together using a wavelength-division multiplexer. A multiplexer is used to combine several signals for transmission over a single medium.

The multiplexed signals are then passed onto a switch. In general, the switch is used for routing individual channels. The switch function can also include adding and dropping some of the channels from the main group of signals. Such an operation is known as an *optical add-drop multiplexing*. Such add-drop lines can be coupled to other nodes or subsystems.

A splitter may be used to tap a small part of the signal and send it to the system management module for wavelength and power monitoring. Upon traveling through a functional module, the signals may be boosted by an amplifier. There can be a number of stages of amplification.

Before the signals are demultiplexed, an attenuation or compensation module can be provided. Alternatively, this module may be placed at the inputs to the switch. The purpose of this module is to control the optical characteristics of each channel to enable better optical performance. Typically, this can involve adjusting the power levels by attenuation to compensate for differences in gain between the channels. It can also involve dispersion compensation, polarization compensation, and other types of compensation for degradations that may be a function of wavelength.

At the receiver end, the signals are separated using a wavelength-division demultiplexer. Individual channels may be monitored by a set of detectors. The receiver uses either a PIN (positive, intrinsic, and negative layers of doped p-type and n-type semiconducting materials) photodiode (a semiconductor device that converts light to electrical current) or an avalanche photodiode (APD — a photodiode that exhibits internal amplification of a photocurrent through avalanche multiplication of carriers in the junction region) to receive the optical signal and to convert it back into an electrical signal.

As the optical gain provided by the optical amplifier and the attenuation and compensation may need to be optimized, the system management module can be implemented to maximize system performance characteristics.

Other functions such as optical transmitters, optical regenerators, and data modulators, among others, can also be included into a photonic processor. Other types of optical switches may prove to be preferable for particular applications if they can be made more compact or more economical or operated at higher speeds. A processor management module is required to optimize performance of the optical amplifiers, attenuators, and compensators.

In summary, a basic processor operation entails, as schematically depicted in Figure 4.5, the following operations:

- Multiplexing
- Adding (sourcing) an optical signal

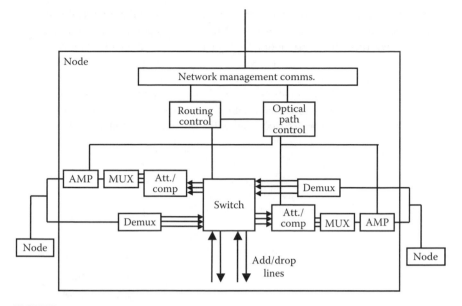

FIGURE 4.5 Transparent optical processor.

- Dropping (receiving) selected data
- Switching
- Amplifying and regenerating the data being transmitted
- Distributing data and control signals to the various components comprising the processor and the system
- Redundant signal rerouting in case of component failure
- Compensate various dispersion mechanisms

Table 4.1 lists the most common subsystems that may be implemented into a photonic node. The subsystems listed in Table 4.1 have been arranged according to their functionality. The following functions have been listed. (1) signal generation, (2) propagation, (3) multiplexing, (4) detection, and (5) maintenance.

Figure 4.5 illustrates a schematic of an advanced optical ("transparent") system that is based on tunable transmitters and tunable receivers. At the central node, signals from an array of tunable transmitters each composed of a tunable laser and modulator, are combined and transmitted through a link with adjacent nodes. Signals arriving at the central node may be locally terminated or passed through to other nodes. Locally added or passed-through signals can be dynamically switched to any other transparent node. Tunable filters can be used in the reconfigurable optical add-drop multiplexers (ROADM) that are the central switching element of transparent optical nodes. At the central node, the signals are demultiplexed and detected by an array of tunable receivers, each composed of a tunable filter and photoreceiver.

It has been estimated that a dynamically controlled "transparent network," that is, one that permits nonterminating channels to pass through nodes in optical

TABLE 4.1
Subsystems of a Photonic Node

Subsystem	Function
	Generation
Transmitters	Inject data flow into the propagation channel with multiple laser sources
Modulators	Imprint streaming digital data into optical channels
	Propagation
Passive components	Direct light from one port to the next in a sequential fashion
	Prevent backward propagation of optical signals
	Select transmission wavelengths
Switches	Direct traffic between channels
	Multiplexing
Multiplexers	Combine two or more signals that are transmitted over a single communication channel
	Separate two or more signals previously combined by multiplexing
	Add or remove a single wavelength from multiplexed channels
	Detection
Receivers	Convert demultiplexed optical channels into distinct electrical signals
	Monitor channel intensities
	Maintenance
Channel compensators	Correct for absorption losses
	Compensate for velocity differences that arise in multiplexed channels having slight wavelength differences
	Compensate for velocity differences that arise in multiplexed signals having slight polarization state differences
	Control relative levels of multiplexed channels

form, can eliminate more than half the OEO conversions found in an opaque network [4.2]. End-user requirements change in a dynamic fashion because of both traffic growth and bandwidth demand patterns in a telecom network. In fixed-wavelength networks, the traffic demand results in stranded and unusable bandwidth. This situation demands that networks be upgraded to transparent and optically switched designs, i.e., designs that do not require OEO conversions.

Such a transparent network provides a very attractive solution for carriers seeking to maximize revenue and minimize both capital investments and operating expenses. There are many possible combinations of processing functions, which can be integrated into a photonic microprocessor. There can be many devices included in a single processor and many different ways of dividing the functions.

Table 4.2 summarizes some of the components and devices that constitute a typical photonic node. The devices listed in Table 4.2 are described in the following chapters.

TABLE 4.2
Devices of a Photonic Node

Subsystem	Devices
Transmitter	Light-emitting diodes (LEDs)
	Laser diodes (LDs)
	Vertical-cavity surface-emitting lasers (VCSELs)
	External-cavity lasers (ECLs)
	Tunable laser modules
	Optical pulse generators
	Fiber lasers
	All-silicon lasers
	Wavelength lockers
Modulator	External modulators
	Lithium niobate modulators
	Waveguide-based modulators
	Si-based waveguide modulators
Passive components	Splitters and combiners
	Couplers
	Circulators
	Isolators
	Gratings
	Waveguide collimators
	T junctions
Switches	Optical switches
	MOEMS-based switches
	Waveguide switches
	SOA switches
	Evanescent switches
	Optical cross connects
	Hybrid PBG / MOEMS switches
Multiplexers	Time-division multiplexers
	Coarse wavelength-division multiplexers
	Dense wavelength-division multiplexers
	Optical add-drop multiplexers
	ROADMs (reconfigurable optical add-drop multiplexers)
Receivers	PIN photodiodes
	Avalanche photodiodes
	Light emitters or detectors
	Si-based photodetectors
Channel compensators	SOAs (semiconductor optical amplifiers)
	EDFA (erbium-doped fiber amplifiers)
	EDWA (erbium-doped waveguide amplifiers)
	ROAs (Raman optical amplifiers)
	DGE (dynamic gain amplifiers)
	Dispersion compensators
	Wavelength converters

REFERENCES

1. Salib, M., Liao, L., Jones, R., Morse, M., Liu, A., Samara-Rubio, D., Alduino, D., and Paniccia, M., Silicon photonics, *Intel Technol. J.*, 8, 144–160, 2004.
2. Berger, J.D. and Anthon, D., Tunable MEMS devices for optical networks, *Optics and Photonics News,* 43–49, 2003.

5 Transmitters

5.1 TRANSMISSION SYSTEMS

Optical transmission uses the same basic elements as traditional copper-based communication systems: a transmitter, a receiver, and a medium by which the signal is passed from one to the other. Figure 5.1 illustrates these basic elements.

An *optical transmitter* is a device that includes an optical source and driving electronics. It functions as an electrical-to-optical converter. The transmitter uses an electrical interface to encode the data through an optical modulator. An optical modulator modulates the optical source in order to transmit the data. In most cases the optical sources use wavelengths of 850, 1310, or 1550 nm. A waveguide or optical fiber connects the transmitter and the receiver.

The receiver uses either a PIN photodiode (the name comes from positive, intrinsic, and negative layers of doped p-type and n-type semiconducting materials) or an avalanche photodiode (APD) to receive the optical signal and convert it back into an electrical signal. A demodulator converts the data back into its original electrical signal.

Many long-haul transmission systems require signal regenerators, signal repeaters, or optical amplifiers such as erbium-doped fiber amplifiers (EDFAs) to maintain the optical signal quality. System drop, repeat, or add functions, such as those in

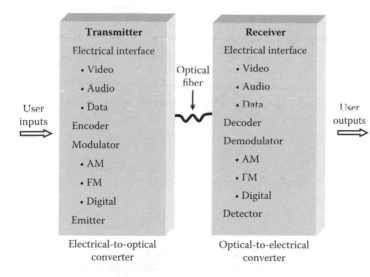

FIGURE 5.1 Data transmission system.

multichannel networks, further augment the optical transmission system, incorporating optical add-drop multiplexers, couplers, splitters, dispersion management devices, and other signal processing devices. A *transceiver* is a device that performs, within one chassis, both transmitting and receiving functions.

5.2 OPTICAL SOURCES

Optical sources are the key element in any photonic system. These components convert the electrical signal into a corresponding optical signal that can be injected into the photonic circuit. The optical source is often the most costly element, and its characteristics strongly influence the overall performance of a given system.

The optical sources must operate in the low-loss transmission windows of glass fiber or waveguides. There are two main types of optical sources: light-emitting diodes (LEDs) and lasers. LEDs are typically used at the 780-, 850-, and 1310-nm transmission wavelengths, whereas lasers are primarily used at 1310 and 1550 nm.

5.2.1 LEDs

LEDs use semiconductor p–n junctions in the same way as they are used in electronic diodes and transistors to generate light due to the recombination of electron–hole pairs. The generation of electron–hole pairs requires an energy supply. The recombination of these particles leads to emission of energy in the form of light (see Figure 5.2). The energy is typically below 1 eV; therefore, visible or infrared light is usually emitted. The conversion process is quite efficient as it generates little heat compared to other sources, such as incandescent lights.

LEDs are of interest for microphotonic applications because of their performance characteristics, which are listed as follows:

Peak wavelength: This is the wavelength at which the optical source emits peak power. This wavelength should be matched to the wavelengths that are transmitted with the least attenuation through the transmissive medium.

Spectral width: The range of the optical signal centered on the peak wavelength is called the spectral width of the optical source.

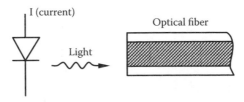

FIGURE 5.2 Schematic of LED.

Emission pattern: The pattern of emitted light affects the amount of optical energy that can be coupled into the optical channel. The size of the emitting region should be similar to the diameter of the fiber core or the optical waveguide.

Power: The key requirement is that the output power of the source be strong enough to provide sufficient power to the detector at the receiving end, considering transmission attenuation, coupling losses, and other system constraints. In general, LEDs are less powerful than lasers.

Switching speed: A source should switch on and off fast enough to meet the bandwidth requirements of the system. The switching speed is measured as a source's rise or fall time that is required to go from 10% to 90% of peak power. LEDs have slower rise and fall times than lasers.

Linearity: Linearity represents the degree to which the optical output is directly proportional to the electrical current input. Analog applications require linear response. Nonlinearity in LEDs may cause distortion in the analog signal that is transmitted over an optical link.

Applications of LEDs are limited to lower-data-rate and shorter-distance multimode systems because of their inherent bandwidth limitations and lower output power. They are used in applications in which data rates are in the hundreds of megahertz as opposed to gigahertz data rates, which require the use of lasers.

Another limiting factor is related to the LED's numerical apertures. LEDs typically have large numerical apertures, which makes light coupling into a single-mode channel difficult. For this reason LEDs are most often used with multimode fibers.

Figure 5.3 shows a graph of typical output power vs. drive current for LEDs and laser diodes. An LED's optical output is approximately proportional to the driving current. Other factors, such as temperature, also affect the optical output.

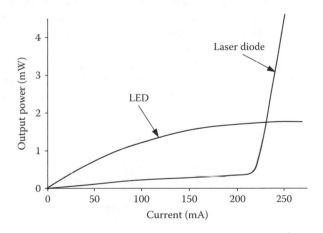

FIGURE 5.3 LED characteristics.

TABLE 5.1
Light-Emitting Materials

Material	Wavelength (nm)
GaP (gallium phosphide)	550
AlAs (aluminum arsenide)	590
GaAs (gallium arsenide)	870
AlGaAs (aluminum gallium arsenide)	770–870
InP (indium phosphide)	930
InGaAsP (indium gallium arsenide phosphide)	1100–1670

As the LED is driven to higher currents, the LED chip gets hotter, causing a drop in conversion efficiency. LEDs are typically operated at currents up to approximately 100 mA peak. Only specially customized devices operate at higher current levels.

Table 5.1 lists some common light-emitting materials and their emission wavelength. As can be seen in Table 5.1, GaP and AlAs are used to make emitters in the visible portions of the electromagnetic spectrum. GaAs, InP, and AlGaAs are used to make emitters in the near-infrared region that is often referred to as the *first window* in optical fiber. InGaAsP is used to make emitters in the infrared portion of the spectrum referred to as the *second and third windows* in optical fibers.

As shown in Figure 5.4 two basic structures for LEDs are used in optical systems: surface emitting and edge emitting. Edge emitters offer high output power levels and high-speed performance, but they are more complex and expensive devices. The output power density is high because the optical beam diameter is a relatively small, i.e., typically 30 to 50 μm. Such a small optical beam diameter allows for a good coupling efficiency to optical fibers and waveguides.

Another variant of the edge emitter is the superradiant LED. These devices are a cross between a conventional LED and a laser. They usually have a very high power density and possess some internal optical gain similar to a laser, but their optical output is noncoherent. Superradiant LEDs have very narrow emission spectra, typically 1 to 2% of the central wavelength, and offer power levels

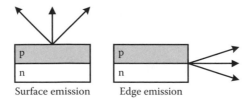

Surface emission   Edge emission

FIGURE 5.4 LED structures.

comparable to a laser diode. These devices are popular for fiber-optic gyroscope applications.

The second type of LED is the surface emitter. Surface emitters have a comparatively simple structure, are relatively inexpensive, offer low-to-moderate output power levels, and are capable of low-to-moderate switching speeds. The total output power is as high as or higher than the edge-emitting LED, but the emitting area is large, causing poor coupling efficiency to the optical channels. Adding to the coupling efficiency deficit is the fact that surface-emitting LEDs emit light in all directions. Thus, very little of the total light may be coupled into a waveguiding channel.

5.2.2 LASER DIODES

A laser diode (LD) is based on three main components (1) an active material that is able to generate and to amplify light by stimulated emission of photons, (2) an optical cavity that provides the optical feedback to sustain the laser action, and (3) a pumping mechanism that is able to excite the active material such that population inversion can be achieved (see Figure 5.5).

In an injection diode laser, the pumping mechanism is provided by carrier injection through a p–n junction and the optical feedback is provided by a Fabry–Perot (FP) cavity (see Figure 5.6). The use of electrical injection makes the device particularly interesting for integration with microphotonic devices.

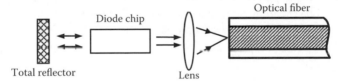

FIGURE 5.5 Schematic of a laser diode.

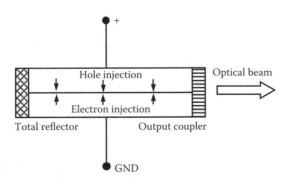

FIGURE 5.6 Schematic of an injection laser based on p–n junction.

An LD has a much higher output power than an LED; it is thus capable of transmitting information over longer distances. Consequently, and given the fact that the LD has a much narrower spectral width, it can provide high-bandwidth communication over long distances. The LD's smaller numerical aperture (NA) also allows it to be more effectively coupled with single-mode fibers. The difficulty with LDs is that they are inherently nonlinear, which makes analog transmission more difficult. They are also very sensitive to fluctuations in temperature and driving current, which causes their output wavelength to drift. In applications such as wavelength-division multiplexing, in which several wavelengths are being transmitted down the same fiber, the stability of the source becomes critical. This usually requires complex circuitry and feedback mechanisms to detect and correct for drifts in wavelength. The benefits, however, of high-speed transmission using LDs typically outweigh the drawbacks and added expense.

Table 5.2 summarizes the characteristics of commercially available LDs operating at different wavelengths between 980 nm and 1550 nm. As shown, the power conversion efficiency and the available optical output power decrease substantially as the operating wavelength of the LD is increased. This is due to the smaller energy band gap of the underlying semiconductor material for lasers operating at longer wavelengths and the correspondingly higher thermal recombination and lasing inefficiency. Even moderate power LDs operating near 1550 nm require thermoelectric coolers (TECs). In contrast, relatively-high-power 980-nm LDs can be operated with just a heat sink. The operating wavelength of LDs can be tuned using an external cavity.

It should be mentioned that a LD source can be used to directly generate the desired wavelength for the optical carrier.

TABLE 5.2
Characteristics of Laser Diodes

Parameter	Unit	LD-980-30	LD-980-100	LD-980-150	POM-14-1	POM-18-1	POM-18-2
Wavelength	nm	980	980	980	1300	1550	1550
Spectral width	nm	1.5	3	3	3	3	3
Output power	mW	30	100	150	1.5	1.5	10
Threshold current	mA	20	30	35	15	25	25
Operating current	mA	100	250	300	40	50	175
Operating voltage	V	2.0	1.6	1.6	1.3	1.3	1.7
Efficiency	%	15.0	25.0	31.2	2.9	2.3	3.4
TEC power	W	< 0.4	< 0.4	< 0.4	1.5	1.5	2.0

The typical edge-emitting diode laser actually operates in several fundamental modes, especially when operated using its own facets as the cavity. External cavities are employed to select the required single-mode operation from an edge-emitting diode laser.

LDs operating near 980 nm are usually based on GaInAs/AlGaAs quantum-well structures grown on GaAs substrates. Single-stripe LDs have a lifetime rating up to 100,000 h, and can typically be operated up to 60°C with no TEC cooling. Output powers exceeding 150 mW are commercially available. LDs operating near 1550 nm are usually based on InGaAsP quantum wells deposited on InP. Low-power telecom signal sources are available with reliabilities to Belcore standards. Higher-power 1550 LDs (10 mW) require TEC cooling because of inefficient electrical to optical power conversion.

LDs require great care in their drive electronics to prevent a critical failure. There is a maximum current that must not be exceeded for even a microsecond, and this depends on the particular device as well as junction temperature. It is not sufficient in most cases to just use a constant current power supply. This sensitivity to high currents is due to the very large amount of positive feedback that is present when the LD is lasing. Damage to the end mirrors can occur nearly instantaneously from the concentrated electromagnetic fields of the laser beam. Closed-loop regulation using optical feedback to stabilize beam power is usually implemented to compensate for device and temperature variations.

LDs can be divided into two generic types depending on the method of confinement of the lasing mode in the lateral direction:

Gain-guided LDs work by controlling the width of the drive-current distribution; this limits the area in which lasing action can occur. Because of different confinement mechanisms in the lateral and vertical directions, the emitted wavefront from these devices has a different curvature in the two perpendicular directions. This astigmatism in the output beam is one of the unique properties of LD sources. Gain-guided injection LDs usually emit multiple longitudinal modes and sometimes multiple transverse modes.

Index-guided LDs use refractive index steps to confine the lasing mode in both the transverse and vertical directions. Index guiding also generally leads to both single-transverse-mode and single-longitudinal-mode behavior. Typical linewidths are on the order of 0.01 nm. Index-guided lasers tend to have less difference between the two perpendicular divergence angles than gain-guided lasers.

Single-frequency LDs are another interesting member of the LD family. These devices are now available to meet the requirements for high-bandwidth communication links. Other advantages of these structures are lower threshold currents and lower power requirements.

There are two basic types of LD structures: Fabry–Perot (FP) and distributed-feedback (DFB).

The DFB LD is shown in Figure 5.7. With the introduction of a corrugated structure into the cavity of the laser, only light of a very specific wavelength is diffracted and allowed to oscillate. This yields output wavelengths that are

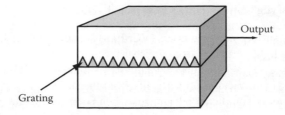

FIGURE 5.7 Distributed-feedback laser.

extremely narrow. This is the basic beam characteristic that is required for dense wavelength-division-multiplexing systems, in which many closely spaced wavelengths are transmitted through the same fiber. DFB lasers have been developed to emit light at fiber-optic communication wavelengths between 1300 and 1550 nm.

Of the two types of lasers, FP lasers are the most economical, but they are generally more noisy and slower. DFB lasers are less noisy devices (e.g., with higher signal-to-noise ratio), have narrower spectral widths, and are usually faster.

DFB lasers offer the higher performance levels and also have the higher cost of the two types. They are nearly monochromatic, whereas FP lasers emit light at a number of discrete wavelengths. DFB lasers tend to be used for high-speed digital applications and for most analog applications because of their greater speed, lower noise, and superior linearity.

FP lasers further break down into buried hetero- (BH) and multi-quantum-well (MQW) types. BH and related types have been used for many years, but now MQW types are more often applied. MQW lasers offer significant advantages over all former types of FP lasers. They offer lower threshold current, higher slope efficiency, lower noise, better linearity, and much greater stability over temperature. As an additional bonus, laser manufacturers have been able to obtain better yields; the manufacturing cost is thus greatly reduced. However, one disadvantage of MQW lasers is their tendency to be more susceptible to back reflections.

LDs require a method of stabilizing the threshold current to achieve maximum performance. As operating temperature changes, several effects can occur. First, the threshold current may change. The threshold current is always lower at lower temperatures, and vice versa. The second change that can be important is the slope efficiency. The slope efficiency is the number of milliwatts or microwatts of light output per milliampere of increased drive current above the lasing threshold. Most lasers show a drop in slope efficiency as temperature increases. Often, a photodiode is used to monitor the light output on the rear facet of the laser. The current from the photodiode changes with variations in light output and provides feedback to adjust the laser drive current.

5.2.3 VERTICAL-CAVITY SURFACE-EMITTING LASERS

Vertical-cavity surface-emitting lasers (VCSELs) are new types of laser structures that emit laser light vertically from their surfaces.

VCSEL is a semiconductor vertical-cavity surface-emitting micro-LD that emits light in a cylindrical beam vertically from the surface. It is fabricated from a wafer, and offers significant advantages when compared to the edge-emitting lasers currently used in the majority of fiber-optic communication devices. These lasers emit at 850 nm and have rather low thresholds.

They are very fast and can give milliwatts of power that is coupled into a 50-μm core optical channel. VCSELs can be tested at the wafer level (as opposed to edge-emitting lasers, which have to be cut and cleaved before they can be tested) and hence are less expensive. In fact, VCSELs can be fabricated efficiently on a 3-in.-diameter wafer. A schematic of a VCSEL is shown in Figure 5.8. The principles involved in the operation of a VCSEL are very similar to those of regular lasers. As shown in Figure 5.8, there are two special semiconductor materials sandwiching an active layer in which the lasing action takes place. There are several layers of partially reflective mirrors above and below the active layer, instead of the conventional cavity reflectors. These mirrors are created from layers of semiconductors of different compositions. Each mirror reflects a narrow range of wavelengths back into the cavity to ensure the light emission at a selected wavelength.

VCSELs are typically MQW devices with lasing occurring in 20- to 30-atoms-thick layers. The laser mirrors are formed from Bragg reflectors. There may be as many as 100 or more layers that form the laser mirrors.

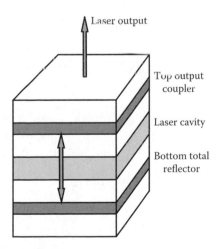

FIGURE 5.8 Structure of VCSEL.

There are many advantages to VCSELs. Their small size and high-efficiency mirrors produce a low threshold current, usually below 1 mA. Their unique feature is that they are stable over a wide temperature range. This feature makes the VCSEL to be an ideal device for applications that require an array of sources.

As has been mentioned previously, these lasers are very sensitive to back reflection. This may limit their usefulness in such applications as, for example, the add-drop multiplexers (ADMs). Some recently designed ADMs have used modified lasers. These modified lasers have a short length of single-mode fiber that provides the interface to an optical connector. This simple technique has not only enhanced the laser stability but has also solved the back reflection problem.

5.2.4 EXTERNAL-CAVITY LASERS

External-cavity lasers (ECLs) are very often used for test and measurement applications. They contain a gain chip and gratings that form a lasing cavity. Usually these lasers are too bulky to be used for DWDM applications. The actual configuration of the cavity may vary from manufacturer to manufacturer. In some commercial designs a rotating grating is used. In other designs a translating mirror is added.

A single-mode tunable ECL was designed by researchers at Intel [5.1]. In this design an anti-reflection (AR) coated III-V semiconductor LD was coupled to a silicon-based waveguide Bragg grating. The lasing wavelength was selected by the grating, and it could be tuned by using the thermo-optical effect by heating the grating. The observed tuning rate was 12.5 nm/100°C.

The Bragg grating used in this design was fabricated by etching a set of microtrenches into a 4 μm-thick silicon-on-insulator (SOI) wafer. These trenches were then filled with polysilicon. The polysilicon was then polished to obtain a planar surface. A 3.5-μm-wide and 0.9-μm-deep rib was then patterned by using lithography and etching. The last step in the fabrication of the laser was to deposit a 0.5-μm-thick oxide layer to provide the necessary upper cladding for the rib waveguides [5.1]. A schematic illustration of the Bragg grating is shown in Figure 5.9.

The laser was formed by coupling a single-angled facet (SAF) gain chip to a waveguide containing the polycrystalline or crystalline silicon Bragg grating. An SAF gain chip features a reflective facet coating on one end and an angled facet with a simple antireflection coating at the other. The angled facet of the waveguide is tilted away from the normal to reduce the occurrence of reflection back into the device [5.1].

The laser cavity designed at Intel was formed between the Bragg grating as one end mirror and a 90% high-reflection coating of the gain chip as the other mirror. The output of the laser was taken from the 90% high-reflectivity coated side of the LD with a conical polished lensed single-mode fiber (see Figure 5.10). The purpose of this lensed optical fiber was to increase the coupling between the laser and the optical fiber.

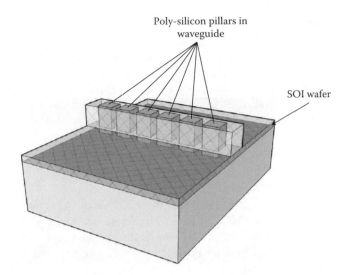

FIGURE 5.9 Schematic of polycrystalline silicon grating in SOI.

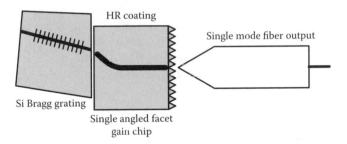

FIGURE 5.10 Schematic of the external-cavity laser.

With the SAF gain chip butt coupled to the Bragg grating, the ECL could run a single mode with a linewidth of 118 MHz. The optimized output power of the laser was 450 μW [5.1]. By altering the period of the grating, different wavelengths could be fed back into the gain chip. The refractive index of silicon can be altered by using the thermo-optical effect. This means that the Bragg wavelength of the polycrystalline or crystalline silicon grating could be tuned by heating the silicon substrate. This was done by placing the silicon die on a thermoelectric cooler and monitoring its temperature with a thermocouple.

For microphotonic integrated circuit applications, the ECL architecture designed by Intel, in which the tuning is done on one side of the ECL and the output taken from the other side, involves coupling to both sides of the gain medium. This is complicated from a device assembly point of view, and different laser geometries are being investigated in which both tuning and the laser output can be done by coupling to just one side of the ECL without degrading its output performance.

5.2.5 TUNABLE LASER MODULES

Networks very often use tunable laser modules. These modules have a similar package structure, i.e., combining a tunable laser, free-space isolator, and wavelength locker on a TEC in a butterfly package with a polarization-maintaining (PM) fiber pigtail and drive electronics [5.2]. Some tunable laser modules also incorporate a semiconductor optical amplifier (SOA) to reach higher output power.

Tunable laser transmitters combine a tunable laser module with an external Mach–Zehnder or an integrated electroabsorption (EA) modulator for 10-Gb/sec data rates, and may use either an EA modulator or direct modulation of the laser diode current for applications at 2.5 Gb/sec.

A novel approach to the design of the tunable laser modules includes micro-electromechanical-system (MEMS)-based elements in the ECL (MEMS-ECL). The MEMS-ECL takes advantage of dramatically reduced cost and complexity at the LD level compared to other tunable laser technologies. The MEMS-ECL is based on a simple, high-power FP LD. The cornerstone of the MEMS-ECL is the silicon MEMS actuator [5.2].

The optical performance of the MEMS-ECL matches that of the fixed-wavelength DFB lasers currently being used in metro and long-haul systems. The MEMS-ECL tunes over 42 nm in the C or L band, with fiber-coupled output powers of up to 40 mW. The laser can be locked sequentially to 100 L-band channels spaced by 25, 50, or 100 GHz. The MEMS-ECL can be directly modulated with low chirp at data rates of up to 2.7 Gb/sec by use of an appropriate low-capacitance LD.

5.2.6 OPTICAL PULSE GENERATORS

High-repetition-rate wavelength-tunable pulses are required in high-speed optical time-division multiplexing. One optical technique is to use a mode-locked fiber ring laser incorporating an SOA as shown in Figure 5.11.

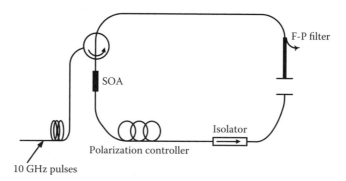

FIGURE 5.11 Optical pulse generator.

The FP laser is gain-switched by sinusoidal modulation at 10 GHz. After transmission through a high dispersion fiber, each spectral mode in the gain-switched pulses is delayed with respect to its nearest mode. So each dispersed pulse creates a sequence of pulses of different wavelengths. If the time separation is in the order of 25 psec, then an effective repetition rate will be 40 GHz. The pulses can be converted to those of the same wavelength by using a fiber ring laser with an SOA as the active element. The 40-GHz optical pulse stream is inserted into the ring laser by a circulator, causing the gain of the SOA to be optically modulated. By choosing an appropriate modulation frequency of the laser, it is possible to mode-lock the laser at the required frequencies. The output wavelength of the fiber laser is selected using a fiber FP filter. The output power is coupled from the fiber ring using a fiber coupler [5.3].

5.2.7 FIBER LASERS

A schematic of the basic structure of an optical fiber laser is shown in Figure 5.12. The fiber laser employs a cavity consisting of a length of Er- or Er-Yb-doped silica fiber or channel waveguide sandwiched between two Bragg reflectors. The Bragg reflectors have a reflection bandwidth near the desired operating bandwidth. One end of the fiber laser cavity has a Bragg reflector with a high reflection coefficient at the desired lasing wavelength. The other end of the cavity has a Bragg grating of FP narrowband filter that is partially reflecting. This allows the stimulated optical signal to be emitted by the device. Fiber laser sources employ a LD pump that operates at a shorter wavelength than the desired operating wavelength. The pump wavelength is transmitted by the Bragg reflector. The optical pump is used to invert the levels within active optical fiber or waveguide. Er-doped fibers and waveguides facilitate the use of a 980-nm laser pump to generate a signal near 1550 nm. The optical signal travels through the fiber and is reflected by the Bragg reflectors at the two ends of the fiber or waveguide cavity. Stimulated emission is reinforced by the optical signal reflected by the narrowband Bragg reflector.

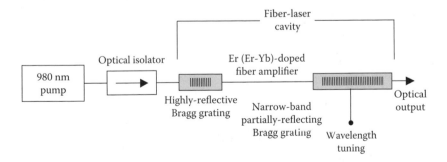

FIGURE 5.12 Schematic of a fiber laser.

The advantages of using a fiber laser light source for microphotonic systems include:

- 980-nm high-reliability pumps that can operate typically without thermoelectric coolers
- High output power capability
- Very stable, narrow linewidth operation for Doppler applications (< 0.005 nm)
- Redundancy in terms of diode pumps

The currently available commercial lasers have demonstrated spectral linewidths below 30 kHz.

5.2.8 ALL-SILICON LASERS

There has been a significant amount of research in silicon photonics that aims at the goal of producing low-cost microphotonic devices. A wide variety of passive devices have been developed in the 1990s. Recent activities have focused on the development of active devices. The main emphasis has been on the development of light amplifiers and lasers in silicon waveguides. One approach that has been investigated for light generation and amplification is based on the Raman effect. This approach relies on the fact that the Raman gain coefficient in silicon is rather strong (more than 100 times higher than in fiber), making it possible to achieve gain over the length scales of an integrated waveguide [5.4].

A Raman laser in silicon was demonstrated for the first time in 2004 [5.4]. This laser consists of a silicon gain medium that is incorporated in a fiber loop cavity. A tapered silicon-on-insulator (SOI) rib waveguide of approximately 2-cm length has been used as a gain medium. It is pumped with 30-psec pump pulses at a 25 MHz repetition rate and is centered at 1540 nm. The laser produces output pulses at the Stokes wavelength of 1675 nm. A lasing threshold has been observed at 9 W peak pulse power. A slope efficiency of 8.5% above the threshold has been recorded. The setup for demonstration of the silicon Raman laser is shown in Figure 5.13.

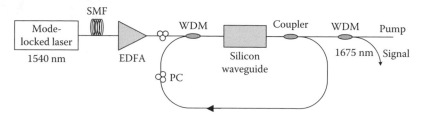

FIGURE 5.13 All-silicon Raman laser.

Pump pulses are coupled into the laser cavity by using a wavelength-division-multiplexing (WDM) coupler. The laser cavity is formed using a fiber ring configuration. Following the silicon waveguide, a tap coupler is used to extract 5% of the power as the output. The remaining 95% output of the tap coupler is looped back into the WDM coupler to form the ring cavity. The residual pump power is blocked by the WDM coupler. Two polarization controllers (PCs) are inserted on the pump arm and in the cavity to adjust the relative polarizations of the pump and the laser. The total cavity loss, including that of the silicon waveguide, measured at the Stokes wavelength (1675 nm) was found to be 3.7 dB. A second wavelength-division multiplexer is used at the laser output to separate the pump and signal wavelengths. An optical spectrum analyzer (OSA) is used to measure the output.

The Raman laser is tunable in the midinfrared range. The laser may be used for optical communication as well as for sensing and biochemical detection.

5.3 MODULATORS

An optical modulator is needed to encode data onto continuous optical waves. The operational characteristics of modulators define their usability. One of the most important characteristics is the modulator bandwidth. The bandwidths that are required for various data transmission applications are listed as follows:

- 1 Mb/sec (10^6 bit/sec) for digital stereo sound
- 100 Mb/sec (10^8 bit/sec) for digital TV
- 1 Gb/sec (10^9 bit/sec) for high-resolution TV
- 100 Gb/sec (10^{11} bit/sec) for three-dimensional TV and teleconferencing

There are basically two methods that can be used to modulate lasers and LEDs: (1) direct modulation and (2) external modulation.

In direct modulation (see Figure 5.14) the output power of the device varies directly with the input current. Both LEDs and lasers can be directly modulated using analog and digital signals. The benefit of direct modulation is that it is simple and cheap. The disadvantage is that it is slower than indirect modulation, i.e., it is limited to 3 GHz.

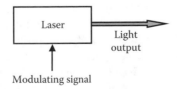

FIGURE 5.14 Direct modulation.

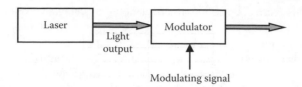

FIGURE 5.15 External modulation.

In external modulation (see Figure 5.15) an external device is used to modulate the intensity or phase of the optical source. Modulators with speeds greater than 1 GHz are typically fabricated from either the electro-optical crystal $LiNbO_3$ or III-V semiconductor compounds such as GaAs/AlGaAs and InGaAsP/InP. These devices have demonstrated modulation frequencies in excess of 40 GHz.

The optical source remains turned on while the external modulator acts similar to a very fast shutter that is controlled by the data transmitter. External modulation is typically used in high-speed applications such as long-haul telecommunication or cable TV. The benefits of external modulation are that it is much faster and can be used with higher-power laser sources. The disadvantage is that it is more expensive and requires complex circuitry to handle high-frequency RF modulation signals.

5.3.1 EXTERNAL MODULATORS

When the required data rates were in the low gigabit range, transmitters were based on directly modulated lasers. However, as data rates and span lengths grew faster, external modulators had to be used.

The external modulators may be divided into the following three types:

- Electro-optical modulators
- Electroabsorptive modulators
- Acousto-optical modulators

Electro-optical modulators are based on changes in the refractive indices that are induced by the external electrical field. The electro-optical effect provides a convenient and widely used means of modulating the intensity or phase of the optical beam. Special designs may be used (e.g., Mach–Zehnder interferometer) to convert induced optical phase changes into amplitude modulation.

Material or structural changes that affect absorption are used in electroabsorptive modulators. Such changes are induced by an external electric field through the Franz–Keldysh effect or quantum-confined Stark effect. Acoustooptical modulators use high frequency sound waves to diffract optical beams. A principle of operation of an acousto-optical modulator is shown in Figure 5.16.

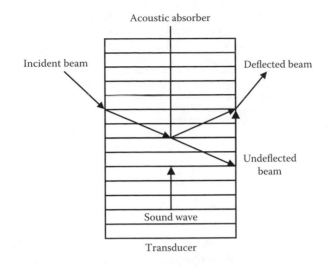

FIGURE 5.16 Principle of operation of acousto-optical modulator.

5.3.2 WAVEGUIDE-BASED MODULATORS

The demand for low-cost solutions has prompted research on waveguide modulators. These are attractive from a cost standpoint because mature processing technology and manufacturing infrastructure already exist and can be used to build cost-effective devices in large volumes. In addition, this approach provides the possibility of monolithically integrating optical elements and advanced electronics on a single chip. Figure 5.17 illustrates basic waveguide structures that are used for modulators.

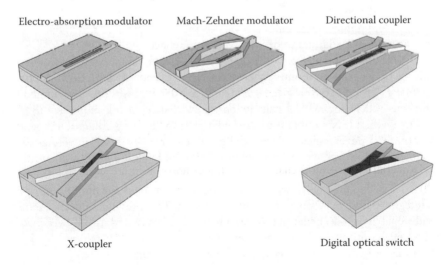

FIGURE 5.17 Waveguide modulator structures.

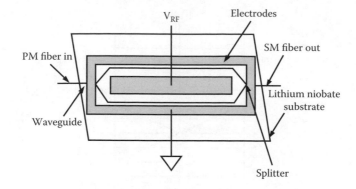

FIGURE 5.18 Lithium niobate intensity modulator.

Figure 5.18 shows one of the possible configurations of a modulator. After the light enters the waveguide, an optical coupler splits it optically into two paths. These two waves travel a distance before another coupler recombines them. If the optical waves are in phase, they will add constructively. If they are out of phase, destructive interference yields little or no output. The two optical waves travel between a set of electrodes arranged so that they have opposite effects on the two paths.

Lithium niobate ($LiNbO_3$) is one of the best-suited materials that is used as an active medium in external modulators. The effectiveness of this material results from its low optical loss and high electro-optical coefficient. The refractive index of $LiNbO_3$ changes in response to an applied electric field. The electric field causes light to travel at a variable speed that is inversely proportional to the value of the refractive index. By applying an external voltage, the refractive index of one path will rise while the refractive index of the other path falls. This causes the optical amplitude at the output to vary as the wave from the two paths moves from constructive addition to destructive interference.

$LiNbO_3$ modulators are well developed. These modulators are available at both the 1310- and 1550-nm wavelengths. High-speed $LiNbO_3$ electro-optical modulators used to convert electrical signals to optical pulses are now limited to a bit rate ~20Gb/sec. Model calculations show that replacing the bulk $LiNbO_3$ substrate with a $LiNbO_3$ thin film will reduce the modulating voltage sufficiently to permit the development of the next generation of 40-Gb/sec traveling-wave modulators. However, the technology of $LiNbO_3$ on Si still requires some further development.

External modulation in waveguide structures is usually accomplished using an integrated optical modulator that incorporates a Mach-Zehnder interferometer. Traditional Mach–Zehnder interferometers are constructed from two 2 × 2 waveguide couplers as shown in Figure 5.19. As each coupler is typically 1- to 2-cm long, the device typically requires considerable substrate area.

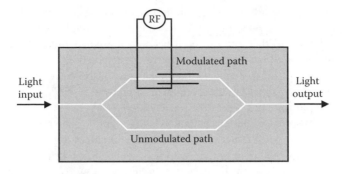

FIGURE 5.19 External Mach–Zehnder modulator.

In a Mach–Zehnder interferometer the light entering the modulator is split into two paths. One path is unchanged. The other path has a modulating element that will cause a phase delay. When the light is recombined, the two waves interfere with one another. If the two waves are in phase, the interference is constructive and the output is in the "on' position. If the two waves are out of phase, the interference is destructive and the waves cancel each other.

A waveguide prototype of Mach–Zehnder modulator consisting of two back-to-back T-bar splitters is shown in Figure 5.20. The first splitter divides the signal into the two arms of the interferometer, and the second splitter recombines the resulting signal from the two arms of the interferometer. As each splitter is only a few 100 μm in dimension, the substrate required for the T-splitter interferometer is almost an order of magnitude smaller than that for the traditional interferometer.

The relative loss through the Mach–Zehnder interferometer, which contained two back-to-back T-bar splitters, is only 1 dB greater than the typical output measured for the 10-μm-wide waveguides of about 24 dB. Therefore, even for the nonoptimized T-bar splitters and resultant interferometer, the intrinsic insertion losses are quite low.

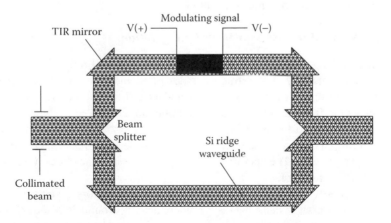

FIGURE 5.20 Mach–Zehnder modulator based on two T-bar splitters.

Building a waveguide in the substrate makes the device suitable for use in microphotonic devices. As with optical fibers, this is accomplished by introducing dopant materials into the area that will become the waveguide. Doping affects the refractive index of the waveguide relative to the surrounding substrate while maintaining optical transparency. If the dimensions of the waveguide are consistent with the dimensions of the core of a single-mode fiber, then light will efficiently couple into and out of the waveguide.

Another type of waveguide modulators is that based on the microoptoelectromechanical system (MOEMS). MOEMS-based modulators use micromechanical beam deflectors to modulate optical beams in waveguides structures. However, these modulators have severe limitations that do not allow for their use with fast communication systems.

5.3.3 SI-BASED WAVEGUIDE MODULATORS

All-optical modulators have been demonstrated with III-V compound semiconductors.

SOI optical modulators can operate with a modulation bandwidth of only 20 MHz. Consequently, these devices are not suitable for today's high-speed communication networks. Achieving high-speed switching in silicon is quite a challenge. The linear electro-optical Pockels effect is absent in the centrosymmetric crystalline Si. The Franz–Keldysh effect and the Kerr effect are small in Si. This leaves two candidate effects for modulation in Si: the plasma dispersion effect and thermal modulation.

Passive photonic structures that can bend, split, couple, and filter optical beams have been demonstrated in silicon, but the flow of light in these structures is predetermined and cannot be readily modulated during operation. All-optical switching in silicon has only been achieved by using extremely high powers in large or nonplanar structures, in which the modulated light is propagating out of plane. Such high powers, large dimensions, and nonplanar geometries are inappropriate for effective on-chip integration.

5.3.3.1 Modulator Based on MOS Configuration

Si modulators may be based on the variation of the index of refraction of Si with the free-carrier density. The refractive index of Si increases with the injected-carrier concentration. A metal oxide semiconductor (MOS) structure can be used to modulate the free-carrier concentration of the underlying Si in the cavity of a FP filter, as shown in Figure 5.21.

This causes a change in the resulting optical interference, resulting in a change in the output intensity. The operation is analogous to that of the channel formation in a MOSFET transistor.

Until recently, waveguide-based silicon optical modulators have been limited to relatively moderate speeds, the fastest reported being around 20 MHz. Recently, researchers at Intel made a breakthrough by moving away from the conventional

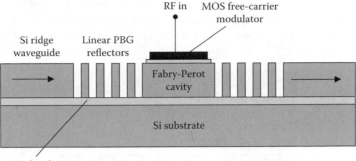

FIGURE 5.21 Si FP-MOS modulator.

current injection-based devices to a novel MOS-capacitor-based architecture [5.1]. This modulator has been the first high-speed optical active device on silicon with a bandwidth that exceeds 1 GHz.

This development of GHz modulator means that silicon could quickly become a viable alternative to the more conventional III-V or LiNbO$_3$ modulators currently being used for optical communications.

Intel's group has achieved the high-speed modulation by using a novel phase-shifter design based on a MOS capacitor embedded in a passive silicon waveguide Mach–Zehnder interferometer (MZI) [5.1]. A schematic representation of the MZI-MOS modulator is illustrated in Figure 5.22. The light wave coupled into the MZI is split equally into the two arms, each of which may contain an active section that converts an applied voltage into a small modification in the propagation velocity of light in the waveguide. Over the length of the active sections, the velocity differences result in a phase difference between the two waves. Depending on the relative phase of the two waves after passing through the arms, the recombined wave will experience an intensity modulation.

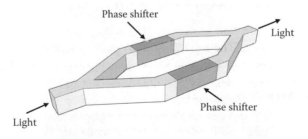

FIGURE 5.22 Schematic of a MZI-MOS modulator.

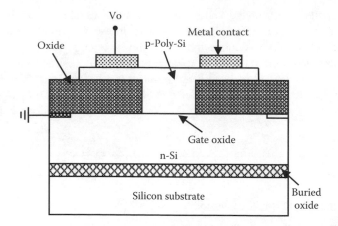

FIGURE 5.23 MOS-based waveguide phase shifter in SOI.

The novel component of Intel's silicon MZI modulator is the MOS capacitor phase shifter. Figure 5.23 shows a schematic of its cross-sectional view. It comprises a n-type doped crystalline silicon slab (the silicon layer of the SOI wafer) and a p-type doped polysilicon rib with a gate oxide sandwiched between them [5.1].

Aluminum contacts were deposited on top of this polysilicon layer, as shown in Figure 5.23. The oxide regions on either side of the rib maintain horizontal optical confinement and prevent the optical field from penetrating into the metal contact areas [5.1].

The n-type silicon in the MOS capacitor phase shifter is grounded and a positive drive voltage, V_D, is applied to the p-type polysilicon, causing a thin charge layer to accumulate on both sides of the gate oxide. The voltage-induced charge density change N_e (for electrons) and ΔN_h (for holes) is related to the drive voltage by [5.1]:

$$\Delta N_e = \Delta N_h = \frac{\varepsilon_0 \varepsilon_r}{e t_{ox} l} \left(V_D - V_{FB} \right)$$

where ε_0 and ε_r are the vacuum permittivity and low-frequency relative permittivity of the oxide, e is the electron charge, t_{ox} is the gate oxide thickness, l is the effective charge layer thickness, and V_{FB} is the flat band voltage of the MOS capacitor. Because of the free-carrier plasma dispersion effect, the accumulated charges induce a refractive index change in the silicon. The change in refractive index results in a phase shift $\Delta\varphi$ in the optical mode and is given by

$$\Delta\phi = \frac{2\pi}{\lambda} \Delta n_{eff} L$$

where L is the length of the phase shifter, λ is the wavelength of light in free space, and Δn_{eff} is the effective index change in the waveguide, which is the difference between the effective indices of the waveguide phase shifter before and after charge accumulation. Because majority carriers govern charge transport in the MOS capacitor, the relatively slow carrier recombination processes of PIN diode devices do not limit device bandwidth. As a result, this capacitor-based design has allowed a bandwidth that was unprecedented in a silicon-based modulator [5.1].

5.3.3.2 Modulator Based on Ring Resonators

The difficulty of modulating light using silicon structures arises from the weak dependence of the refractive index and absorption coefficient on the free-carrier concentration. For example, for a 300-mm-long 1.55-mm Mach–Zehnder modulator based on rib waveguides with a mode-field diameter of about 5 mm, a minimum optical pump pulse energy of 2 mJ is needed to modify the real part of the refractive index by $\Delta n = 10^{-3}$ in order to achieve 100% modulation [5.5]. The absorption due to free carriers under such high powers is also small, which demands a very long waveguide (as long as tens of centimeters) in order to achieve the required modulation depth.

Highly confined resonant configurations for low-power light modulation have been proposed in order to overcome these aforementioned limitations of silicon photonic structures.

An example of such configuration is illustrated in Figure 5.24. This configuration is based on a ring resonator that is coupled to a waveguide [5.5].

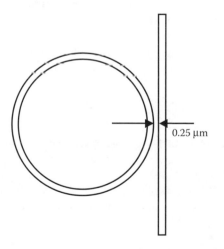

FIGURE 5.24 Ring resonator coupled to a waveguide.

The transmission of a ring resonator, coupled to a waveguide, is highly sensitive to the signal wavelength and is greatly reduced at wavelengths at which the ring circumference corresponds to an integral number of guided wavelengths. Figure 5.24 shows a schematic of a SOI ring resonator with 10-mm diameter. Both the silicon waveguide and the ring resonator were channel waveguides with 450-nm-wide by 250-nm-high rectangular cross sections [5.5].

Changing the effective index of the ring waveguide can modify the resonance wavelength. The modulator could also be used as a switch or router with a response time as high as 100 psec.

5.3.3.3 Spatial Light Modulators

Spatial light modulators (SLMs) play an important role in several technical areas in which the control of light on a pixel-by-pixel basis is a key element, such as optical processing, for inputting information on light beams, and in displays [5.6]. For display purposes, it is required to have as many pixels as possible in as small and as cheap a device as possible. For such applications, designing silicon chips for use as spatial light modulators has been effective. The basic idea is to have a set of memory cells laid out on a regular grid. These cells are electrically connected to metal mirrors such that the voltage on the mirror depends on the value stored in the memory cell. A layer of optically active liquid crystal is sandwiched between this array of mirrors and a piece of glass with a conductive coating. The voltage between individual mirrors and the front electrode affects the optical activity of the liquid crystal. Hence, by being able to individually program the memory locations, one can set up a pattern of optical activity in the liquid-crystal layer. Figure 5.25 shows a spatial light modulator based on the static random access memory (SRAM) that also includes an embedded complementary metal oxide semiconductor (CMOS) circuit.

SLMs are used in optical data storage applications. These devices are used to write data into the optical storage medium at high speed.

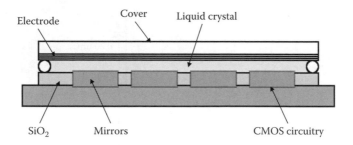

FIGURE 5.25 Spatial light modulator.

REFERENCES

1. Salib, M., Liao, L., Jones, R., Morse, M., Liu, A., Samara-Rubio, D., Alduino, D., and Paniccia, M., Silicon photonics, *Intel Technol. J.*, 8, 144–160, 2004.
2. Berger, J.D. and Anthon, D., Tunable MEMS devices for optical networks, *Optics and Photonics News,* 43–49, March 2003.
3. Connelly, M., Semiconductor optical amplifiers and their applications, http://www.ece.µl.ie/research/ocrg/OPTOEL2003%20Semiconductor%20optical%20amplifiers.pdf (May 2005).
4. Boyraz, O., and Jalali, B., Demonstration of a silicon Raman laser, *Opt. Express*, 12, No. 21, 5269–5273, 2004.
5. Almeida, V.R., Barrios, C.A., Panepucci, R.R., and Lipson, M., All-optical control of light on a silicon chip, *Nature*, 431, 1081–1084, 2004.
6. Goswami, D., Optical computing, *Resonance*, 56–71, 2003.

6 Couplers and Switches

6.1 COUPLERS AND SPLITTERS

Couplers and splitters are the passive devices that are used in every photonic system. They are used to split, combine, and route signals within systems.

Some of the most common applications for couplers and splitters include:

- Local monitoring of a light source output
- Distributing a common signal to several locations simultaneously

6.1.1 SPLITTERS

An optical splitter is a passive device that splits the optical power carried by a single fiber into two or more outputs. These devices have at least three ports but may have more than 32 in more complex configurations. Figure 6.1 illustrates a simple three-port device, also called a *tree splitter*.

One port is called the input port, whereas the other two are called output ports. The splitter manufacturer determines the ratio of the distribution of signal between the output ports. Most common splitting ratios are: 50%/50%, 90%/10%, 95%/5%, and 99%/1%.

An optical combiner is a passive device that combines the optical signals carried by two or more inputs into a single output. Figure 6.2 illustrates the transfer of optical signal in a combiner.

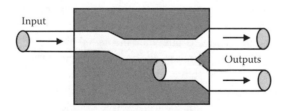

FIGURE 6.1 Three-port splitter.

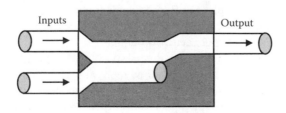

FIGURE 6.2 Optical combiner.

6.1.2 COUPLERS

Optical couplers either split signals into multiple paths or combine multiple signals on one path. Optical signals are more complex than electrical signals, making optical couplers more difficult to design than their electrical counterparts. Similar to electron flow in electrical circuits, a flow of photons comprise the optical beam. However, an optical beam does not flow through the receiver to the ground. Rather, at the receiver, a detector absorbs the signal flow. Multiple optical receivers, connected in a series, would receive no signal past the first receiver, which would absorb the entire signal. Thus, multiple parallel optical output ports must divide the signal between the ports, reducing its magnitude.

Optical couplers prevent the transfer of optical beam from one input to another input. Directional couplers prevent this transfer of optical signal between inputs. Optical couplers can be designed in such a way that they transmit the same amount of power when the input and output are reversed. These devices are known as the symmetrical couplers.

Some optical data links require more than simple point-to-point connections. These data links may be of a more complex design that requires multiport or other types of connections. In many cases these types of systems require optical components that can redistribute (combine or split) optical signals throughout the system. In applications that require links other than point-to-point links, optical couplers find the widest use. Such applications may include (1) bidirectional links and (2) local area networks (LANs). This is schematically illustrated in Figure 6.3, where the use of couplers is shown in examples of ring, bus, and star topologies.

One example of an optical coupler is a fiber coupler. A fiber coupler is a device that can distribute the optical signal from one fiber to two or more fibers. A fiber-optic coupler can also combine the optical signal from two or more fibers into a single fiber. Fiber couplers attenuate the signal much more than a connector or splice because the input signal is divided among the output ports.

Couplers can be either active or passive devices. The difference between active and passive couplers is that a passive coupler redistributes the optical signal without optical-to-electrical conversion. Active couplers are electronic devices that split or combine the optical signal electrically and use optical detectors and optical sources as inputs and outputs.

The so-called X coupler (or 2×2 coupler) combines the functions of the optical splitter and combiner. An X coupler combines and divides the optical power between the two inputs and two outputs.

Figure 6.4 illustrates the design of a passive $N \times M$ optical coupler. The letter N represents the number of inputs, and M represents the number of outputs. An $N \times M$ coupler has N input and M output ports. The values for N and M can be from 1 to 64. The number of inputs and outputs may vary depending on the intended application.

Other types of multiport couplers are star and tree couplers.

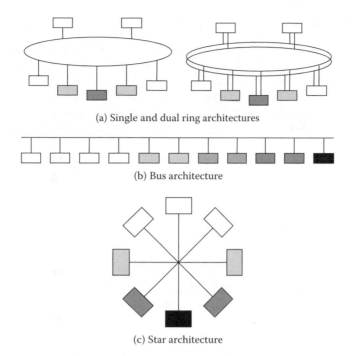

(a) Single and dual ring architectures

(b) Bus architecture

(c) Star architecture

FIGURE 6.3 Ring, bus, and star network topologies.

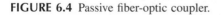

N input ports   M output ports

FIGURE 6.4 Passive fiber-optic coupler.

A star coupler is a passive device that distributes optical signals from two or more input ports to several output ports. Figure 6.5 shows the multiple input and output ports of a star coupler. The star coupler divides all outputs allowing every station to communicate to every other station. Star couplers may have many ports, and couplers with 32 or 64 ports are quite common.

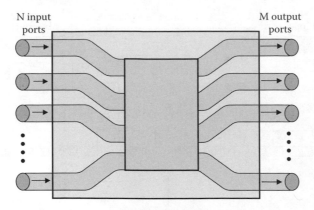

FIGURE 6.5 Star coupler.

The bus topology utilizes a tree-type coupler to connect a series of stations that are connected to a single backbone of cable. In a typical bus network, a coupler at each node splits off part of the power from the bus and carries it to a transceiver. In a system with N terminals, a signal must pass through N – 1 couplers before arriving at the receiver. Its loss increases linearly as the number N increases.

A bus topology may operate in a single direction, bidirectional, or duplex transmission configuration. In a single directional setup, a transmitter at one end of the bus communicates with a receiver at the other end. Each terminal also contains a receiver. Duplex networks add a second bus or use an additional directional coupler at each end and at each terminal. In this way, signals may flow in both directions.

A tree coupler is a passive device that splits the optical power from one input fiber to two or more outputs. A tree coupler may also be used to combine the optical power from more than two inputs into a single output. Figure 6.6 illustrates these two types of tree coupler configurations.

Fiber-optic coupler fabrication technique is a rather complex one. This process may involve beam splitting using microlenses or graded-refractive-index rods as well as beam splitters and optical mixers. The beam splitters are used to divide the optical beam into two or more separated optical signals.

The most common type of coupler in use is the fused coupler. A fused fiber coupler is mostly used with 1 × 2 and 2 × 2 configurations. In this coupler, two (or more) fibers are twisted together and then spot-fused under tension to form an elongated biconical taper structure (see Figure 6.7).

The most interesting application of fused biconical taper (FBT) couplers is for wavelength-division-multiplexing (WDM) systems. This application is only possible with single-mode fibers. There is a coherent interaction between optical waves in the fibers that form the taper (see Figure 6.8). This interaction leads to a periodic variation in the splitting ratio as a function of the taper length and of the wavelengths. By adjusting the taper length it is possible to select two

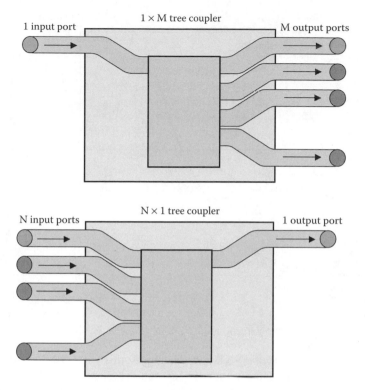

FIGURE 6.6 Tree coupler.

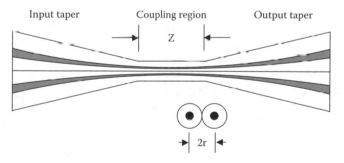

FIGURE 6.7 Fused coupler.

wavelengths of interest. In other words, the pull length is used to control the coupling for the selected wavelengths [6.1]. This is illustrated in Figure 6.8. For example, point A in Figure 6.8 marks a coupler with a coupling ratio of 50% at 1550 nm, and point B marks the taper length required for a 50% coupler at 1310 nm.

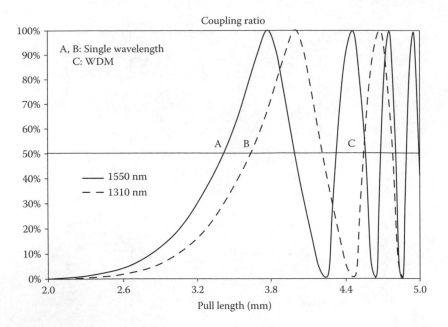

FIGURE 6.8 Wavelength interaction in FBT coupler.

Point C marks a specific WDM application. At this point, 100% of the signal at 1550 nm and 0% of the signal at 1310 nm are coupled [6.1]; this coupler separates two wavelengths such that each output carries only one wavelength. As a result, the fused taper operates as wavelength-division multiplexers for these two specific wavelengths.

The FBT couplers are characterized by the following properties:

- Low insertion loss
- Wavelength-selective or broadband properties
- High thermal and mechanical stability
- Applicable to any coupling ratio (1%–50%).

Another technique that is used for making couplers is planar lightwave circuit (PLC) technology. PLC enables optical waves to pass through a wafer structure in much the same way they do through fiber. PLC has been used to manufacture multiplexer or demultiplexer modules based on arrayed-waveguide gratings (AWGs) for WDM systems. This technology allows for manufacture of 1×8, 1×16, 1×32, and even higher ratio configurations.

Couplers may be used to separate (or combine) signals transmitted at different wavelengths. Essentially, the transmitting coupler is a mixer and the receiving coupler is a wavelength filter. For example, demultiplexing 1310-nm and 1550-nm channels is a relatively simple operation and can be achieved using bulk optical dichroic filters. Wavelengths in the 1550-nm range that are spaced at

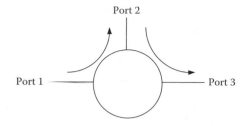

FIGURE 6.9 Optical circulator.

greater than 1 to 2 nm can be resolved using WDM couplers and/or AWGs. Fiber Bragg gratings are used to separate more closely spaced wavelengths in a dense-wavelength-division-multiplexing (DWDM) system if the separation is in the range of 0.8 nm or smaller.

6.1.3 CIRCULATORS

Optical circulators are passive devices that transport an optical signal from one port to the next port only in one direction, i.e., 1 to 2 or 2 to 3 (see Figure 6.9). They may be used to separate forward- and backward-propagating signals with high degree of isolation and a minimum interference between ports.

Separation of the signals by circulators is based on propagation direction. There are no additional losses that are imposed on transmitted signals, as is observed in the case of directional couplers.

6.2 OPTICAL ISOLATORS

Laser oscillators are based on optical feedback, and they are sensitive to weak back reflections. Back reflections can cause the frequency to drift and the power to vary owing to changes in the laser's gain profile.

Any two optical surfaces can form a Fabry–Perot cavity (see Chapter 3) that could cause spectral modulation. When surface reflection is high enough, modulation amplitude can reach beyond 50% (see Figure 6.10).

For example, the so-called DFB (distributed-feedback) lasers are particularly sensitive to the back reflections. Without isolators the DFB lasers will almost certainly fail to operate properly.

A common device that is used to overcome this problem is the optical isolator, which transmits the laser beam but rejects its reflections. One example of the optical isolator is shown in Figure 6.11. This isolator is based on the Faraday effect. The Faraday isolator is usually constructed by placing a block of glass with a high magneto-optical coefficient in an axial magnetic field.

Before entering the Faraday rotator, which is usually an yttrium–iron–garnet (YIG) material, the laser beam passes through the input polarizer P_1 and is linearly polarized. This linearly polarized beam enters then the Faraday rotator rod. The plane of polarization rotates as the light propagates along the axis of the rod. The

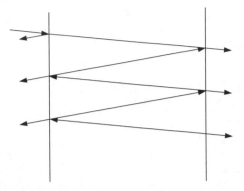

FIGURE 6.10 Fabry–Perot cavity back reflections.

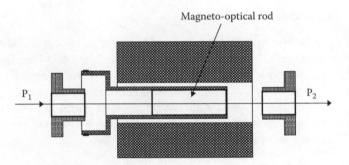

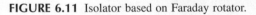

FIGURE 6.11 Isolator based on Faraday rotator.

Faraday rotator is placed in such a way that it rotates the plane of polarization by 45°. The light then passes through the output polarizer P_2, whose transmission axis is also at 45°. In this way most of laser beam passes through the output of the isolator. Any backward-reflected beam will reenter the isolator through the output polarizer (P_2) and becomes polarized at 45°. It then passes through the Faraday rotator, which induces additional 45° rotation and is now polarized at 90° as opposed to the input polarizer. Therefore, the input polarizer stops the back-reflected laser beam. Thus, the laser itself is isolated from its own reflections.

The amount of rejection offered by an optical isolator may improve problems caused by back reflections, but often will not eliminate them. This is schematically illustrated in Figure 6.12. Figure 6.12a shows a laser waveform with no back reflection, and Figure 6.12b shows a laser waveform with a strong back reflection.

It should be noted that optical isolators are not a substitute for properly polished, low-back-reflection connectors. In order to address the back reflection problem, the fiber-optic industry introduced a number of specially prepared connectors, such as (1) physical contact (PC) and (2) angled physical contact (APC) polished connectors. The APC connectors especially help in eliminating any concern regarding back reflections from the optical connectors.

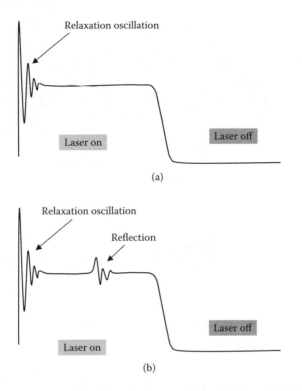

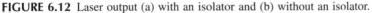

FIGURE 6.12 Laser output (a) with an isolator and (b) without an isolator.

6.3 GRATINGS

There are two current technologies used for splitting or combining the optical beams into rays of waves with different wavelengths that are required for the WDM systems. These are (1) arrayed-waveguide gratings and (2) echelle gratings. These devices accept a single input fiber that contains a wavelength-multiplexed signal corresponding to N separate information channels. Wavelength-dependent filtering or diffraction is employed to separate the signal into its N channel components and direct them to N output fibers.

Arrayed-waveguide devices (see Figure 6.13) consist of an array of curved channels with a fixed optical path difference between adjacent waveguides. The array of channel waveguides is connected to cavities at the input and output. The input optical signal is distributed among the waveguides. The difference in the optical path length of each waveguide introduces controlled phase delays. This results in interference at the output cavity, with maximal interference due to different input wavelengths at different physical positions along the output cavity. The points of maximum signal are subsequently coupled to an array of output fibers. Arrayed-waveguide devices are large, typically one device occupying an entire 4-in.-O.D. optical substrate.

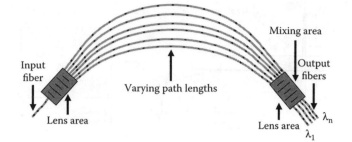

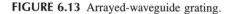

FIGURE 6.13 Arrayed-waveguide grating.

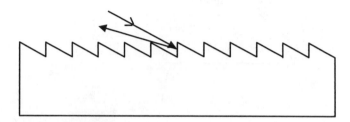

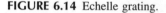

FIGURE 6.14 Echelle grating.

An echelle grating is a blazed grating with a relatively high (on the order of 65°) blaze angle (see Figure 6.14).

Lithographic techniques facilitate the planar patterning of optical structures on a suitable slab waveguide to yield a fully integrated, guided-wave echelle-grating-based spectrometer. WDMs based on the echelle grating are largely based on variations of the Rowland spectrometer geometry. A single blazed concave grating provides both wavelength dispersion and output signal focusing onto an array of output channel waveguides. This has significant advantages over AWGs in terms of device size and potential wavelength resolution. Several spectrometers can be patterned on a single 4-in.-O.D. substrate. The main technical challenge is the patterning of the grating elements. These involve submicron features and require optically smooth, perpendicular sidewalls to minimize optical signal scattering.

Compact high-resolution echelle grating WDM multiplexers can be achieved by using high-order diffraction (m = 6 to 24). This technology has now been under development for over 10 years. The main technical constraint has been the reactive ion etching (RIE) aspect ratio required to define the grating elements. Aspect ratios of 20 to 200 have been achieved, facilitating relatively-high-quality grating faucets. This technical breakthrough in RIE also facilitates sharp bends in high-index waveguides such as silicon on insulator (SOI) and InP. An additional benefit of high-aspect RIE technology development is the possibility of right-angle bends in high-index waveguides such as SOI and InP.

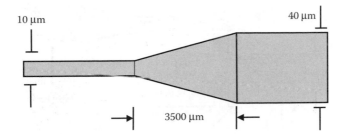

10 µm   40 µm

3500 µm

FIGURE 6.15 Tapered waveguide beam collimator/expander.

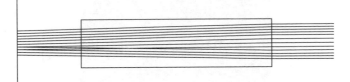

FIGURE 6.16 Zeemax ray tracing for tapered waveguide.

6.4 WAVEGUIDE COLLIMATORS

Tapered waveguide collimators are used to shape the optical beam in the plane of the waveguide. The tapering reduces the numerical aperture (NA) of the beam in the horizontal direction, providing near collimation of the optical signal. This becomes the basis of various devices, including the T-junction signal splitters, Mach–Zehnder interferometers, and 2 × 2 microelectromechanical systems (MEMS) switches. This waveguide structure may also be used to improve input/output (I/O) signal coupling to the SOI platform.

Figure 6.15 illustrates the tapered waveguide collimator. The basic part of the device is a tapered waveguide that is capable of collimating a laser beam. One method of achieving this is to gradually expand the width of the channel waveguide to reduce the effective NA of the optical beam in the horizontal direction. This was simulated using nonsequential Zeemax ray tracing, as shown in Figure 6.16.

The beam collimation was analyzed in terms of the beam expansion factor, from 1 to 5, and the beam expansion rate, from $100w$ to $1000w$, where w is the initial width of the channel waveguide. It was found that a beam expansion factor of about 4 is sufficient.

6.5 TOTAL INTERNAL REFLECTION T JUNCTION

Figure 6.17 presents an optical beam splitter based on (1) a collimator, (2) beam splitter mirrors, and (3) a tapered waveguide. The main advantage of this design

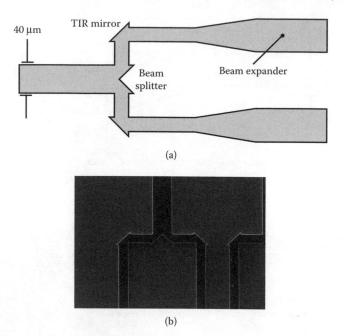

(a)

(b)

FIGURE 6.17 Total internal reflection waveguide (a) schematic, (b) SEM micrograph.

is precise control of the beam splitting ratio, feasibility of 1 × N splitters or combiners, and the substantially smaller substrate area required than that offered by fused optical couplers.

6.6 OPTICAL SWITCHES

The rapid and global development of the Internet and broadband network services is accelerating the growth of optical communication networks. The newest branch of optical communications, i.e., photonic network, is based on WDM systems, and they are playing a key role in increasing their capacity and flexibility. An optical switch is a crucial component of any WDM system. The basic function of an optical switch is to allow the signal on any one of the inputs to be redirected to any one of the outputs in the manner configured by the user.

Most networking equipment today are still based on electronic signals, meaning that the optical signals have to be converted to electrical ones to be amplified, regenerated, or switched and then reconverted to optical signals. This is generally referred to as an *optical-to-electronic-to-optical* (OEO) conversion and is a significant bottleneck in transmission. The basic premise of optical switching is that by replacing existing electronic network switches with optical ones, the need for OEO conversions is removed.

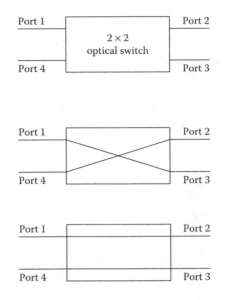

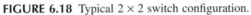

FIGURE 6.18 Typical 2 × 2 switch configuration.

Optical switches are very attractive for commercial implementation because they are transparent to the data rate of the signal. This not only reduces the system cost but also facilitates future upgrading of the system's data capacity. To achieve this, it is important for the optical switch to be wavelength independent, polarization independent, and scalable. With the aforementioned advantages, the optical switch is independent of the system upgrading and network evolution from single-wavelength to multiwavelength systems.

The number of input and output ports, expressed as an N × M configuration, characterizes a switch (similar to a coupler). Switches can be made in any configuration, but they commonly use multiples of two.

A digital optical switch has a matrix arrangement consisting of 1 × 2 or 2 × 2 unit switches. Each switch has two states and acts similar to an on–off shutter. An input optical signal can be switched to any desired output port by operating a unit switch at a specific intersection. Figure 6.18 illustrates the most common switch based on a 2 × 2 configuration.

The design and fabrication of optical switches can be categorized into two types depending on whether they use free-space transmission or optical waveguides as the propagation medium.

The main technologies for optical switches include: optomechanical switches, electro-optical switches, liquid crystals, bubble switches, thermo-optical switches, holographic switches, acousto-optical switches, total internal reflection (TIR) switches, semiconductor optical amplifier (SOA) switches, and MEMS mirror-based switches.

6.6.1 Optomechanical Switches

Optomechanical switches are the oldest type of optical switches. These devices work by moving optical elements by means of stepper motors or relay arms. This causes them to be relatively slow with switching times in the 10 to 100 msec range.

The most common configurations of optomechanical switches are the (1) gate switch and (2) cross-connect switch (see Figure 6.19). Optomechanical switches demonstrate excellent reliability combined with low insertion loss and no cross-talk interferences.

6.6.2 Electro-Optical Switches

Electro-optical devices are typically built on a lithium niobate ($LiNbO_3$) substrate. These switches offer the best switching times but consume the most substrate real estate on a per channel basis. The structure of a 1×2 electro-optical switch is shown in Figure 6.20.

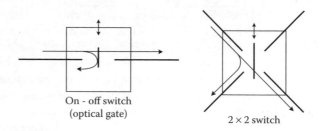

On - off switch
(optical gate)

2×2 switch

FIGURE 6.19 Optomechanical switches.

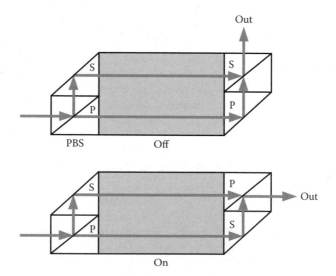

FIGURE 6.20 Electro-optical switch.

A randomly polarized optical beam is incident on the switch via a polarizing beam splitter (PBS). The splitter divides the optical beam into two perpendicularly polarized components. The S-polarized component is reflected, whereas the P-polarized component is transmitted. A reflecting prism is used to direct S-polarized component into the electro-optical crystal. If the electro-optical crystal is in the off state (i.e., no voltage is applied to the electrodes), the two components maintain their polarization during propagation. At the crystal output, a reflecting prism and a splitter combine the two components and a single beam emerges from the switch at the output in the vertical direction. When the electro-optical crystal is in the on state, i.e., a suitable voltage is applied to the crystal electrodes, then the polarization of the two components are subjected to a rotation. The combiner at the output recombines the two components, and the optical beam emerges at the output in another direction. In this way a 1×2 optical switching is realized.

The LiNbO$_3$-based switch has drawbacks of long device length and high operating voltage (20 to 60 V). It is difficult to make high-performance, low-cost, and wafer-size LiNbO$_3$ switches. In addition, a large area is required for each switching node, in excess of about 1 cm^2. Also, the switching voltage has to be tailored for the individual characteristics of a node to obtain high optical transfer efficiency. Therefore, the control methodology becomes very complex if several nodes are involved.

6.6.3 LIQUID CRYSTALS

Liquid-crystal switches have been borrowed from laptop screen technology. Electric currents alter the properties of liquid crystals in such a way that light passing through them is polarized in different ways. These optical devices can then be used to steer each wavelength of light depending on its polarization.

The liquid-crystal state is a phase that is exhibited by a large number of organic materials over certain temperature ranges. In the liquid-crystal phase, molecules can take up a certain mean relative orientations due to their permanent electrical dipole moments. It is thus possible, by applying a suitable voltage across a cell filled with a liquid-crystal material, to act on the orientation of the molecules.

A 1×2 liquid-crystal optical switch structure is shown in Figure 6.21. The birefringent plate at the input port manipulates the polarization states to the desired ones. Without applying a bias, the input signal passes through the liquid-crystal cell. By applying a voltage on the liquid-crystal spatial modulator, molecules rotate the polarizations of the signal passing through them. With sufficient voltage, the signal polarizations rotate to the orthogonal ones, and the polarization beam splitter reflects the signal to the other output port.

Liquid-crystal switches are wavelength selective, i.e., they can switch signals depending on their wavelength. This is a very attractive feature, as it allows adding and dropping single wavelengths from a multiwavelength beam.

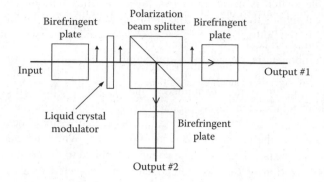

FIGURE 6.21 Liquid-crystal switch.

Another version of these switches are liquid-crystals-in-polymer switches. These switches are built by filling an active cell with a mixture of liquid crystals and a particular monomer. This mixture undergoes a process of polymerization that produces a stable structure, the process being characterized by alternating polymer and liquid-crystal microdroplets layers. The refractive index of the polymeric layers normally differs from that of the liquid-crystal layers. By applying a suitable driving voltage, the orientation of the optical axis of the liquid-crystal microdroplets can be changed. This variation can be made such that the refractive index of the polymeric layers match the liquid-crystal microdroplets layers. In this case, the cell is transparent to the light beam (see Figure 6.22a). If there is no driving voltage applied, the difference of the refractive indexes makes the active cell work as a Bragg grating. In this case the signal is deflected to another output port (see Figure 6.22b).

6.6.4 BUBBLES

Bubble switches are based on tiny bubbles that act as mirrors, reflecting light onto intersecting paths as they traverse microscopic channels carved in silica. The bubbles are generated using ink-jet printer technology.

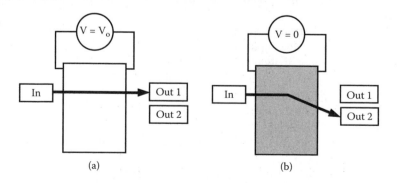

FIGURE 6.22 Liquid-crystal-in-polymers switch.

A bubble switch is made up of two layers: a silica bottom layer, through which optical signals travel, and a silicon top layer, containing the ink-jet printer technology. In the bottom layer, two series of waveguides intersect each other at an angle of around 120°. At each cross point between two guides, a tiny hollow is filled with a liquid has the same refractive index of silica in order to allow propagation of signals in normal conditions. Thus, a light beam travels straight through the guide, unless the guide is interrupted by a bubble placed in one of the hollows at the cross points. In this case, light is deflected into a new guide crossing the path of the previous one. Bubbles are generated by means of tiny electrodes placed in the top silicon layer, which heat the liquid until it gasifies.

The technology competes with MEMS — arrays of microscopic mirrors — and has the advantage of not having moving parts to seize up or wear out. Bubble switches, unlike some MEMS devices, are less prone to vibration and realignment problems.

6.6.5 HOLOGRAPHIC SWITCHES

The principle of holographic switches is based on dynamically controlled wavelength-specific reflective gratings. The grating structure in these devices is in the form of a hologram. Holograms are stored as spatial distribution of charge in crystals. The application of a driving voltage is used to activate prestored holograms in order to deflect light beams. The holograms can be energized by means of externally applied electric field.

A schematic diagram of a holographic switch is presented in Figure 6.23. As shown in Figure 6.23, if there is no voltage applied the electrodes, then the crystal is transparent to the optical beams. If a suitable driving voltage is applied, then the optical beams crossing the crystal are deflected. As it is possible to store several holograms in the same crystal, these devices can be used to drop even a single wavelength or groups of wavelengths from a WDM signal. The biggest advantage of holographic switches is that there are no moving parts.

6.6.6 ACOUSTO-OPTICAL SWITCHES

Acousto-optical devices are commonly called *Bragg cells*. These optical-switching devices employ the effects of sound to cause varying mechanical stresses in

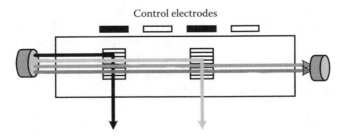

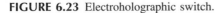

FIGURE 6.23 Electroholographic switch.

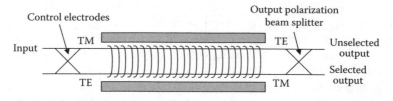

FIGURE 6.24 Acousto-optical switch.

the optical medium, which in turn causes change in its refractive index. A piezo-electric transducer produces the sound wave. Acousto-optical devices offer moderate switching speeds in the microsecond range. These devices are used to produce the acousto-optical tunable filters (AOTFs) that provide wavelength-selective switching.

The principle of operation of a polarization-insensitive acousto-optical switch is schematically illustrated in Figure 6.24. The input signal is split into its two polarized components by a polarization beam splitter. These two components are directed to two parallel waveguides. A surface acoustic wave is generated, and this wave travels in the same direction as the optical beams. Through an acousto-optical effect in the material, the equivalent of a moving grating is formed, which can be phase-matched to an optical beam at a selected wavelength. This allows for control of the optical beam's polarization.

It is possible with these devices to switch several different wavelengths simultaneously. The switching speed of acousto-optical switches is limited by the speed of sound and is in the order of microseconds.

Acousto-optical switching can also be used to provide $1 \times N$ switching in one node by using the acousto-optical signal as a Bragg reflector. However, this requires guiding the optical signal in a slab waveguide by using planar optics. This is very difficult to achieve without substantial optical losses.

6.6.7 TOTAL INTERNAL REFLECTION SWITCHES

TIR switches employ a prism interface, as schematically illustrated in Figure 6.25. At the boundary between two media, characterized by different values of their refractive indices, a beam of light may be reflected or can go through. The last case will occur if the two prisms are forced together. Otherwise, there will be an air gap that will form the condition for TIR.

Another version of the TIR switches are analogous to MEMS micromirror arrays but employ prealigned mechanically translated channel waveguides with an insertable index-matching microbubble at the crossing junctions. These switches based on controlling active optical amplification in III-V compound semiconductors can facilitate high functional density.

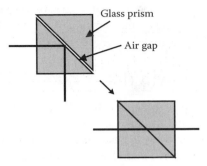

FIGURE 6.25 TIR switch.

6.6.8 THERMO-OPTICAL SWITCHES

The operation of thermo-optical switches is based on the variation of the refractive index of a dielectric material due to temperature variation of the material itself. There are two categories of thermo-optical switches: interferometric and digital.

Interferometric switches are usually based on Mach–Zehnder configurations. These devices consist of a coupler that splits the signal into two beams. The beams then travel through two distinct arms of equal length and through a second coupler that combines the signal again. Heating one arm of the interferometer causes its refractive index to change and varies the phase difference between the beams, causing constructive or destructive interference at the output.

Digital optical switches are normally based on waveguides made in polymers or silica. Their operation relies on the change of refractive index with temperature created by a resistive heater placed above the waveguide. Heating the waveguide locally can increase the index of refraction of Si or SiO_2 waveguides. This is done in order to match the refractive index change between two adjoining waveguides so that light can be switched from one branch of a y junction to another branch of the y junction. A schematic diagram of a 2 × 2 digital thermo-optical switch is shown in Figure 6.26.

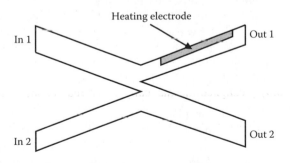

FIGURE 6.26 Thermo-optical digital switch.

Thermo-optical switches contain no moving parts and can therefore be highly reliable and repeatable. However, the switching speed is relatively slow, approximately 2 msec.

6.7 MOEMS-BASED SWITCHES

Microoptoelectromechanical systems (MOEMS) are miniature devices that perform optical, electrical, and mechanical functions. They have been made using batch process techniques derived from microelectronic fabrication processes. MOEMS offer attractive performance characteristics with very low cross talk, wavelength insensitivity, polarization insensitivity, and scalability. MOEMS-based switches may be manufactured as free-space switches or waveguide switches.

MOEMS-based switches consist of mirrors no larger in diameter than the human hair that are arranged on special pivots such that they can be moved in three dimensions. Several hundred such mirrors can be placed together on mirror arrays no larger than a few square centimeters. Light from the input is aimed at a mirror, which then directs the light to another mirror on a facing array. This mirror then reflects the light towards the desired output.

There are two types of MOEMS approaches for optical switching, i.e., (1) two-dimensional, or digital and (2) three-dimensional, or analog.

In two-dimensional MOEMS, the switches are digital as the mirror position is bistable (on or off), which makes driving the switch very straightforward. Figure 6.27 shows a top view of a two-dimensional MOEMS device with the mirrors arranged in a crossbar configuration to obtain cross-connect functionality. Collimated light beams propagate parallel to the substrate plane. When a mirror is activated, it moves into the path of the beam and directs the light to one of the outputs. This arrangement also allows light to be passed through the matrix without hitting a mirror. This additional functionality can be used for adding or dropping optical channels. The trade-off for the simplicity of the mirror control in a two-dimensional MOEMS switch is optical loss. The optical loss grows rapidly with the number of ports. The two-dimensional MOEMS architectures are found to be impractical beyond 32-input and 32-output ports. Although multiple stages of 32 × 32 switches can theoretically form a 1000-port switch, high optical losses also make such an implementation impractical.

Considerable work has been done on three-dimensional MOEMS switches. These devices are used to bend the optical beam path from an input optical fiber and direct it through free space onto a selected output optical fiber by using miniature mirrors fabricated on planar substrates (see Figure 6.28). This technology requires additional components in the form of (1) planar lens arrays to collimate the optical signal from the input fibers and (2) a second lens array to focus the resulting deflected optical signal into the desired receiving fiber. Such a system is mechanically complex and requires the fabrication of several wafers that must be aligned with micron precision in three-dimensional space. Because of the tolerance specifications for optical alignment between inputs and outputs,

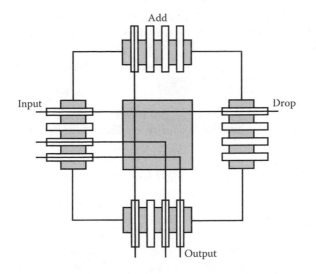

FIGURE 6.27 Two-dimensional MOEMS switch.

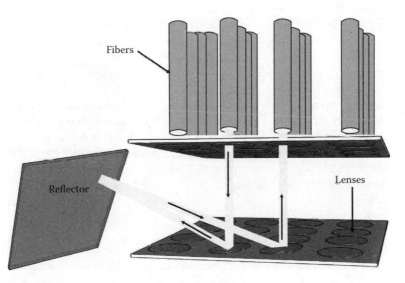

FIGURE 6.28 Three-dimensional MOEMS switch.

active alignment is employed during switching. Switching speeds are generally greater than 15 msec. The technology is currently being employed for scanning systems for high-definition projectors in HDTV systems.

Another approach has been based on the use of high-displacement MOEMS actuators to move optical fibers as a means of providing optical switching. This approach requires lateral displacement of fibers by several hundred microns for

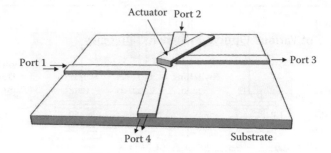

FIGURE 6.29 MOEMS-based 2 × 2 optical cross connect.

a 1 × N device. To achieve this displacement range, several centimeter-long cantilevers are needed that can be actuated by using microheaters.

A MOEMS cantilever may be employed to form an optical cross connect (OXC) switch. As shown in Figure 6.29, the cross connect directs the optical signal from port #1 to either port #3 or port #4. The actuation can be electrostatic, thermomechanical, or electromagnetic.

In the design shown in Figure 6.29, a 2 × 2 optical switch consists of ridge waveguides and an electrostatically actuated flip-up mirror.

The MOEMS cantilever can be driven by the electrostatic linear motor [6.3]. Two sets of fixed electrodes may be used, one at a potential V+ and the other at V–. When the potential of the movable slider is at V–, the actuator is drawn towards the V+ electrode. This inserts the Si slider into the gap between the channel waveguides. This action reduces the difference in refractive index between the gap and the Si core of the waveguides. As a result, the light from port #1 travels straight through to port #3 with minimal reflection at the intersection. If the potential to the actuator is reversed such that it is at V+, it will be repelled from the V+ fixed electrode and attracted towards the V– electrode. This will cause the Si slider to retract from the waveguide intersection. This will result in a large refractive index difference between the Si waveguide (n = 3.45) and the air gap (n = 1). As a result, light traveling into port # 1 will be reflected at the intersection and directed into port #4. Similarly, light traveling from port #2 will be directed to port #3 [6.3].

The cantilever can be coated with a reflective metallic layer, resulting in light from port #1 being directed to port #4 when the actuator is inserted between the waveguides and in passing from port #1 to port #3 when the actuator is in a retracted position [6.3]. In this case the waveguides and actuator are fabricated on separate substrates, and the cantilever actuator requires a large gap in the waveguide, which compromises the attainable optical performance. This entails careful alignment and positioning of the two substrates and accounts for the relatively high insertion loss.

Table 6.1 summarizes the main characteristics of the various technologies used for switches.

TABLE 6.1
Summary of Various Optical Switch Technologies

Technology	Substrate	Switching Speed	Channel Capacity	Size (cm²)	Functional Density (channels/cm²)
Electro-optical	LiNbO₃	< nsec	6 × 6	350	0.03
Thermo-optical	Silica, SOI	2 msec	16 × 16	110	2.4
Total internal reflection switch	Silica, SOI	5–10 msec	32 × 32	6	170
2 × 2 MOEMS switch	SOI	ms to μs		0.2 × 0.3	120 (limited by MEMS actuator size)

6.8 WAVEGUIDE SWITCHES

Currently, optical waveguide switches use a number of effects as their principle of operation, i.e., optical gain variation, interferometric effect, internal reflection effect, etc.

Figure 6.30 illustrates a waveguide switch based on the Mach–Zehnder interferometer and the total internal reflection T junction (the same configuration may be used for a modulator — see Figure 5.20 in Chapter 5). It consists of two back-to-back T junctions. One of T junction serves to divide the signal; the other T junction recombines the two arms of the interferometer.

A microheater can be applied to one arm of the interferometer in order to control the refractive index of the Si waveguide core. Depending on the temperature of the heated arm of the waveguide interferometer, the transmittance can vary from 0 to 100% (see Figure 6.31).

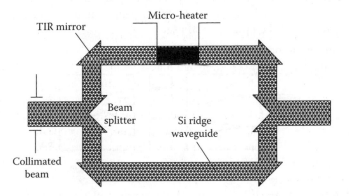

FIGURE 6.30 Waveguide Mach–Zehnder switch.

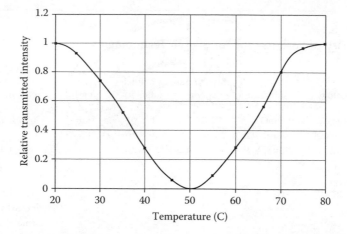

FIGURE 6.31 Transmitted intensity at 1550 nm vs. microheater temperature.

This type of device is limited to a simple two-port device in a single stage. However, higher channel capacities may be obtained by using a cascade of the primal 1×2 or 2×2 waveguide switching nodes. Considerable substrate area per switching node is required because of an active interaction length of 5 to 10 mm. This limits the practical switch channel capacity N because the device size increases as N^2. Various cascade architectures are being developed to reduce the number of switching layers and nodes.

6.9 SOA SWITCHES

Optical switches can be constructed using SOAs. The simplest method to control an SOA switch is by turning the device current on or off. The great advantage of SOA switches is that they can be integrated to form switch arrays. In the 2×2 switch shown in Figure 6.32, incoming input beams can be routed to any output port by switching on the appropriate SOA. The switching time of a current switched SOA is of the order of 100 psec.

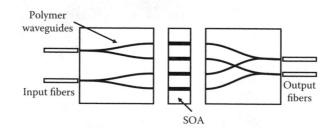

FIGURE 6.32 SOA switch.

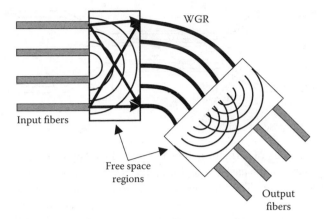

FIGURE 6.33 Optical cross connect based on WGR.

6.10   WAVEGUIDE GRATING ROUTERS

Among the most promising switches with many input ports to many output ports is the generalized Mach–Zehnder waveguide grating router (WGR).

The WGR is an interferometer consisting of an array of I/O waveguides, a waveguide grating, and two free-space regions that connect the I/O ports to the waveguide grating (see Figure 6.33). Optical signal from one of the input ports is coupled through the free-space region to the grating arms. A path length difference between neighboring arms causes a wavelength-dependent linear phase shift between the individual grating arms. The optical signal at a certain wavelength will predominantly couple to only one output port.

The WGR was first demonstrated in glass-based materials. It is now commercially available and deployed in long-haul WDM systems. Its advantages include scalability to a large number of channels, highly reproducible optical characteristics, low insertion loss, and N × N connectivity.

The WGR can also be fabricated on InP. Although WGRs on InP have shown excellent performance, it is generally believed that they cannot outperform their glass-based counterparts in passive demultiplexing applications. The advantage of InP-based WGRs is the possibility to monolithically integrate them with components such as semiconductor amplifiers, detectors, or modulators.

6.11   EVANESCENT SWITCHES

These types of switches are based on the evanescent coupling between waveguides (see Chapter 3). A schematic of an evanescent switch is shown in Figure 6.34.

Waveguides can exchange energy by means of evanescent waves that exist in the region that separates them. Evanescent coupling is a function of the length of the interaction and of the optical properties of the materials in which the waves

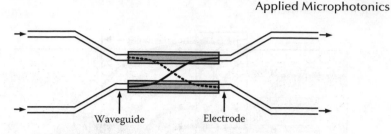

FIGURE 6.34 Evanescent switch.

propagate. The index of refraction may also be changed by a number of externally controlled effects. In the case of electro-optical switching, the device is controlled by an electrical signal. The voltage is used to modify the refractive index of the waveguide along the path of the beam. By modifying the refractive index, the optical signal's propagation speed is changed to provide matching to an adjacent waveguide. In this way it is possible to modify and control the coupling and energy exchange between the two waveguides. For example, identical waveguides with complete coupling from one waveguide to the other (cross state) can be placed in the bar state when a voltage is applied resulting in changing the substrate's index of refraction. The bar state is created by altering the propagation constants in the waveguides in order to ensure that there is no coupling between the two waveguides. Therefore, the change in the index of refraction allows directing the optical beam to the desired output port.

In addition to being used as a switch, such a device may be used as a modulator and as an element of a multiplexer.

Another configuration of an evanescent switch is shown in Figure 6.35. This switch contains two movable parallel waveguides that can be electrostatically actuated in such a way as to vary the spacing between them [6.2].

If these two single-mode waveguides are brought close together, then the compound waveguide structure can support two modes that travel along the waveguides. These modes travel at different velocities along the waveguide. Both modes will contribute to the optical power measured at the output. The resulting

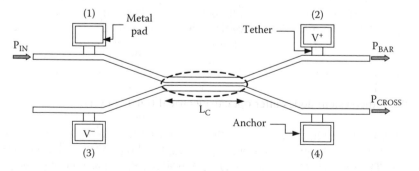

FIGURE 6.35 Movable evanescent switch.

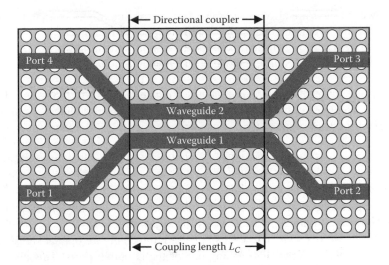

FIGURE 6.36 PBG-based evanescent switch.

output will depend on the phase shift between the modes. Therefore, there will be a constantly varying exchange of power between bar and cross waveguides throughout the interaction length. In this case, the coupling coefficient depends exponentially on the waveguide gap, the waveguide geometry, and refractive index, as well as on the optical polarization of the modes [6.2].

Photonic-band-gap (PBG) structures may also be used for the fabrication of evanescence switches that will allow miniaturizing the microphotonic integrated circuits (micro-PICs) to a scale comparable to the wavelength of light.

Figure 6.36 illustrates a switch that is similar to that shown in Figure 6.30. The difference is that, instead of conventional waveguides, the structure is based on PBG waveguides.

The optical properties of the PBG switch may also be controlled by the use of external electrical or optical signals.

6.12 OPTICAL CROSS CONNECTS

Channel connecting is a key function in most communication systems and the optical switches are the basic components of OXCs. The first WDM networks were point-to-point transmission systems and the signals were processed electrically in the system nodes after optical–electrical (OE) conversion.

Currently, very attractive new-generation photonic networks are implemented. They are based on many point-to-point DWDM transmission systems. In these networks, optical signals are processed without OEO conversion at the nodes. They require OXC and optical–add-drop-multiplexing (OADM) systems. Key devices for these systems are large-scale arrays of optical switches. As large as 1000×1000 optical switch arrays have already been proposed.

Switch arrays are also needed to protect the transmission. For example, when one fiber link is cut, then the signal can be redirected to another fiber link. There is also a demand to form multipoint-to-multipoint optical networks. The switch array is the key enabling component that is required for forming such optical networks.

In electronic systems, a digital cross connect (DXC) is constructed with massive integrated circuitry that is capable of interconnecting thousands of inputs with thousands of outputs. Cross connection in an optical domain is much more difficult to implement. It may be accomplished by converting optical data streams into electronic data by using electronic cross-connection technology and then converting electronic data streams back into an optical stream. This is known as the hybrid, or opaque, OXC.

Another way is to cross connect optical channels directly in the photonic domain with all-optical cross-connects, i.e., transparent OXC.

Optical M × N switch arrays can be used to reconfigure optical signals from any one of the M input ports to any one of the N output ports. This allows for information sharing at different terminals. Also, it allows for N input optical ports to be mapped onto N output ports in a programmable manner.

A typical array consists of basic 1 × 2 or 2 × 2 switching nodes. Switch sizes larger than 2 × 2 can be realized by appropriately cascading small switches. A simple 1 × 8 switch array is illustrated in Figure 6.37.

Photonic OXCs can be (1) free-space optical-switching devices, (2) optical solid-state devices, and (3) electromechanical mirror-based devices [6.3]. Some examples of free-space switching arrays are schematically illustrated in Figure 6.38.

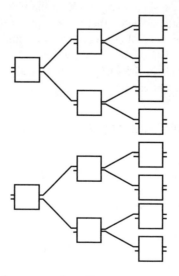

FIGURE 6.37 1 × 8 switch array configuration.

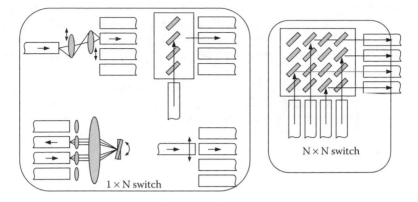

FIGURE 6.38 Switching arrays.

In terms of the switching function achievable, switches are of two types: blocking or nonblocking. A switch is nonblocking if an unused input port can be connected to any unused output port. Thus, a nonblocking switch is capable of realizing every interconnection pattern between the inputs and the outputs. On the other hand, if some interconnection pattern cannot be realized, then the switch is blocking. Most applications require nonblocking switches. Two examples of switching arrays are shown in Figure 6.39, i.e., they are 4×4 and 8×8 arrays based on small 2×2 switches.

Figure 6.40 shows the basic arrangement of an OADM system, which consists of optical AWGs and a series of 2×2 optical switches.

In an OADM system, the number of 2×2 switches corresponds to the number of WDM transmission wavelengths.

6.13 HYBRID PBG/MOEMS SWITCHES

A hybrid MOEMS/PBG switch consisting of an actuator and channel waveguides can be fabricated simultaneously using planar processing. The X-bar structure fabrication of the switch can be reduced to a single wafer, as shown in Figure 6.41, if a sliding actuator replaces the classical MEMS cantilever.

The actuator is made from Si that is index-matched to the channel waveguide. In the unenergized state, the actuator is withdrawn from the air gap (bar state). It is fabricated in this position to enable simultaneous fabrication of the waveguides and MOEMS device on the same substrate. The channel waveguides are slightly tapered at the junction to collimate the optical signal in the horizontal plane within the gap between the channel waveguides, improving optical coupling in the cross state. The large index-of-refraction difference between air and Si causes the interface to act as a mirror, reflecting the input signals (bar state). When the MOEMS actuator is energized, it slides into the gap between input and output channels (cross state). An index-matching optical grease can be used to further reduce losses at the waveguide or actuator air gap in the cross state. An

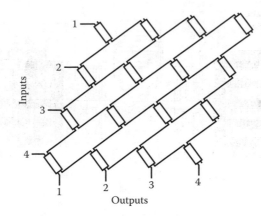

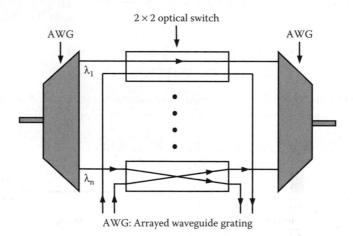

FIGURE 6.39 Switching arrays (4×4 and 8×8) built from 2×2 switches.

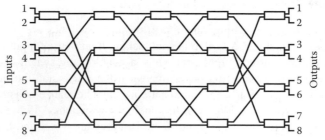

AWG: Arrayed waveguide grating

FIGURE 6.40 OADM structure.

additional innovative feature of the design is that the Si actuator has a top and bottom cladding to provide vertical optical confinement of the optical signal in the cross state. This will provide a significant improvement in the optical through-put compared to current integrated MOEMS switches.

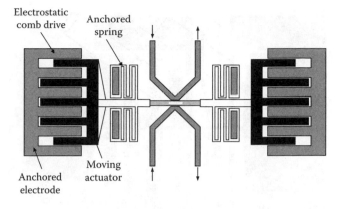

FIGURE 6.41 Schematic of a hybrid MOEMS/PBG 2 × 2 switch.

This hybrid MOEMS/PBG switch is based on the concept of high-order PBG structures (see Chapter 10). In the crossing state of the junction, multiple interference at the two air–Si gaps result in high transmittance across the junction. In the bar state, a short PBG structure is used to provide high reflectance with practically 0% transmittance across the junction (see Figure 6.42). Depending on the exact structure of the PBG in the switching junction, the optical switch can either be narrowband wavelength selective or wideband.

The MOEMS/PBG X-bar switch can be driven by various actuators, such as the interdigitated electrostatic linear motor. There are two sets of fixed electrodes, one at a potential V+ and the other at V−. When the potential of the movable slider is at V−, the actuator is drawn towards the V+ electrode. This inserts the Si slider into the gap between the channel waveguides. If the potential to the actuator is reversed so that it is at V+, it will be repelled from the V+ fixed

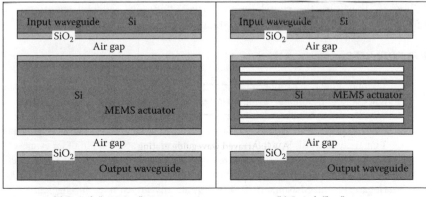

(a) Switch "crossing" state   (b) Switch "bar" state

FIGURE 6.42 Detail of the MOEMS/PBG switching junction in the two states.

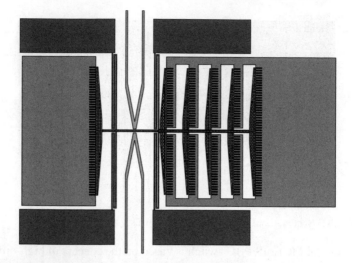

FIGURE 6.43 Layout of MOEMS/PBG switch with a multifinger electrostatic comb drive.

electrode and attracted towards the V– electrode. This will cause the Si slider to retract from the waveguide intersection.

The active drive side has five sets of interdigitated electrodes to provide the required force (see Figure 6.43). Four Si springs provide a restoring force to return the switch back to the nominally off bar state. These are assisted by a single set of electrodes for the actuator return. By powering both sets of actuators, it is possible to control the tension in the MOEM actuator to compensate for any strain effects and to reduce stiction.

A SEM photograph of the MOEMS/PBG switch is shown in Figure 6.44 [6.4]. The photo in Figure 6.44a shows the switch in the bar state; the photo in Figure 6.44b shows the switch in the cross state. The switch can effectively function as a digital light valve for optical signals.

REFERENCES

1. Tekippe, V.J., Passive fiber optic components made by the fused biconical taper process, *Fifth National Symposium on Optical Fibers and Their Application*, Warsaw, 1085, 1989.
2. Pruessner, M.W., Amarnath, K., Datta, M., Kelly, D.P., Kanakaraju, S., Ho Ping-Tong, and Ghodssi, R., Optical and mechanical characterization of an evanescent coupler optical switch, *Proceedings of Solid-State Sensor, Actuator and Micro-systems Workshop*, Hilton Head Island, South Carolina, 238–244, June 6–10, 2004.
3. Hogari, K. and Matsumoto, T., Electrostatically driven micromechanical 2 by 2 optical switch, *Appl. Opt.*, Vol. 30, 1253–1257, 1991.
4. Kruzelecky, R. and Haddad, E., Advanced Optical MEMS Components for Integrated Optical Micromachines, CSA's Report # DSS9F028-024104/004/MTB, 2004.

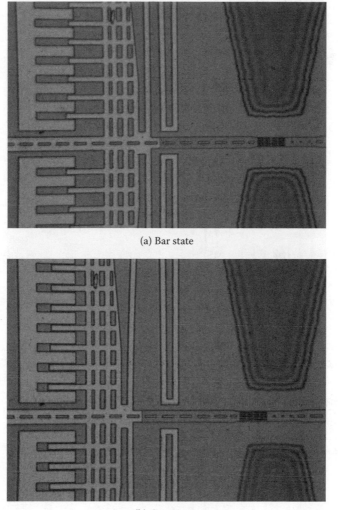

(a) Bar state

(b) Cross state

FIGURE 6.44 Microscopic photographs of the MOEMS/PBG switch in two states: (a) bar state, (b) cross state.

7 Multiplexers

Multiplexing is defined as the process by which two or more signals are transmitted over a single data channel. The purpose of multiplexing is to allow for more efficient use of the available transmission capacity so that a single channel may be used for several sources of data.

Since the very beginning of fiber-optic telecommunication systems, the main effort was focused on increasing transmission capacity. This was done first by increasing the signal modulation speed from 155 Mb/sec to 622 Mb/sec, 2.5 Gb/sec, 10 Gb/sec, and recently to 40 Gb/sec. The total available bandwidth of standard optical fibers is about 20 THz. As it is impossible for a single-wavelength laser to utilize such a large bandwidth, schemes based on multiplexing have been developed. There are several techniques of multiplexing. The two techniques applied in photonic processors are (1) time-division multiplexing (TDM) and (2) wavelength-division multiplexing (WDM).

Initially, digital systems used TDM, which broke different voice or data signals into pieces and sent them in alternating slots in one stream. TDM has enabled a single fiber strand to carry up to 32,000 voice calls simultaneously.

WDM, which was introduced in 1995, splits light waves into different frequencies, with each frequency capable of transmitting data at high speeds. One of the most critical technologies enabling the capacity expansion of fiber-optic systems is dense wavelength-division multiplexing (DWDM), which can exponentially increase the bandwidth of a fiber-optic strand.

7.1 TDM

TDM is a technique whereby several low-speed channels are multiplexed into a high-speed channel for transmission. Each low-speed channel is allocated a specific position based on time.

In TDM, time on the channel is shared among many data sources. The time-division multiplexer is a type of a high-speed switch that connects each of the multiple inputs to the communication channel for a fixed period of time. After each channel has been connected, the process repeats itself. One complete cycle is called a *frame*. Start and stop markers are added to the frame to ensure that each channel on the input is connected to its corresponding channel on the output. A schematic illustration of a TDM frame is shown in Figure 7.1.

Optical time-division demultiplexers and add-drop multiplexers are key components required by optical-time-division-multiplexed network nodes. In an add-drop multiplexer one channel is dropped from an incoming TDM data stream, leaving the other channels undisturbed. A new channel can be added by inserting data pulses into the vacant time slot.

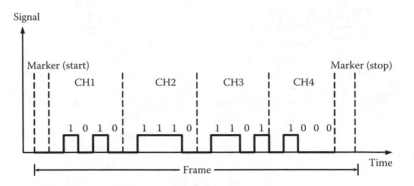

FIGURE 7.1 TDM frame of data transmission.

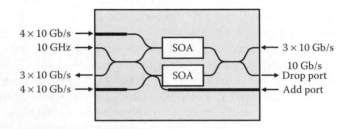

FIGURE 7.2 SOA add-drop multiplexer.

Many configurations are possible to implement an add-drop-multiplexing (ADM) scheme into a link. One possible design is based on Mach–Zehnder interferometer (MZI) switches incorporating semiconductor optical amplifiers (SOAs) [7.1]. A schematic of MZI-SOA add-drop multiplexer is shown in Figure 7.2.

In the configuration shown in Figure 7.2, the input data signal at 40 Gb/sec, is split into two drive signals. One of the drive signals is delayed by half a bit period. The interferometer is configured such that when a nondelayed signal pulse is present in the upper arm of the interferometer, then a 10 GHz input pulse is directed to the drop port. At the same time the 3×10 GHz pulse stream is directed to the through port. When the delayed signal pulse is present in the lower arm of the interferometer, then the data is directed away from the drop port [7.1].

7.2 WDM

The first commercially implemented optical links were based on a single-wavelength transmission. As requirements for more data transmission capacity increased, the industry developed the so-called WDM, in which each data channel is transmitted using a different wavelength. In this way many channels can be transmitted through the same optical link simultaneously. The biggest advantage

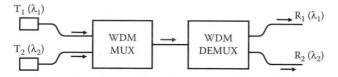

FIGURE 7.3 Two-channel WDM system.

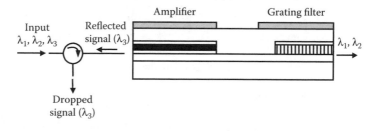

FIGURE 7.4 ADM system.

of the WDM method has been its ability to upgrade the capacity of the existent optical communication networks. In its first version, two separate channels were transmitted simultaneously, thus doubling the link capacity. WDM may be applied to four- or even more-wavelength systems. A schematic illustration of a two-channel WDM system is shown in Figure 7.3.

The ability to add and drop wavelength channels in WDM networks is useful for wavelength routing. The function of a wavelength add-drop multiplexer is to separate a particular wavelength channel without interference from adjacent channels. This can be achieved by a wavelength demultiplexer or by using an integrated SOA with a grating filter [7.1] as shown in Figure 7.4.

The filter characteristics can be tuned by changing its current. The selected wavelength channel is reflected by the filter, amplified a second time by the amplifier, and extracted to a drop port using a circulator. The remaining channels pass through the filter section.

7.2.1 Coarse Wavelength-Division Multiplexing

Coarse wavelength-division multiplexing (CWDM) was developed as the result of growing demand for optical networks. It was an intermediate technology that offered greater capacity than WDM. CWDM allows for a modest number of channels, typically eight or less, corresponding to the optical window in the C band (the region around 1550 nm; see Figure 3.13 in Chapter 3). The biggest advantage of this technique is the dramatically lower cost as less expensive uncooled lasers may be used. These lasers have a central line stability tolerance of ± 3 nm and they can be used for spacings of 20 nm. CWDM transmission is usually applied at one of eight wavelengths: 1470, 1490, 1510, 1530, 1550, 1570, 1590, and 1610 nm (see Figure 7.5).

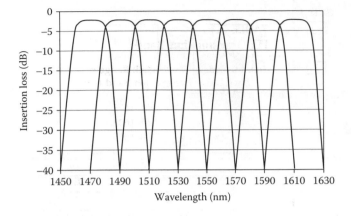

FIGURE 7.5 CWDM eight-channel transmission band.

CWDM technology employs a wider wavelength spacing and less expensive components than does long-haul DWDM. DWDM's narrower channel spacing requires the use of thermoelectric coolers to stabilize the laser emissions. On the other hand, the wide wavelength spacing of CWDM can accommodate the wavelength fluctuation of uncooled laser diodes. However, owing to the output power limitations of the uncooled laser diodes, the loss budget of CWDM systems can be limited to < 30 dB. Thus, the typical transmission distance is between 40 and 80 km.

7.2.2 Dense Wavelength-Division Multiplexing

DWDM is a later version of the WDM method. DWDM refers to a larger number of closely spaced optical signals through the same optical link.

DWDM has revolutionized data transmission technology by increasing the signal capacity of the existing optical channels (Figure 7.6). This technique has allowed for multiple data channels to be transmitted over one optical link while maintaining the overall system performance. These multiplexing systems are capable of working with 60 wavelengths, where each wavelength carries 40 Gb/sec data. It has been demonstrated that the total capacity of a single-mode fiber can be more than 2.5 Tb/sec.

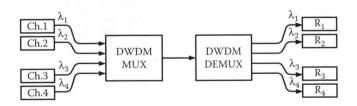

FIGURE 7.6 DWDM system.

TABLE 7.1
ITU Transmission Bands

Transmission Band	Wavelength Range of Transmission Band (nm)
U	1625–1675
L	1565–1625
C	1530–1565
S	1460–1530
E	1360–1460
O	1260–1310

DWDM systems operate in the 1550-nm window because of the low attenuation characteristics of glass at 1550 nm. The International Telecommunications Union (ITU) has regulated the spacing between transmission channels, i.e., standards developed by the ITU define the exact optical wavelength used for DWDM applications.

The ITU has specified six transmission bands for fiber-optic transmissions. These bands are expressed in nanometers (nm). The six bands are the O band (1260 to 1310 nm), E band (1360 to 1460 nm), S band (1460 to 1530 nm), C band (1530 to 1565 nm), the L band (1565 to 1625 nm), and U band (1625 to 1675 nm) (see Table 7.1). A seventh band, not defined by the ITU, but used in private networks, is at around 850 nm. It is interesting to note that, typically, higher transmission window corresponds to lower attenuation (or signal degradation) — see Figure 7.7. At the same time, the bands with lower attenuation require more expensive electronics.

The standard channel spacing on the ITU grid has been selected as 200 and 100 GHz. A wavelength spacing of 200 GHz corresponds to about 1.6 nm; a 100

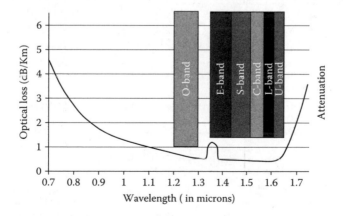

FIGURE 7.7 Transmission bands.

TABLE 7.2
A Sample of ITU Channel Frequency Grid (C Band)

Center Wavelength (nm, in vacuum)	Optical Frequency (THz)	Center Wavelength (nm, in vacuum)	Optical Frequency (THz)
1530.33	195.9	1546.92	193.8
1531.12	195.8	1547.72	193.7
1531.90	195.7	1548.51	193.6
1532.68	195.6	1549.32	193.5
1533.47	195.5	1550.12	193.4
1534.25	195.4	1550.92	193.3
1535.04	195.3	1551.72	193.2
1535.82	195.2	1552.52	193.1
1536.61	195.1	1553.33	193.0
1537.40	195.0	1554.13	192.9
1538.19	194.9	1554.93	192.8
1538.98	194.8	1555.75	192.7
1539.77	194.7	1556.55	192.6
1540.56	194.6	1557.36	192.5
1541.35	194.5	1588.17	192.4
1542.14	194.4	1558.98	192.3
1542.94	194.3	1559.79	192.2
1543.73	194.2	1560.61	192.1
1544.53	194.1	1561.42	192.0
1545.32	194.0	1562.23	191.9
1546.12	193.9	1563.05	191.8
1546.92	193.8	1563.86	191.7

GHz spacing corresponds to about 0.8 nm (see Figure 7.8). The standard spacing of 100 GHz allows for transmission of 45 channels on one optical link (e.g., on a single fiber). A 45-channel system spaced at 100 GHz would cover an optical span of 35 nm. A sample of the ITU grid frequencies is listed in Table 7.2. A comparison between CWDM and DWDM is summarized in Table 7.3.

TABLE 7.3
Performance Comparison between CWDM and DWDM

Parameter	CWDM	DWDM
Transmission band	O + S + C + L	C + L
Number of channels (maximum)	Up to 18	Up to 80
Channel spacing (minimum)	6 nm	0.8 nm
Channel capacity (maximum)	< 2.5 Gb/sec	10 Gb/sec (and higher)
Applicable distance (maximum)	< 80 km	Up to 1500 km (and more)

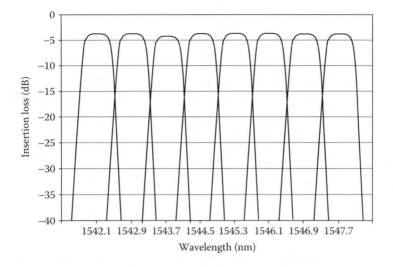

FIGURE 7.8 100-GHz channel spacing.

Recently a lot of effort has been put toward the implementation of 50-GHz communication links; 50 GHz corresponds to 0.4-nm channels spacing. In the near future a further decrease in channel spacing down to 25 GHz is expected. As the channel spacing decreases, the number of channels that can be transmitted increases, thus further increasing the transmission capacity of the system.

Channel spacing is the minimum frequency separation between two adjacent multiplexed signals. Channel spacing imposes certain requirements on the optical signal characteristics; the requirements are such that they allow for each set of adjacent channels to be clearly identified by the demultiplexer. Figure 7.9 illustrates the typical DWDM characteristics.

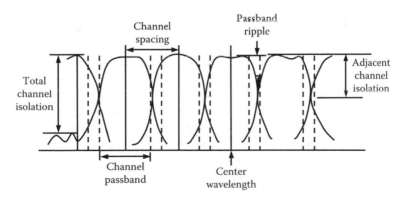

FIGURE 7.9 Optical characteristics of DWDM channels.

7.3 FILTERS

The recent explosion of DWDM technology forced manufacturers to develop dense wavelength-division multiplexers that can handle closely spaced optical wavelengths. These designs require a narrow passband, usually 0.4-nm wide, with temperature-stable operation. Recently, multiplexers have gained versatility, moving beyond the wideband wavelengths and into densely packed wavelengths that can be integrated into multiple high-frequency (191 to 195 THz) transmission systems (see Table 7.2). This type of system can maintain up to 16 channels.

Demultiplexers need to eliminate channel interference. Wavelength-division multiplexers use several methods to separate wavelengths, the method used being dependent on the wavelength spacing. Couplers and filters, both passive devices, are the most favorable demultiplexers today.

7.3.1 DICHROIC FILTERS

Dichroic filters are those with wavelength dependence; they are used to reflect and transmit parts of the incoming signal. In a wavelength-division multiplexer coupler, a number of cascaded dichroic filters may be used. A dichroic filter is placed at the center of the wavelength-division multiplexer. The long-wave pass (LWP) filter allows wavelengths longer than λ_t to pass through, whereas shorter wavelengths are reflected.

For example, as the signal hits the filter, λ_3 wavelength ($\lambda_3 > \lambda_t$) passes through whereas λ_1 and λ_2 wavelengths are reflected onto the short-wave pass (SWP) filter. The SWP filter allows λ_2 to pass through while reflecting λ_1. Thus, the information on the three wavelengths can be independently decoded.

7.3.2 FIBER BRAGG GRATINGS

Fiber Bragg gratings are passive devices that use diffraction to separate wavelengths, similar to a diffraction grating. They are of critical importance in DWDM systems, in which separation of multiple closely spaced wavelengths are required. An optical beam entering a fiber Bragg grating may be diffracted by periodic variations in the index of refraction. Each variation in refractive index partially reflects light with a 360° phase shift. Spacing the periodic variations at multiples of the half wavelength to be reflected causes constructive interference of a very specific wavelength while allowing other wavelengths to pass. Fiber Bragg gratings are available with bandwidths ranging from 0.05 nm to > 20 nm.

Fiber Bragg grating are typically used in conjunction with circulators, which are used to drop single or multiple narrowband WDM channels and to pass (express) other channels (see Figure 7.10).

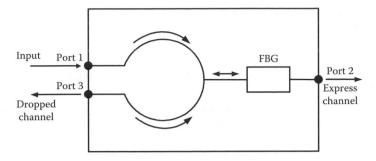

FIGURE 7.10 Wavelength selector.

7.4 RECONFIGURABLE OPTICAL ADD-DROP MULTIPLEXERS

Reconfigurable optical add-drop multiplexer (ROADM) has been promoted as the solution offering greater flexibility and lower operational costs. Unlike their conventional predecessors (i.e., add-drop multiplexers), ROADMs improve network flexibility by allowing switching of optical paths remotely instead of installing a new fixed switching device. So far the technology has been used primarily in long-haul networks, in which its costs are more easily justified.

The capability of ROADMs surpasses that of a simple optical switch. Depending on the technology used for their switching mechanism, ROADMs may offer additional features such as add-drop capability for all wavelengths, variable optical attenuation, dynamic power equalization, and remote monitoring of chosen wavelengths wherein data may be checked or modified before it travels through the network. Most ROADMs can perform these functions on single wavelengths without affecting other wavelengths.

Figure 7.11 shows a schematic of ROADM configuration. Arrays of 1×2 add-drop switches in ROADM may be used to increase their flexibility. The

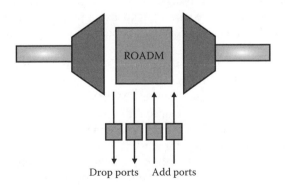

FIGURE 7.11 ROADM configuration.

switches may be arranged as 1×2 switches to drop channels or as 2×1 switches to add channels.

There are several ways of making ROADMs. The competing technologies include microelectromechanical systems (MEMS), planar lightwave circuits (PLCs), wavelength blockers, and wavelength-selective switches.

PLC provides the advantage of monolithic integration and potentials of increased performance that is matched by low-cost mass production. PLCs are typically thermal switches that split incoming light along the two paths of a MZI. Heating one path elongates it and alters the phase of the traversing beam. The two beams recombine and follow a path according to how their phases interact. PLCs allow for fine adjustments to channel intensity. Their phase-based switching mechanism, however, can incur polarization-dependent losses. Also, the minimum bend radius of PLC is a limitation. Therefore, integrating conventional PLCs into an ROADM may be limited to one or two devices per 6-in. wafer.

Another architecture features wavelength blockers, which uses either arrayed waveguides or free-space gratings to demultiplex light into discrete channels. Signals are then selectively expressed or dropped using an array of liquid-crystal shutters. Wavelength blockers are particularly attractive at nodes when dropping to a small number of ports. Cascading arrays of these devices can create a simple, inexpensive switching architecture. This approach can be expensive for a system with a large number of ports or when wavelength designations need to be dynamically changed. Also, such arrays can be susceptible to wavelength dependencies causing bandwidth narrowing that could induce network loss. More stable lasers make this less of a problem, but wavelength blockers still pose multiplexing and demultiplexing problems that might inhibit their overall performance [7.2].

Operating on principles similar to wavelength blockers, wavelength-selective switches redirect the beam into another channel. A highly flexible and powerful technology, these switches can dynamically drop multiple channels per port. Their scalable architecture also allows for future expansion. Wavelength-selective switches currently available can add or drop 8 out of 40 channels and must be cascaded in multiple units in order to provide large add-drop capacity. Adding or dropping more channels requires an expansion port that can direct signals to a cascaded unit. But this sort of configuration is more expensive and increases loss.

Microoptoelectromechanical systems (MOEMS) technology attracted a great deal of interest for both small- and large-scale optical-switching applications. Of all the available switching architectures, MOEMS technology strikes the best balance between add-drop performance for all wavelengths and scalability. Modular MOEMS arrays of 1×2 or 2×2 functioning as add-drop switches provide simple and inexpensive upgrade for existing multiplexed systems.

ROADMs offer additional features such as variable optical attenuation, dynamic power equalization, and remote monitoring of chosen wavelengths (see Figure 7.12). All these functions can be performed simultaneously.

Adding wavelength selectivity that is offered by ROADMs will open up the mass market to a wide variety of interactive communications for both consumers

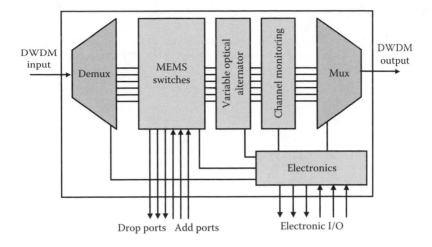

FIGURE 7.12 MEMS-based ROADM configuration.

and businesses. It will allow implementing the so-called "last mile" of optical links (see Figure 7.13) and will allow for the implementation of fiber-to-the-curb (FTTC), i.e., fiber-optic service to a node connected to several nearby homes (typically on a block). Also, it will allow for a full implementation of fiber-to-the-home (FTTH) configurations, i.e., fiber-optic service to a node located inside an individual home.

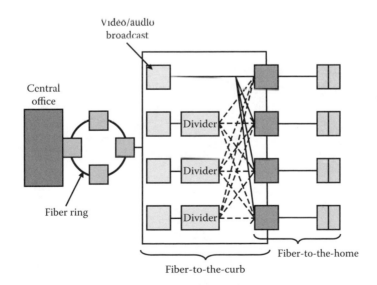

FIGURE 7.13 Schematic illustration of FTTC and FTTH.

REFERENCES

1. Conelly, M., Semiconductor optical amplifiers and their applications, http://www. uc3m.es/uc3m/dpto/IN/dpin08/Semiconductor_optical_amplifiers.pdf.
2. Auyeung, J., *Lightwave,* November 15–17, 2004.

8 Receivers

8.1 DETECTORS

Detectors convert optical signals into electrical impulses that are used by the receiving end of an optical data link. The most common detector is the photodiode.

It operates on the principle of the p–n junction. There are two main categories of photodiodes: PIN (positive, intrinsic, negative) photodiodes and avalanche photodiodes (APDs), which are typically made of InGaAs or Ge.

The process by which light is converted into an electrical signal is the opposite of that which produces light. Light striking the detector generates a small electrical current that is amplified by an external circuit. Absorbed photons excite electrons from the valence band to the conduction band, resulting in the creation of an electron–hole pair. Under the influence of a bias voltage these carriers move through the material and induce a current in the external circuit. For each electron–hole pair created, the result is an electron flowing in the circuit. Typical current levels are small and require the use of electronic amplifiers (see Figure 8.1).

The choice of a detector for a photonic system depends on several factors including the following:

- Wavelength sensitivity
- Responsivity
- Quantum efficiency
- Dark photocurrent
- Rise time

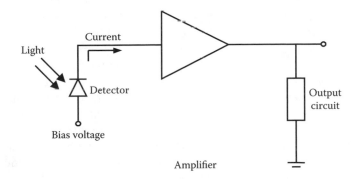

FIGURE 8.1 Typical detector–amplifier circuit.

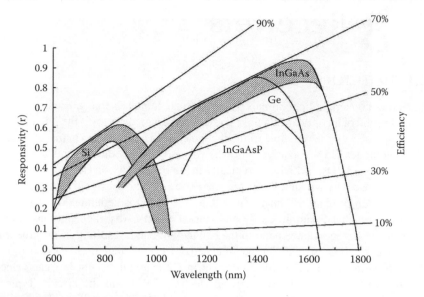

FIGURE 8.2 Detector spectral response.

Wavelength sensitivity is determined by the material composition of the photodiode. In general, silicon photodiodes are used for detection in the visible portion of the spectrum, InGaAs crystals are used in the near-infrared portion of the spectrum between 1000 nm and 1700 nm, and germanium detectors are used between 800 nm and 1500 nm. Figure 8.2 depicts the various types of detectors and their wavelength dependence.

Responsivity is the ratio of the current output to the optical input. A photodetector's responsivity is directly related to the absorption coefficient. The theoretical value for the responsivity is 1.05 A/W at a wavelength of 1310 nm. High responsivity of a detector implies high sensitivity. Responsivities for commercial III-V photodetectors are typically around 0.8 to 0.9 A/W.

Quantum efficiency is defined as the ratio of the number of electrons generated by the detector to the number of photons incident on the detector.

Dark photocurrent is a small current that flows through the photodetector even when there is no external light source. The dark photocurrent is induced by the intrinsic resistance of the photodetector and the applied reverse voltage. The magnitude of the dark photocurrent is sensitive to temperature and contributes to noise in the electrical output of the detector. Because the output electrical current of a photodiode is typically in the range of microamperes, a transimpedance amplifier (TIA) is used to amplify the electric current to the level of a few milliamperes.

Rise time is the time it takes for the detector reading to rise to a value equal to 63.2% of its final steady-state reading.

Table 8.1 gives some typical characteristics of photodetectors.

TABLE 8.1
Typical Photodetector Characteristics

Photodetector	Wavelength (nm)	Responsivity (A/W)	Dark Current (nA)	Rise Time (nsec)
Silicon PN	550–850	0.41–0.7	1–5	5–10
Silicon PIN	850–950	0.6–0.8	10	0.070
InGaAs PIN	1310–1550	0.85	0.5–1.0	0.005–5
InGaAs APD	1310–1550	0.80	30	0.100
Germanium	1000–1500	0.70	1000	1–2

8.2 PIN PHOTODIODES

A p–n diode's limitations are related to the small active detection area. The small detection area leads to many electron–hole pairs that recombine before they can create a current in the external circuit. In a PIN photodiode, the active detection area can be made much larger. A lightly doped intrinsic layer separates the more heavily doped p types and n types. The diode's name comes from the layering of these materials positive, intrinsic, negative, i.e., PIN.

Figure 8.3 shows the cross section of a PIN photodiode.

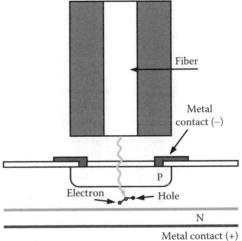

FIGURE 8.3 PIN photodiode.

8.3 AVALANCHE PHOTODIODES

APDs provide much more gain than PIN photodiodes (Figure 8.4). However, they are much more expensive and require a high-voltage power supply to operate them. APDs are also more sensitive than PIN photodiodes to temperature changes.

The APD is based on the free electrons and holes that are created by absorbed photons. They act as very fast carriers. A collision of these fast carriers with neutral atoms causes the accelerated carriers to use some of their own energy to break out the bounded electrons of the valence shell. As a result, free-electron–hole pairs are created. They are called *secondary carriers*. The secondary carriers in turn create new series of carriers. This process is known as *photomultiplication*.

APDs require high-voltage power supplies for their operation. The voltage can range from 30 or 70 V for InGaAs and over 300 V for Si. This high-voltage requirement adds to the circuit complexity. The fact that APDs are very temperature sensitive further complicates circuit requirements. In general, APDs are only useful for digital systems because they exhibit a poor linearity.

Because of the added circuit complexity and the high voltages, APDs are usually less reliable than PIN detectors. The performance of PIN detectors at lower data rates is similar to APDs, and therefore PIN detectors are usually the first choice for most deployed low-speed systems. However, APD performance is superior at multigigabit data rates.

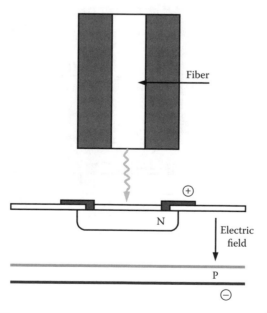

FIGURE 8.4 APD.

8.4 LIGHT EMITTERS

Light emitters such as LEDs and lasers may also be used as optical detectors. This allows for a unique application of light emitters in half-duplex fiber-optic communication devices. These devices allow simultaneous bidirectional transfer of data. They involve using an LED or laser alternately as a light emitter and as an optical detector. Such a scheme allows the transmission of information in either direction over the fiber. The key parameter is the coupling efficiency between the light emitter and the transmission channel. High coupling efficiency allows good performance in both modes of the operation.

A good InGaAs detector may have a responsivity of 0.8 A/W at a wavelength of 1310 nm. An LED operating as a detector may provide a responsivity in the range of only 0.08 A/W. The main reason for this much lower response is the fact that an LED operating as a detector has a relatively narrow spectral response spectrum that does not fully overlap with the LED emission spectrum [8.1]. Figure 8.5 shows the spectral response as well as the emission spectrum of a typical InGaAs detector.

It can be seen from Figure 8.5 that an InGaAs detector has a very broad spectral response, i.e., from 800 nm to beyond 1600 nm. Because of the wide response, the detector responds to all photons emitted by the LED. The spectral emission of the LED is a relatively narrow spectrum, nearly 60 nm wide, and is centered around 1310 nm. The spectral response of the same LED operating as a detector is shifted towards the shorter wavelengths, i.e., centered around 1270 nm. The overall response as a detector is a slightly wider than the emissions as an LED [8.1]. Laser diodes exhibit characteristics similar to the LED shown in Figure 8.5.

8.5 SILICON-BASED PHOTODETECTORS

Most communication-grade semiconductor lasers operate in the near-infrared wavelength region of 0.850, 1.310, and 1.550 μm (a region in which silicon is a

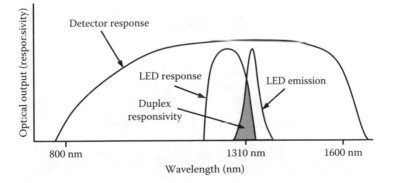

FIGURE 8.5 Full-Duplex LED.

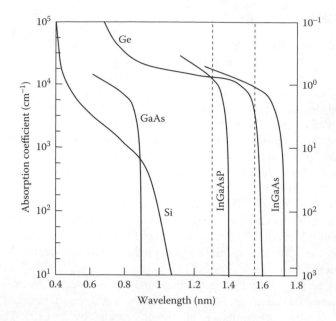

FIGURE 8.6 Absorption coefficient and penetration depth.

poor detector). In order to improve the performance of silicon-based detectors, the most common approach is to introduce germanium to reduce the band gap and extend the maximum detectable wavelength. As seen in Figure 8.6, the InGaAs material is most suitable for detectors used at 1.310 and 1.550 μm. The vertical dash lines in Figure 8.6 mark the communication wavelengths region of 1.310 and 1.550 μm [8.2].

A cross section of a silicon detector structure that was designed at Intel is shown in Figure 8.7. In this design, an SiGe layer was deposited directly on top

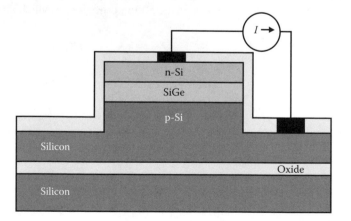

FIGURE 8.7 SiGe waveguide-based photodetector on a SOI wafer.

of a silicon rib waveguide [8.2]. SiGe was used to shift the detector absorption towards the infrared region of interest. The amount of Ge required for efficient photodetection is dependent on the wavelength. If detection at 1.310 or $1.550 \, \mu m$ is desired, then very high (> 40%) Ge concentrations are needed [8.2].

This detector had a dark current of less than $1 \, \mu A$ (<1 nA/μm^2) at 3 V, which is acceptable for most applications (for comparison, InGaAs PIN photodetectors typically have dark currents close to 1 nA).

REFERENCES

1. http://www.fiber-optics.info/articles/detector.htm.
2. Salib, M., Liao, L., Jones, R., Morse, M., Liu, A., Samara-Rubio, D., Alduino, D., and Paniccia, M., Silicon photonics, *Intel Technol. J.*, 8, 144–160, 2004.

9 Amplifiers and Compensators

9.1 AMPLIFIER SUBSYSTEMS

Optical amplifiers enhance optical signals without converting the signal into electrical form. As the demand for wider bandwidth grows, there is a requirement for more efficient and reliable optical amplifiers. This is the same requirement that is in demand for long transmission lengths, i.e., amplifiers developed to support dense wavelength-division multiplexing (DWDM). Thus, optical amplifiers have become an essential component in both (1) long-haul and (2) higher-data-transmission systems.

There are various types of optical amplifiers that are used in photonic systems. Optical amplifiers can be divided into two classes: (1) optical fiber amplifiers (OFAs) and (2) semiconductor optical amplifiers (SOAs). OFAs have been used in conventional system applications, such as in-line amplification, to compensate for fiber losses. Erbium-doped fiber amplifiers (EDFAs) and Raman optical fiber amplifiers (ROAs) belong to this category.

SOAs are showing great promise of use in evolving optical communication networks. These devices, in addition to being used for signal enhancement, lessen the effects of dispersion and attenuation.

The amplifiers may be utilized in a number of ways:

1. Power boosters: Many tunable lasers provide low optical power levels that must be amplified.
2. In-line amplifiers: Signals are amplified within the signal path.
3. Wavelength converters: The wavelength of an optical signal may be changed.
4. Receiver preamplifiers: They can be placed in front of detectors to enhance the incoming signal.

Figure 9.1 illustrates the use of amplifiers as boosters, in line amplifiers, and preamplifiers. Figure 9.2 illustrates the gain bandwidth of various optical amplifiers that might be used for coarse wavelength-division multiplexing (CWDM) systems. EDFAs are common elements in optical networks; SOAs can cover a wide gain band; ROAs can extend the usable optical bandwidth by optimizing the pumping spectrum.

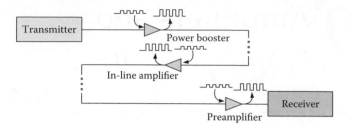

FIGURE 9.1 Various types of amplification.

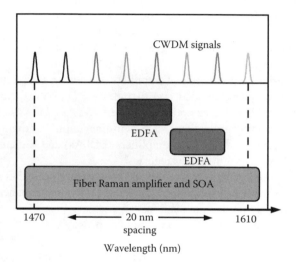

FIGURE 9.2 Gain bandwidth of optical amplifiers.

9.2 SOAs

SOA works in a manner similar to a basic laser. SOAs are essentially laser diodes without cavity mirrors. They amplify any optical signal that passes through the cavity. The signal may come from either end. The signal is then transmitted and amplified (see Figure 9.3).

SOA is driven by an electrical current. An electrical current is used to excite electrons that can then produce photons by the stimulated emission effect. There are two types of SOAs — Fabry–Perot amplifiers (FPAs) that are based on the Fabry–Perot cavity and traveling-wave amplifiers (TWAs) [9.1].

SOAs can work for 1310- and 1550-nm systems. Their drawbacks include high insertion loss, polarization dependence, and a higher noise figure [9.3]. A comparison between OFA and SOA is given in Table 9.1.

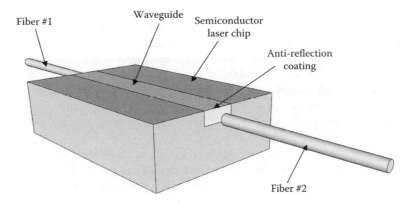

FIGURE 9.3 Semiconductor optical amplifier.

TABLE 9.1
Comparison between SOA and OFA

Characteristics	SOA	OFA
Pump source	Electrical	Optical
Insertion loss	10 dB	2.0 dB
Gain bandwidth	30 dB	50 dB
Polarization dependence	Yes	No
Noise figure	12 dB	5 dB
Operational bandwidth	50 nm	30 nm
Saturation	20 dBm	15 dBm

9.3 ERBIUM-DOPED AMPLIFIERS

Erbium ions may be used as a dopant for optical fibers or waveguides. Thus, there are two types of erbium-doped amplifiers, EDFA and erbium-doped waveguide amplifiers (EDWA). Doped glasses are highly transparent at the erbium lasing wavelength, in the spectral region from 2 to 9 μm. Erbium ions are pumped to an upper energy level by the absorption of light from the pump source, usually at 980 or 1480 nm. The transition to the ground state generates a photon either by spontaneous or stimulated decay. Signal photons in the erbium-doped amplifiers stimulate depopulation of the excited state, amplifying the signal. The relatively long lifetime of the excited state (approximately 10 msec) minimizes noise emitted by spontaneous emission; most erbium ions will amplify signals by stimulated emission. Therefore, this naturally occurring coincidence makes it possible to use doped glasses both as optical signal transmitters and as amplifiers.

Optical amplification is similar to electronic amplification in which the transistor converts the applied DC bias power to amplify a desired AC signal. Optical amplification is very useful in telecommunications applications because it avoids the need for an optical signal to be converted to electronic signal. Such conversion from optical signals to electrical signals is expensive as it requires photodetectors and high-frequency electronics. Before the invention of EDFA, optical losses were compensated every tens of kilometers by an electronic repeater. Drawbacks of electronic repeaters are their high cost and that they work only for a designed bit rate and at a single carrier wavelength.

There are two commercial applications of optical amplifiers. The first application is related to regeneration of optical signals. The optical signal usually passes through various integrated optical components and micro-PIC optical processors. Each of these components will induce some insertion loss. This loss has to be compensated if the integrity of the data is to be maintained. Even in short link spans, the signal passes through several splitting and branching devices that can substantially reduce the signal strength.

A second application is related to various optical logic circuits and processors. Optical amplifiers can be used to provide a number of digital logic functions that can be used in optical processors.

9.3.1   EDFAs

The idea of erbium-doped fiber lasers and amplifiers began in 1964 with the first amplification experiments in rare-earth-element-doped fibers. The first diode-pumped fiber lasers were demonstrated in 1974. Erbium-doped fiber lasers and amplifiers provide a significant advantage for transmission systems that operate in the 1.5-μm region.

EDFAs provide advantages over standard regenerative repeaters. Regenerative repeaters require high-speed electronics that increase significantly in cost and complexity as the bit rates increase. In addition, regenerative circuits are required for each optical wavelength in wavelength-division-multiplexing (WDM) applications.

EDFAs provide a significant amplification with low insertion loss, no polarization sensitivity, and broad bandwidth. Erbium-doped materials are more efficient than the traditional semiconductor laser and ROAs. The EDFA gain is a function of pump power, signal wavelength, fiber length, and erbium doping concentration. The pump efficiency is a function of the pump band, the signal wavelengths, and the fiber length. For example, a 3-m long erbium-doped fiber can provide a gain of up to 28 dB.

EDFAs are silica-based optical fibers that are doped with erbium (see Figure 9.4). A pump signal is used (usually at 980 nm) to inject energy into the doped fiber. When a 1550-nm signal enters the fiber, it stimulates the rare earth atoms to release their stored energy at 1550 nm. This process continues as the signal passes along the fiber. Figure 9.5 illustrates a schematic of an EDFA.

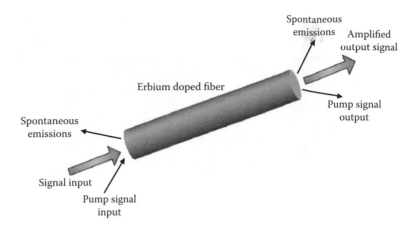

FIGURE 9.4 Erbium-doped optical fiber amplifier.

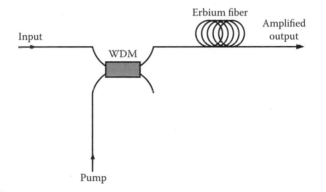

FIGURE 9.5 Schematic of EDFA.

As can be seen in Figure 9.5, an EDFA consists of a short length of optical fiber whose core has been doped. A laser diode beam is used as an optical pump. In the case of the EDFAs, the optical gain does not change with bit rate. Therefore, these amplifiers can be used to provide simultaneous amplification of all of the transmitted wavelengths in WDM links (see Figure 9.6).

The following are some applications whose performance can be improved with the use of EDFA:

Boosters: The so-called booster amplifiers are placed directly after an optical source. This application requires EDFA to accept a relatively high input signal and then to enhance it to the maximum output level. In other words, the function of a booster amplifier is to increase the input signal prior to its transmission. Boosting laser power in an optical transmitter enables the construction of medium-haul links with increased

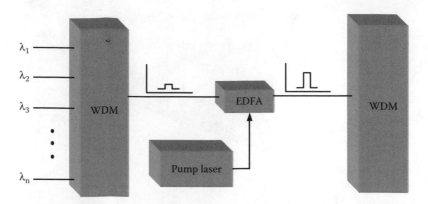

FIGURE 9.6 WDM system with EDFA.

transmission distance. Such links simply consist of an optical fiber between the transmitter and receiver. As this involves no active components in the transmission link, the reliability and performance of such a link are greatly improved. In long-haul links, the use of a booster amplifier can increase the link power budget and reduce the number of the required in-line amplifiers or regenerators. Booster amplifiers are also useful in distribution networks in which there are additional losses associated with the use of fiber splitters and taps. Booster amplifiers are also needed when it is required to simultaneously amplify a number of input signals at different wavelengths, as is the case in WDM links.

In-line amplifiers: These are used to boost signals before they are retransmitted along the fiber. The use of EDFA in this application improves performance and reduces the noise introduced by various components.

Preamplifiers: The function of an optical preamplifier is to increase the power level of an optical data signal before detection and demodulation. The increase in power level can increase receiver sensitivity. This allows construction of longer unrepeated links. Such an improvement may be achieved by placing an EDFA prior to the receiver.

Loss compensators: Inserting an EDFA before an $1 \times N$ optical splitter increases the power and allows each of the output signals to provide an output that is almost equal to the original transmitter power (see Figure 9.7). This can greatly increase the overall efficiency of the transmission link.

There are various configurations of commercially implemented EDFA systems. Among those are (1) hybrid EDFA, (2) dual-pumped EDFA, and (3) multistage EDFA:

Hybrid EDFA: The usable bandwidth of an EDFA is about 40 nm (1525 to 1565 nm). Optical communication systems carrying 100 or more

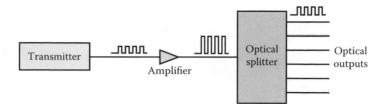

FIGURE 9.7 Loss compensator.

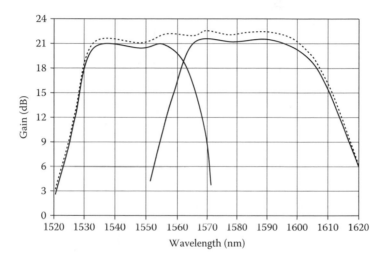

FIGURE 9.8 Gain spectrum of a hybrid EDFA.

optical wavelengths require an increase in the bandwidth of the optical amplifier to nearly 80 nm. The usable bandwidth of the EDFA may be extended by implementation of the so-called hybrid EDFA. A hybrid EDFA consists of two separate optical amplifiers that allow for expanding the amplification band. This may be achieved by selecting the amplifiers in such a way that one amplifier works for the lower 40-nm band and the second amplifier works for the upper 40-nm band. Figure 9.8 illustrates the optical gain spectrum of such a hybrid EDFA. The solid lines illustrate the response of two individual amplifier sections. The dotted line shows the gain spectrum of the combined hybrid amplifier [9.1].

Dual-pumped EDFA: A schematic of a dual-pumped EDFA is shown in Figure 9.9. Wavelength-division multiplexers are used as a means to inject the 980-nm pump signals into the erbium-doped fiber, as shown in Figure 9.9. The second pump laser's function is to increase the gain and output power of the transmission system [9.1].

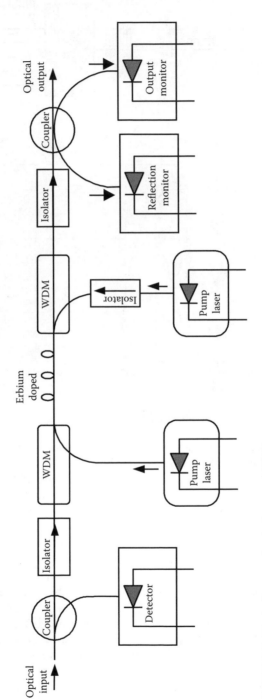

FIGURE 9.9 Block diagram of a dual-pumped EDFA.

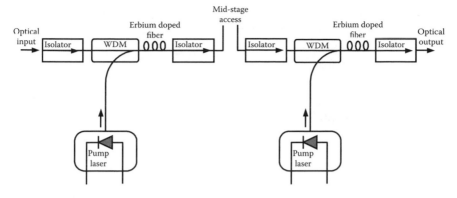

FIGURE 9.10 Two-stage EDFA.

Multistage EDFAs: Figure 9.10 shows an example of a multistage EDFA. In this particular case two single-stage EDFAs are packaged together. This configuration includes a midstage user access to the system [9.1]. This midstage access allows for adding some compensation devices that may improve the efficiency of the system. For example, dispersion compensating fiber can be added at the midstage to reduce the overall dispersion of the system.

9.3.2 EDWAs

The use of optical amplifiers shows a great potential in various fields such as LANs and access networks, in which they can compensate for signal attenuation resulting from distribution or from component insertion losses. In these cases, EDWAs promise additional advantages over EDFAs, such as integration of passive components on the same chip. In such applications, EDWAs will complement EDFAs and not replace them.

EDWAs are quite compact. They do not need to accommodate tens of meters of fiber. Most of the advantages of EDWA come from their ability to provide high gain in a very short optical path. For example, a 15-dB gain EDWA may fit into $120 \times 11 \times 6$ mm package.

EDWAs can exhibit similar absorption and gain spectra as the larger EDFAs but at a much lower cost. Moreover, they can offer added benefits due to their potential for integrating additional devices such as gain-flattening filters (GFFs) and dispersion compensating gratings that can enhance the overall device performance. EDWAs can also be fabricated as arrays to provide amplification for a number of separate channels. This area is also gaining importance in current long-haul telecom links.

At first sight, it may seem straightforward to translate the concept of a fiber amplifier to that of an integrated planar waveguide. However, scaling down the size of device from that of EDFA (typically 40-m length) to that of EDWA (several

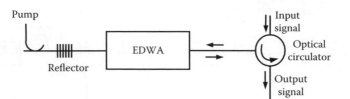

FIGURE 9.11 EDWA-based multipass amplifier.

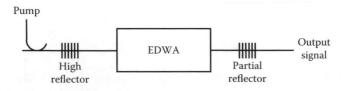

FIGURE 9.12 EDWA-based laser.

centimeters or even less) implies using concentrations of Er^{3+} ions higher by about two orders of magnitude.

Figure 9.11 shows a block diagram of an EDWA-based multiple-pass amplifier. An input signal passes through a port of an optical circulator and then is directed out of the second port of the optical circulator into an EDWA.

In this design, a reflector is used to return the signal back through the EDWA for a second pass and back into the second port of the optical circulator. The signal, having passed through the EDWA two times, is then directed to a third port of the optical circulator and out of the multiple-pass amplifier [9.2].

It is interesting to note that an EDWA may be also used to form a waveguide laser. A laser can be formed using a distributed grating in the doped region of the amplifier (see Figure 9.12).

If the laser is equipped with a tunable feedback device, such as a tunable grating, then a tunable laser can be realized. Such a laser can be operated in both the C and L optical bands.

9.4 ROAs

Both ROAs and EDFAs amplify optical signals in the fiber by transferring energy from a pump to the signal. In ROAs, energy is transferred through an effect in the fiber known as stimulated Raman scattering (SRS). ROAs are very attractive devices because of their low noise figure, low connection loss, high gain, high output power, and broad bandwidth. However, their overall efficiency is lower as compared to that of EDFAs.

When light is transmitted through matter, part of it is scattered in random directions. The scattered-light frequency is different from the frequency of the incident beam. The difference is related to vibration frequencies of the scattered

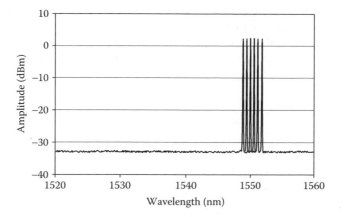

FIGURE 9.13 Spectrum of a six-wavelength DWDM system.

medium. This effect is known as *Raman scattering* (see Chapter 3). In a particular case, when the initial beam is sufficiently intense, a threshold can be reached beyond which signal at the Raman frequencies may be amplified. This is called SRS and is utilized in ROAs.

Figure 9.13 and Figure 9.14 illustrate SRS. Figure 9.13 shows an example of an optical spectrum of a six-channel DWDM system. Figure 9.14 shows the spectrum of the same system affected by SRS. It should be pointed out that all six wavelengths in Figure 9.13 have the same amplitude. One can see that the noise background in Figure 9.14 has increased and the transmitting channel signals have been changed as well. The shorter wavelengths in Figure 9.14 have smaller amplitude than the longer wavelengths; SRS has transferred energy from the shorter wavelengths to the longer wavelengths [9.1]. By applying SRS it is

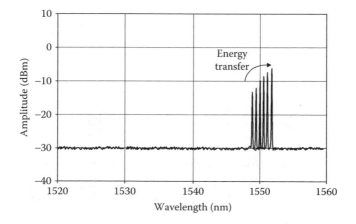

FIGURE 9.14 SRS spectrum of a six-wavelength DWDM system.

obvious that the noise background has increased, making the amplitudes of the six wavelengths different [9.1].

SRS amplification occurs in the transmission fiber itself, and it is distributed along the transmission path. A gain bandwidth for the ROAs is in the range up to 10 nm. The ROA can be used together with EDFAs to expand the optical gain bandwidth.

Figure 9.15 shows a schematic a typical ROA. The ROA consists of the pump laser and the circulator. The circulator is used to inject the signal back into the transmission line [9.1]. Figure 9.16 illustrates the optical spectrum of a forward-pumped ROA. In this case, the pump laser is injected at the transmitting end [9.1]. The SRS makes the amplitude of the six data signals much stronger. The energy from the pump laser is redistributed to the six data signals.

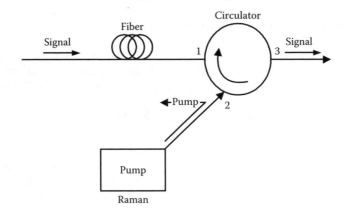

FIGURE 9.15 Schematic of Raman optical fiber amplifier.

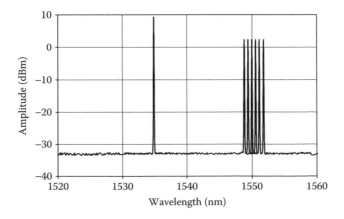

FIGURE 9.16 Transmitted spectrum of Raman optical fiber amplifier.

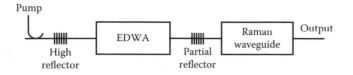

FIGURE 9.17 EDWA-based Raman waveguide amplifier.

Figure 9.17 shows a schematic an EDWA-based Raman waveguide amplifier. The Raman waveguide amplifier is an amplifier comprising a waveguide section constructed from a material with high scattering coefficient and pumped with a waveguide laser [9.2].

9.5 DYNAMIC GAIN EQUALIZERS

The implementation of EDFAs has revolutionized telecommunication systems. However, EDFAs do require some compensation because of their uneven gain profiles. These uneven gain profiles may limit performance specifications. Gain equalization devices are installed to eliminate the uneven gain profile problem.

In the past, gain equalization was addressed by using GFFs that were based on fiber Bragg gratings. These act as static gain filters. They perform the function of flattening the spectrum; however, their correction function is fixed to a predetermined shape of the amplifier gain profile.

An alternative approach led to the design of dynamic gain equalizers (DGEs). DGEs allow for dynamic change of the gain correction (see Figure 9.18). In other words, they adapt their response to changes in the transmission. High accuracy of the DGE allows equalizing the DWDM spectrum with the low residual ripples.

9.6 DISPERSION COMPENSATORS

Standard single-mode fiber used for telecom applications exhibits an approximately linear chromatic dispersion. This dispersion causes the different wavelength components of a data pulse to travel at different group velocities. As a

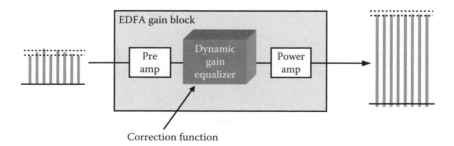

FIGURE 9.18 Dynamic gain equalizer.

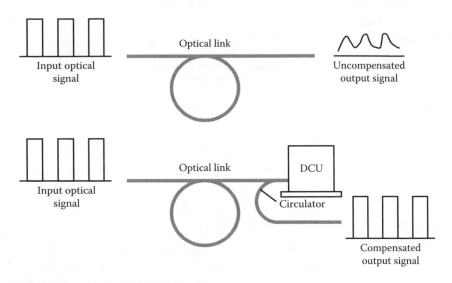

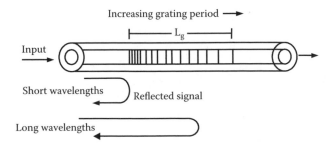

FIGURE 9.19 Dispersion compensation.

result, the pulse peak intensity decreases, and its width broadens with distance along the fiber (see Figure 9.19). This causes an increase in bit error rate for long spans and is currently the main limiting factor in the performance of optical telecommunications networks at higher data rates.

A dispersion compensation unit (DCU) may be used to eliminate dispersion-related errors. A DCU may be based on a linearly chirped fiber Bragg grating. The grating provides a wavelength-dependent time delay in the frequency components of the reflected pulse that can compensate for the chromatic dispersion introduced by the carrier fiber and other network components. Each wavelength is reflected back at different positions along the grating as the optical signal travels through the DCU (Figure 9.20). The maximum time delay depends on the length of the grating and the refractive index of the fiber.

The chirp of the grating must be sufficiently small to provide an adequate compensating dispersion (see Figure 9.21). For broadband WDM applications,

FIGURE 9.20 Linearly chirped fiber grating.

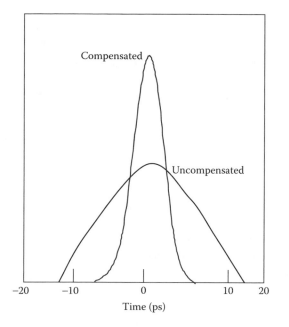

Compensated

Uncompensated

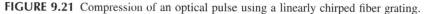

−20   −10   0   10   20

Time (ps)

FIGURE 9.21 Compression of an optical pulse using a linearly chirped fiber grating.

this would require long, linearly chirped gratings, which are tens of centimeters long and hence are technically difficult to produce and impractical to package. Compensating a single WDM wavelength on the ITF grid (see Table 7.2) is a more feasible approach and can be realized using chirped gratings under 11 cm in length. A cascade of such compensators can be employed to provide dispersion compensation for the different wavelengths of the WDM ITU network.

It is possible to use a fiber grating compensator at add-drop nodes to provide both wavelength-selective add-drop and dispersion compensation in one device. In this case, it is important that the central wavelength of the compensator is temperature-stabilized and that the overall bandwidth does not exceed the channel spacing of the ITU grid.

Because the chirp required for dispersion compensation is relatively small, fixed-period gratings can be used as the basis for a dispersion compensation device. A chirp can be induced either mechanically or thermally in the grating.

A DCU can be formed by affixing a fiber with a uniform Bragg grating onto a tapered substrate with a suitable cross-sectional profile. The substrate can then be mechanically elongated along its z axis, as shown in Figure 9.22. As the substrate is stretched, segments of the grating affixed to thinner cross sections of the substrate will be elongated more, increasing the local period of the grating and the refractive index of the fiber because of the corresponding stress. If the elongation plastically deforms the substrate, then the grating will be permanently chirped at a fixed value. If a relatively elastic substrate is used, then a dynamically

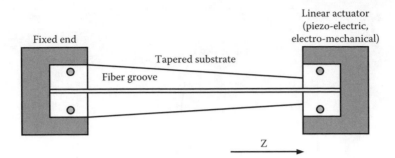

FIGURE 9.22 Mechanically chirping a long uniform grating.

tunable chirp can be obtained that can be varied by controlling the amount of elongation.

A robust thermomechanical actuator may be used to provide dynamic tuning of the grating chirp and central wavelength. Either single or multiple channels (4 to 8) on the ITU grid can be compensated using a single actuator.

9.7 WAVELENGTH CONVERTERS

All-optical wavelength converters play an important role in broadband optical networks. Their most important function is to increase the flexibility and capacity of a network using a fixed set of wavelengths.

One type of wavelength converter is based on nonlinearities observed in SOAs. These nonlinearities are induced by input signals. For example, the gain spectrum of an SOA can be homogenously broadened by a strong signal. The broadening will affect all other weaker input signals that pass through the amplifier. Therefore, it is possible to affect the gain of weaker signals at other wavelengths. This effect is known as *cross gain modulation*.

A schematic illustration of a wavelength converter that uses cross gain modulation is shown in Figure 9.23. A weak continuous wave (cw) optical signal and a modulated pump signal are injected into an SOA [9.3]. The cross gain modulation effect will impose the pump beam modulation on the weaker signal. In this way, the amplifier acts as a wavelength converter.

The same cross gain modulation effect can also be used for wavelength conversion in a Mach–Zehnder interferometer configuration. This is shown in

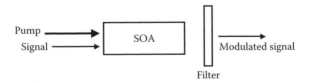

FIGURE 9.23 Cross gain modulation wavelength converter.

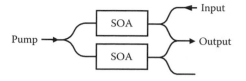

FIGURE 9.24 Wavelength converter based on Mach–Zehnder interferometer.

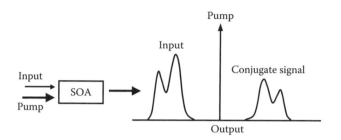

FIGURE 9.25 Wavelength converter based on FWM.

Figure 9.24. A coupler is used to split the input signal and send a portion to each arm of the interferometer [9.3]. The intensity-modulated signal induces different phase shifts in the input signals. The output coupler recombines the signals. They can interfere constructively or destructively when recombined by the output coupler. The state of interference depends on the SOA bias current and input optical power.

Four-wave mixing (FWM) is another nonlinear process that can be used to build wavelength converters. The pump beam and a modulated input beam are injected into an SOA. The injected fields cause the amplifier gain to be modulated at the beat frequency [9.3]. This gain modulation in turn gives rise to a new conjugate signal, as shown in Figure 9.25.

SOA-based devices show great promise for applications in evolving optical communication networks. They can be used as amplifiers, switches, and wavelength converters. All these functions will be required in all-optical systems.

REFERENCES

1. http://www.fiber-optics.info/articles/op-amp.htm (September 2005).
2. Demaray, R.E. and Dawes, D., Planar Optical Waveguide Amplifier with Mode Size Converter, U.S. Patent Application Publication No. 20040081415, April 29, 2004.
3. Conelly, M., Semiconductor optical amplifiers and their applications, http://www.uc3m.es/uc3m/dpto/IN/dpin08/Semiconductor_optical_amplifiers.pdf.

10 New Technologies

Several newly developed technologies offer opportunity for additional miniaturization and functionality of the next generation of microphotonic devices. Four technologies are of particular interest: microoptoelectromechanical systems (MOEMS), photonic-band-gap (PBG) structures, ring resonators (RRs), and smart coatings.

10.1 MOEMS

Microphotonic technology has been greatly advanced by recent developments in miniature optoelectronic systems. These devices couple photonics with electronics and micromachined structures that operate on the micron level. They are called optical MEMS, or MOEMS. These structures merge the functions of sensing and actuation with computation and communication to locally control physical parameters at the micron scale. MOEMS is a portfolio of techniques and processes that are used to design and create miniature systems. The integration of MOEMS with guided-wave optics can have a significant impact in improving the performance and reducing the size of photonic systems.

The advantages of MOEMS are their small size, low cost, and the ease with which these devices may be integrated. These advantages can outweigh disadvantages such as the relatively slow speed of mechanical switches or tuning devices.

MOEMS consist a number of elements that can perform several functions. Figure 10.1 illustrates schematically MOEMS functional interrelation within microoptics, microelectronics, and micromechanics.

MOEMS may include a number of microoptics elements, such as the following:

Lenses — to shape optical beams
Mirrors — to direct optical beams
Gratings — to select specific wavelengths of optical beams
Splitters — to split or combine optical beams
Waveguides — to guide optical beams
Switches — to redirect optical beams

Micromechanics mechanisms allow for these microoptical elements to be linearly translated or rotated. Microelectronics provides (1) signal processing capability and (2) drivers for the mechanical motions.

Current MOEMS technology is capable of producing mechanical devices that are on the scale of 10 to 500 μm. For example, the diameter of microlenses may

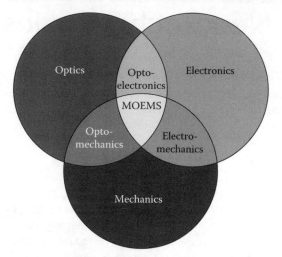

FIGURE 10.1 MOEMS functional interrelation.

be in the range of 50 μm. The speed of moving elements may be as fast as a fraction of a millisecond. There can be as many as 10^6 components on a single chip.

The driving force behind MOEMS is the realization of an all-optical network. An optical signal traveling from its source to destination does not get converted into an electrical signal. This should enable increased network speeds and eventually reduce expenses. A major electrical component to be addressed in current networks is the switch. A MOEMS switch is a device that combines free-space optics and microscopic mirrors to route individual wavelengths of light transmitted simultaneously on an optical fiber. The mirrors are so small they can add or drop specific wavelengths without disturbing the others. It is expected that the application of MOEMS switches will reduce costs and increase efficiency while requiring less space than conventional routers.

An example of a MOEMS-based wavelength-division-multiplexing (WDM) switching system may include a multiwavelength light source from an optical fiber that is imaged through a diffraction grating onto a column of micromechanical tilt-mirrors. The mirrors are positioned so that each is illuminated by a single wavelength, and they are tilted so that individual wavelength signals are either passed into the output fiber or reflected directly back into the input fiber (see Figure 10.2).

The linear array of micromirrors is controlled with simple digital electronics. Each mirror is positioned accurately in one of several highly stable positions with switching times of less than a millisecond.

For integrated optics, switches can be based on the electrostatic deflection of a waveguide. Silica-on-silicon moving waveguide switches have been demonstrated with the input channel guide fabricated on a cantilever suspended over an etched cavity. Using surface electrodes, the cantilever is deflected electrostatically

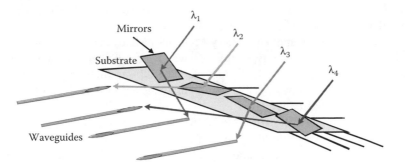

FIGURE 10.2 Array of MEMS mirrors.

from side to side to allow connection to either of the output waveguides. The switching time on the order of milliseconds is long compared to the data rate, but is sufficient for network reconfiguration.

Other MOEMS switches are based on the insertion of small mirrors into the nodes of a cross-connect structure. These devices can be fabricated by deep reactive ion etching (DRIE) of the silicon. DRIE can simultaneously form vertical mirrors, fiber alignment grooves, and a simple electrostatic drive. The electronic drive is used to remove or insert the mirrors into the regions of cross connects. Switching times for these switches are on the order of milliseconds.

Three-dimensional MOEMS may be used to construct larger mirror insertion cross connects. For example, a 64-node cross-point switch may be formed from a 8 × 8 array of small mirrors. The mirrors may be moved between two states by an electrostatic drive. Switches of this type are now well developed and reliable.

Mirror insertion devices are also being used as variable optical attenuators. A small mirror on a cantilever may be inserted into the optical path to block part of a collimated beam (the reflected part is directed onto an absorber). One example of such an application is channel equalization in dense wavelength-division multiplexed systems.

Linear arrays of mirrors have also been used to construct add-drop multiplexers. A lens is first used to collimate the output from a channel carrying a number of different wavelengths. A diffraction grating is used to disperse the wavelengths into a number of beams, which are then passed through another lens to form spatially separated focused spots. The mirror array is placed at the focal plane such that each focused wavelength falls on a separate mirror. Each mirror may then either reflect the beam back through the multiplexer or redirect it to a second multiplexer and a transmission channel. In this way the spectral compositions of two output channels may be arbitrarily selected.

MOEMS technology has attracted a great deal of interest for both small- and large-scale optical-switching devices. At the beginning, optical cross connects (OXCs) and add-drop modules represented the two most promising markets.

However, MOEMS need to overcome challenges such as nonreliability, vibration sensitivity, high voltages, and optical losses.

These challenges applied to early MOEMS devices that were based on thin polysilicon mirrors. Recently, suppliers have begun replacing this material with bulk-machined single-crystal silicon material. Bulk-machined mirrors are much thicker, making them more stable and highly resistant to influences of temperature, vibration, and fatigue [10.1].

The operating voltages for MOEMS are still relative high, but they have dropped considerably from the 250 V of earlier systems. For example, a 16×16 MOEMS switch now requires about 100 V. It should be pointed out that MOEMS do not require electrical currents because they operate on electrostatic forces.

The surface forces are a well-known problem in the fabrication of MEMS and MOEMS devices. This is related to stiction, which occurs when surface adhesion forces are higher than the mechanical restoring force of the microstructure. When a device is removed from solution after wet-etching, the liquid meniscus formed on surfaces pulls the microstructure towards the substrate. Applying a dry-etching technique or using CO_2-drying may avoid this stiction problem. However, a more critical problem is related to in-use stiction, which occurs during the MOEMS operation. In-use stiction may be caused by capillary forces, electrostatic attraction, or direct chemical bonding. To overcome the stiction problem, many MEMS and MOEMS manufacturers have decided to switch to bulk micromachining (see Chapter 11).

Another approach to solve the stiction problem is to provide a low-energy surface coating in the form of an organic passivation layer on the inorganic surface. Such a coating, applied to the semiconducting substrate, can eliminate or reduce capillary forces, eliminate direct chemical bonding, and can reduce electrostatic forces.

Antistiction coatings for MOEMS devices must meet the following characteristics before they can be fully implemented in commercial systems: (1) they should be covalently bonded to the substrate; (2) they should be compatible with dry or aqueous etching processes; and (3) the monolayers should be chemically and mechanically stable under conditions of processing and operation.

Commercial acceptance of MOEMS-based devices will be accomplished when the stiction-related problems are fully resolved.

Nevertheless, it is expected that the significant progress in the performance of MOEMS-based devices and systems will be achieved by merging them with a new cluster of engineered nanostructures. Such a merger will allow for further exploration of MOEMS functionalities that can be derived from manipulation of spectral, polarizing, and spatial properties of optical beams.

There have been many exciting predictions regarding the future of MOEMS. However, despite the demonstration of numerous MOEMS device and product concepts, a relatively small number of MOEMS devices have actually succeeded in the market place. Two prominent examples are MEMS-based devices such as digital mirrors and accelerometers.

10.2 PBG STRUCTURES

Unlike the case for microelectronics, significant changes in microphotonic devices cannot be brought about through direct scaling and reduction of critical dimensions. Rather, changes to the underlying physics of operation will be needed to radically change optical device designs. The recent advances in the development of PBG structures provide the possibility of building devices in the nanoscale.

The operation of active electronic devices such as transistors and diodes are based on the regular periodic structure of crystalline semiconductor materials such as Si and InGaAsP. This periodicity results in an electron energy band gap that separates filled valence states from relatively empty conduction states with an associated periodicity in the electron k space (momentum space). The exact structure of the electron space defines the response of electrons to external stimuli. Defects, such as dopants, can be introduced into the periodic structure to modify the electronic band structure and the resulting optoelectronic characteristics. For example, the inclusion of phosphorus into a silicon matrix results in a shallow band of electron donors within the energy gap just below the edge of the conduction band. This shifts the electron Fermi energy toward the conduction band edge, resulting in an n-type semiconductor with an excess conduction electron population relative to the intrinsic electron concentration of the pure crystal. Similarly, a semiconductor crystal can be doped with acceptor-like impurities such as boron that create excess holes in the valence electric band, resulting in a p type semiconductor. The formation of electrical junctions, such as p n or metal–semiconductor creates voltage-controlled barriers to the motion of electrons. This forms the basis for active devices such as diodes and transistors. Using lithographic patterning techniques and planar fabrication, complex electronic structures can be fabricated on a single semiconductor substrate in a very compact area. This forms the basis of VLSI electronic integrated circuits. The dimensions for guiding the electronic signal are currently in the submicron regime, facilitating high-speed electronic systems (> 1 GHz) with thousands of active devices in very compact areas of a few millimeters.

PBG materials (also called photonic crystals) are periodic dielectric structures fabricated in one to three dimensions in dielectrics and semiconductors that act as optical (electromagnetic) analogs of electronic semiconductor crystals (see Chapter 3). The periodicity of the dielectric constant induces the removal of degeneracies of the free-photon states at Bragg planes. Analogous to electrons in a crystal lattice, under suitable conditions, a photonic crystal will exhibit a frequency band (i.e., PBG) for which electromagnetic waves are forbidden irrespective of their direction of propagation in space.

PBG materials (or photonic crystals) were predicted theoretically by Yablonovich [10.2] and John [10.3]. This prediction has led to two fundamentally new optical principles:

- Complete inhibition of spontaneous emission over a broad frequency range
- Localization and trapping of light in bulk materials

PBG devices are obtained by introducing defects (line or point) into the photonic lattice. This can induce wavelength-specific allowed states within the PBG that are localized to the vicinity of the defect. The PBG outside the defect provides the optical confinement. This is in contrast to traditional optical channel waveguides and fibers that rely on propagation-dependent total internal refraction. Isolated defects can be used to form cavities to effectively store light. Line defects can be used as the basis for channel waveguides. The tight optical confinement provided by PBG structures allows much sharper channel waveguides than is feasible using standard diffused or etched channel waveguides.

These characteristics of PBGs have led to a wide range of novel concepts that have triggered vigorous activities in designing PBG structures with a view to develop a variety of new and improved optical and microwave devices. By introducing a defect in the periodic structure, one can construct such devices as miniaturized waveguides, laser cavities, couplers, beam splitters, filters, etc.

Standard channel waveguide technologies require very gradual waveguide bends to minimize losses. This results in relatively large fan-outs when interfacing multichannel integrated optical devices and input/output channel arrays with a l periodicity of approximately 125 or 250 μm. A high-index substrate such as Si or InP is desirable for PBG structures to provide a useful band gap. PBG techniques can be employed to improve optical confinement in channel waveguide structures to provide a significant reduction in the substrate area required for passive signal distribution.

A review of published results indicates that microphotonics currently has a strong theoretical basis with various concepts and simulation tools being developed for photonic crystal devices. In terms of experimental technical readiness, it is still at the development level of the basic building blocks. The main issues are the accuracy (resolution) of the lithographic patterning to define the PBG structures, optical insertion losses of PBG waveguides, and the input/output coupling between waveguides and fibers.

One-dimensional PBG structures can be written in channel waveguides and optical fibers using the technology developed for writing fiber Bragg gratings (FBGs). Applications of these structures include control of spontaneous emission and wavelength selectivity. Two-dimensional PBG structures have been produced by various groups on Si and InP using high-resolution lithography (e-beam direct-write and deep-UV). The technical challenges currently encountered for PBG devices include:

- Efficient interfacing to optical fibers and traditional channel waveguides
- Fabrication tolerances of air holes in two-dimensional PBG lattices of approximately ± 50 nm in diameter due to optical proximity effects
- Development of active PBG devices

FBGs can be considered as an example of one-dimensional PBG structures. Conventional semiconductor fabrication techniques allow for making one-dimensional and two-dimensional PBG structures.

10.2.1 SILICON PBGs

Ideally, in the future, commercial PBG structures will be made from the same semiconductor materials as computer chips by using common chip-making techniques. They will contain regularly spaced gaps of air or other materials that form boundaries within the crystal that refract, or bend, specific wavelengths of light. Therefore, silicon-based PBG structures are of a great interest to the industry.

Almost all research conducted in two-dimensional planar PBGs are processed using e-beam lithography and are designed for 1.55- and 1.31-μm light [10.4]. Several advantages of e-beam lithography systems are their relatively low cost, ability to pattern features smaller than 20 nm, and their ease of use for processing wafers. Most physical contact (PC) structures (holes), designed for 1.55- and 1.31-μm light, are of similar dimensions and design rules as those for 0.130- and 0.090-μm CMOS transistor designs.

The Interuniversity MicroElectronics Center (IMEC) in Belgium has fabricated PBGs with conventional deep-UV lithography and reactive ion etch techniques [10.4]. Their successes with 0.18-μm technology tools have demonstrated applicability of these techniques to PBG designs. This resulted in increased attention to the possibility of pursuing PBG research in modern high-volume fabrication.

PBG designs can be dramatically expanded by using standard fabrication equipment and adapting lithography and etching techniques. Currently, for both fabrication techniques, the most challenging processing problems that face PBGs are coupling losses and optical losses through the waveguides. Recently, researchers at IBM managed to create PBG waveguides in SOI with transmission losses as low as 20 dB/cm [10.4].

10.2.2 HIGH-ORDER PBGs

One of the most promising PBG designs has PBGs operating in higher-order structures. Such a design allows making the overall device dimensions larger than those for the conventional first-order PBG structures. Manufacturing conventional PBG structures necessitates very sharp control as periodic structures are usually smaller than 1 μm.

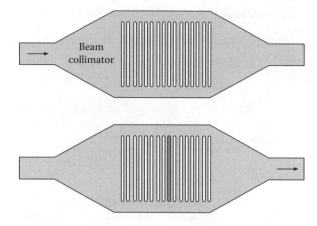

FIGURE 10.3 Design of a high-order PBG structure.

Designing PBG structures at higher orders relaxes technical challenges by permitting larger periodical structures in the range of ~ 1 μm.

The advantages of the higher-order PBG structures are as follows:

- Relaxed fabrication tolerances relative to first-order PBG structures (dimensions are on the micron scale)
- High wavelength selectivity up to 1 pm with a relatively short one-dimensional structure (e.g., with 15 elements, such a resolution would require over 2000 grating periods for the conventional linear Bragg grating structures)
- Possibility of switching by filling in the air gaps
- Possibility of wavelength-selection tuning using microheaters

Figure 10.3 illustrates an example of a high-order PBG structure design. Figure 10.4 shows the SEM view of a microfabricated high-order PBG device with a gap spacing of about 1.4 μm.

Despite using very few grating elements (5 to 9), wavelength selectivity better than 0.5 nm is feasible for filters and WDM applications (see Figure 10.5). These devices have been fabricated in a very compact waveguide structure.

The layer thickness of high-order PBG structures is determined by the value λ/n (e.g., for Si, $n = 3.45$ and $\lambda = 1.55$ μm). As can be seen from Figure 10.4 and Figure 10.5, a structure with only a few periods (7 and 5, respectively) provides a relatively broad spectral response. As the number of layers increases, the spectral bandwidth for such a PBG structure can approach the picometer range, even though the total number of periods is only on the order of 10 to 20. A comparative Bragg grating filter with a similar spectral resolution would have to be almost 10-cm long and contain thousands of grating elements. Thus, this approach to providing wavelength-selective devices is much more efficient than the conventional Bragg grating structures.

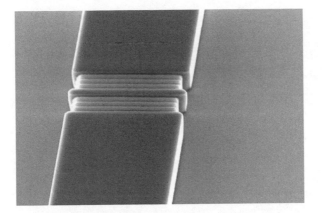

FIGURE 10.4 SEM micrograph of the fabricated P7 high-order linear PBG structure.

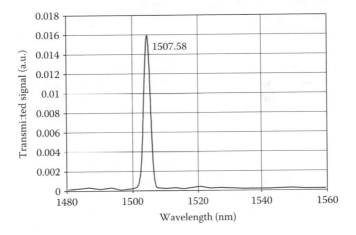

FIGURE 10.5 PBG-5 transmittance characteristic.

10.3 RING RESONATORS

Optical ring resonators (RRs) are components used for wavelength filtering, multiplexing, switching, and modulation.

The basic structure of an optical RR is shown in Figure 10.6. The optical signal is split into two beams. Each beam couples from their respective channel waveguide into the RR, one portion of the beam traveling in a clockwise direction and the other traveling in the counterclockwise direction. External stimuli such as the Sagnac effect can cause a relative phase change between the two counter-spinning optical beams. The Sagnac effect (see Chapter 3) states that light beams propagating in opposite directions in a rotating frame experience a difference in their optical path lengths. Such a design can be used for fabrication of miniature sensors and waveguide optical gyroscopes.

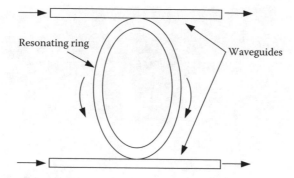

FIGURE 10.6 Example of a ring resonator.

In the case of a RR, the two output frequencies from the resonator can beat together. Resonances occur when the path length within the RR is an integral multiple of the input wavelength.

The key performance characteristics of the RR includes the free spectral range (FSR), the finesse (or Q factor), the resonance transmission, and the extinction ratio. These quantities depend not only on the device design but also on the fabrication tolerance. Although state-of-the-art lithography may not be required for most conventional waveguide designs, RR and the conventional PBG designs require microfabrication resolution on the order of 0.1 μm [10.5].

For such designs, the microfabrication resolution is of great importance to the success of the device. In the case of Si-based RRs, one of the critical parameters to control is the coupling efficiency between the RR and the input or output waveguide. As a compact waveguide (for example, 220×500 nm strip waveguide) is usually used in the RR to obtain a large FSR, the gap between the ring and bus waveguide is in the 100- to 200-nm range. As the device operates through evanescent coupling, the coupling is exponentially dependent on the size of the separating gap. Thus, in order to reliably process high-Q RRs, precise microfabrication control readily achieved by modern 0.18- or 0.13-μm lithography is required [10.5].

10.4 SMART COATINGS

The use of thin-film smart coatings may simplify the design of switches and other microphotonic devices by replacing mechanical hinges with nonmoving elements.

Several material systems have been identified as potential candidates for smart materials, i.e., materials that change one or more of their physical properties with external stimuli. For example, oxides of metals such as V, W, Mn, and L have been found to exhibit a change from metallic to insulator behavior (metal–insulator transition) in response to an external electric or thermal stimulation. There are three groups of these so-called smart coatings that have been extensively investigated in the recent years — $La_{1-x}M_xMnO_n$, WO_3, and VO_n.

10.4.1 $LA_{1-x} M_x MnO_N$

$La_{1-x} M_x MnO_n$ is a class of double-exchange (DE) ferromagnets. DE refers to a double electron exchange between neighboring Mn and O sites. This affects the Mn-O-Mn bond angle and hence the crystal structure. DE ferromagnets differ from the vanadium oxide family in that the DE class of materials is metallic at lower temperatures and semiconducting above the transition temperature. Here, M_x can be Ca, Sr, or Ba. The large change in optical and electrical properties occurs near x = 0.3 of composition, with $LaMnO_3$ behaving as an insulator at one extreme of the compositional variation.

Bulk samples are typically prepared by sintering a powder mixture of the appropriate composition at high temperatures above 1200°C for about 60 h. This method can be used to produce tiles of the material. Thin films can be prepared by multicomponent laser ablation onto a suitable substrate such as ZrO_2 with subsequent high-temperature annealing at about 900°C. In the Sr composition range of 0.16 to 0.30, the material exhibits metallic behavior below the transition temperature and is a paramagnetic insulator above it. Transition temperatures near 250 to 360 K can be obtained. The metal–insulator transition in this system is from a semimetallic reflective state at lower temperatures to an insulating state at higher temperatures. The insulating state is generally opaque with high optical absorption in the UV and visible range.

10.4.2 WO_3

Tungsten oxide (WO_3) is the basis of most electrochromic devices. These films can be prepared by either e-beam evaporation or reactive sputtering using WO_3 targets. Films prepared at lower temperatures tend to be amorphous. WO_3 exhibits high optical transmittance from below 400 nm to beyond 2500 nm. The optical band gap is near 3 eV. The corresponding refractive index is about 2.4. As charge is injected into WO_3, the transmittance in the red spectral range above 500 nm is reduced. The coloration by Li (or H or Na) causes the emergence of an optical absorption band with a peak near 1100 nm that is responsible for the reduction in transmittance above 500 nm.

Electrochromic devices fabricated at lower substrate temperatures are in a disordered (i.e., amorphous) state; thus, optical switching occurs between a highly transmissive state and a highly absorbing state in the near- to midinfrared. The switching mechanism depends on the field-assisted diffusion of ions such as Li or H and, therefore, typically requires several seconds at room temperature. The response time decreases exponentially as the ambient temperature is lowered. Addition of the colorant can impede the reversibility of the process.

Electrochromic devices require the use of an electrolyte to facilitate the migration of the colorant ions (H^+, Li^+, and Na^+) from a reservoir to WO_3. Typical electrolytes for electrochromic devices, such as zirconium phosphate in a gel suspension, tend to be relatively volatile with upper operating temperatures typically below 100°C. As a result, electrochromic devices require hermetic sealing.

The main research endeavor for potential applications has been the development of a solid-state electrolyte such as amorphous Ta_2O_3.

The efficiency of the electrochromic devices can be increased by using a NiO_x layer near the anode to obtain supplemental anodic coloration of the Ni oxide.

10.4.3 VO$_N$

Recently a new method has been proposed for active microphotonic devices based on VO_n smart thin-film materials. VO_2 can pass from the insulator phase (low temperature) to metallic phase at higher temperature (e.g., ~ 65°C). The transition temperature is adjustable and easily controlled. For example, adding Tungsten as a dopant can shift the transition temperature down (see Chapter 3).

There are three methods to control the VO_2 switching transition (i.e., metal–insulator transition):

1. Thermochromic (millisecond to microsecond transition times): By heating the thin film over the transition temperature (65°C), the transition can be as fast as a few microseconds if the triggering impulse is generated by a low-power laser pulse. For slower transitions, a microheater or microcooler can be used.
2. Electrochromic (microsecond to nanosecond transition times): Applying a low voltage (1 to 5 V) through a multilayer structure based on VO_2 and a transparent conductor (ITO) can trigger the transition. In this case the transition can be as fast as 1 nsec.
3. Short laser pulses (< 1 psec): The transition can be in the range of 1 psec if generated by very short laser pulses [10.6].

It is possible to use a laser beam itself as the triggering element in thin-film smart-coating switches. A schematic illustration of the switch operation is shown in Figure 10.7.

A WDM coupler is used to combine 980- and 1550-nm optical beams. The 980-nm beam (or 632 nm) is used to trigger the switching transition in the thin-film VO_2 coating. In the off position, i.e., when the coating is not exposed to the triggering 980-nm optical beam, the coating is transparent to the 1550-nm beam. On the other hand, when the coating is exposed to the 980-nm beam, then the

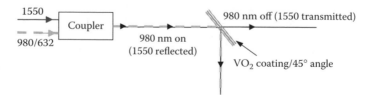

FIGURE 10.7 VO$_2$ optical switch.

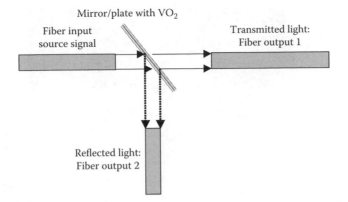

FIGURE 10.8 VO$_2$ switch inserted into an optical path.

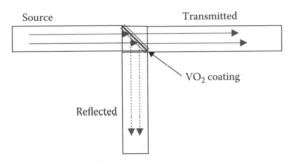

FIGURE 10.9 VO$_2$ switch incorporated within a fiber.

coating switches from the transparent to its metallic or reflective phase. In this case the switch is in on position.

Figure 10.8 and Figure 10.9 show two possible ways to use the VO$_2$ switch. In the first method, the VO$_2$ is deposited on a plate (fixed at 45°) that acts as the switch. Depending on the switch state, the film acts as a mirror or as a transmitter. In the second method, the VO$_2$ is deposited directly on the surface of a fiber.

10.5 HYBRID STRUCTURES

In view of the characteristics that are offered by MOEMS, PBG structures, and smart materials, it is beneficial to consider a hybrid integration of these new technologies with traditional integrated optical structures in order to afford a net reduction of device size and improve device performance. This may include the use of PBG structures to improve optical confinement and reduce the bend radius of channel waveguides. The use of smart coatings may improve the reliability of some of MOEMS-based devices by allowing removal of their moving parts.

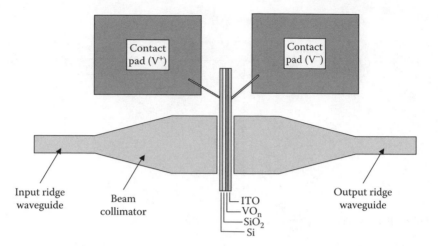

FIGURE 10.10 Hybrid SOI/VO$_n$ optical switch.

Figure 10.10 shows a hybrid scheme of an optical switch based on SOI and VO$_2$. Applying the voltage triggers the insulator to metallic transition and the VO$_2$ passes from the transmitter to reflector [10.5]. Such a transition can be as fast as 1 nsec [10.7].

Coupling of VO$_2$-based smart coatings with SOI technology can provide:

- Variable optical attenuators and phase delay lines
- High-speed optical switching in the nanosecond to picosecond range
- High-speed optical modulators

As VO$_2$ is relatively transparent near 1550 nm, it could be used directly for small sections of active channel waveguides.

REFERENCES

1. Berger, J.D. and Anthon, D., Tunable MEMS devices for optical networks, *Optics and Photonics News*, 43–49, 2003.
2. Yablonovitch, E., Inhibited spontaneous emission in solid-state physics and electronics, *Phys. Rev. Lett.*, 58, 2059–2062, 1987.
3. John, S., Strong localization of photons in certain disordered dielectric superlattices, *Phys. Rev. Lett.*, 58, 2486–2489, 1987.
4. Salib, M., Liao, L., Jones, R., Morse, M., Liu, A., Samara-Rubio, D., Alduino, D., and Paniccia, M., Silicon photonics, *Intel Technol. J.*, 8, 144–160, 2004.
5. Kruzelecky, R., Haddad, E., and Zou, J., Microphotonics for Space Applications, ESA's Report 16112/02/NL/CP (unpublished document), 2003.

6. Cavalleri, A., Toth, C., Siders, C.W., Squier, J.A., Raksi, F., Forget, P., and Kieffer, J.C., Femtosecond structural dynamics in VO_2 during an ultrafast solid-solid phase transition, *Phys. Rev. Lett.*, 87, 7401–7404, 2001.
7. Stefanovich, G. et al., Electrical switching and Mott transition in VO_2, *J. Phys. Condens. Matter*, 12, 8837–8845, 2000.

11 Materials, Fabrication, and Integration

Customer demand for price reduction has put pressure on manufacturers to reduce costs of microphotonic devices while maintaining their high performance and functionality. This market demand has been the main reason the microphotonics industry has embarked on manufacturing processes similar to those used in the semiconductor industry.

The biggest challenge to manufacturers is the transition from custom-designed to standardized microphotonic devices. Even though each system is unique, they all require the same basic functions, many of which occur in the same sequence. Electronic component manufacturers were able to effectively define standardized integrated electronic circuits that were derived from the common elements of application-specific integrated circuits (ASICs). Based on that history, it is logical to assume that component manufacturers of microphotonics will define a host of standardized components from the common elements of application-specific micro-PICs.

Microphotonic device manufacturers will be required to develop a library of standard building blocks similar to those developed by the electronic manufacturers. These standard building blocks will drive micro-PIC standards and serve as the fundamental tools for systems designers. Built with hybrid elements, these devices may include switching matrices, amplifiers, lasers, modulators, detectors, splitters, couplers, and others. Such scalable, recyclable product blueprints will reduce design cost and time to market, allowing equipment manufacturers to update existing product lines or bring new product lines to market with speed and reliability.

The main advantage of this approach is batch fabrication. Batch fabrication includes the following very well-developed manufacturing procedures: (1) bulk processing of substrate material (i.e., cutting, making holes, etc.) and (2) multiple-cycle surface microprocessing (deposition, pattern transfer, and removal).

Batch-manufacturing technology is currently being used to produce a wide array of passive and active optical building blocks such as:

- Waveguides
- Photonic-band-gap (PBG) structures
- Erbium-doped waveguide amplifier (EDWA) structures
- Microoptoelectromechanical systems (MOEMS)

In addition, it is expected that, in the near future, other building blocks such as lasers, acousto-optical (AO) modulators, electro-optical (EO) modulators, and other types of active components will be fabricated in a similar manner.

Batch manufacturing allows for wafer-based manufacturing processes to tailor materials on the submicron scale. These processes enable subwavelength structures to be created in a variety of different shapes on a variety of different substrates. With the appropriate selection of materials, substrates, shapes, and sizes, a wide variety of microphotonic devices may be created.

11.1   MATERIALS

There are three families of materials that are the prime candidates for microphotonic devices. These are (1) silicon, (2) InGaAsP, and (3) LiNbO$_3$. Traditional integrated optical devices have been prepared on all of these substrates. In addition to these, there are several other material systems that are used such as silica-on-silicon (SOS), silicon-on-insulator (SOI), silicon oxynitride, sol-gels, and polymers.

For example, currently commercially available high-speed modulators are based on LiNbO$_3$ and high-channel arrayed-waveguide gratings (AWGs) are based on SOI. PBG structures have been fabricated on both Si and III-V compound substrates.

11.1.1   SILICON

Although micromachining has been demonstrated in glass, ceramics, polymers, and compound semiconductors made of groups III and V, silicon remains the primary material of choice for microphotonic devices. This popularity arises from the large momentum of the electronic integrated circuit industry. The great economic benefit is the extensive industrial infrastructure that has already been established.

Silicon is one of the very few materials that can be economically manufactured in single-crystal substrates. This crystalline nature provides significant electrical and mechanical advantages. The low cost of ultrapure electronic-grade silicon wafers makes them attractive for the fabrication of microphotonic components and devices.

Silicon is an element that exists in any of three forms: crystalline, polycrystalline, or amorphous. Polycrystalline (or simply polysilicon) and amorphous silicon are usually deposited as thin films with typical thickness below 5 μm. Crystalline silicon substrates are commercially available as circular wafers with 100-, 150-, 200-, and 300-mm diameters. Double-sided polished wafers are commonly used for micromachining. The silicon material system encompasses, in addition to silicon itself, a host of materials commonly used in the semiconductor integrated circuit industry. Normally deposited as thin films, they include silicon oxides, nitrides, and carbides, as well as metals such as aluminum, titanium, tungsten, and copper.

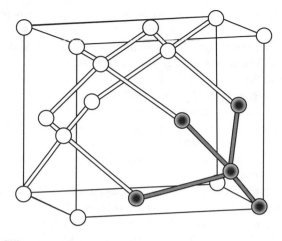

FIGURE 11.1 Silicon crystal structure.

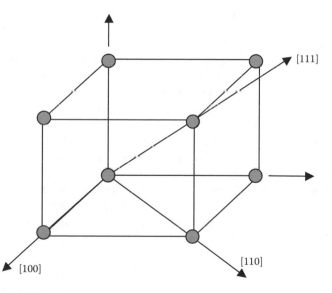

FIGURE 11.2 Miller's indices.

The crystallographic silicon structure is the key to understanding the dependence of material properties on crystal orientation and effects of plane-selective etching. Silicon has a diamond-like lattice crystal structure that can be regarded as simple cubic (see Figure 11.1).

The different planes are designated by Miller's indices, e.g., (100), (110), and (111) (see Figure 11.2). The angle between the (100) and (111) planes in silicon is of particular interest. This is important because many alkaline aqueous solutions, such as potassium hydroxide, can selectively etch the (100) but not the

(111) planes of silicon. The etch results in cavities that are bounded by (111) planes.

Polysilicon deposited onto c-Si may provide an additional layer that is used for making MOEMS actuators. It has been successfully used for fabrication of micromechanical structures and for their integration with electrical interconnects, thermocouples, p–n junction diodes, and other electrical devices.

Surface micromachining based on polysilicon is today a well-established technique for forming thin and planar devices. Mechanical properties of poly-crystalline and amorphous silicon are similar to those of single-crystal silicon except that they exhibit high levels of intrinsic stress that requires annealing at elevated temperatures.

Silicon oxide (SiO_2) is a very stable material, and it is used as an electrical insulator on Si. Various forms of silicon oxides (SiO_2, SiO_x, and silicate glass) are widely used in micromachining because of their excellent electrical and thermal insulating properties. They are also used as sacrificial layers in surface-micromachining processes because they can be preferentially etched in hydro-fluoric acid (HF) with high selectivity to silicon. Silicon nitride (Si_xN_y) is also widely used as an insulating thin film and is effective as a barrier against mobile ion diffusion, in particular, sodium and potassium ions. It can also be used as an effective masking material in many alkaline etching solutions.

The capabilities of MOEMS actuators can be augmented by the possibility of deposition of $LiNbO_3$ on Si. This can be realized by reactive ion sputtering in a clean vacuum system with the substrate heated to 550°C. The first epitaxial growth of $LiNbO_3$ on (111) Si using the chemical beam epitaxy was demonstrated by growing alternating layers of the single-metal oxides of Li_2O/Nb_2O_5 [11.1].

SOS integrated process offers an efficient platform for the implementation of optical power splitters or combiners, couplers, and wavelength-division mul-tiplexers and demultiplexers. The real benefit of SOS, however, is the ability to apply wafer-scale, planar lithography and processing techniques to integrate substantial numbers of functions onto common optical chips, either as arrays of identical devices or in the form of customized circuit configurations. SOS tech-nology has also been used to manufacture lasers, amplifiers, couplers, filters, switches, and attenuators.

SOI technology is a substitute for SOS technology. The starting substrate is a silicon wafer with a buried silica layer.

Silicon oxynitride (SiON) technology uses SiO_2 as a cladding. A core is made from a mix of SiO_2 and silicon nitride (SiN_4). This planar waveguide technology uses a low-pressure chemical vapor deposition (CVD) process.

11.1.2 InGaAsP

InGaAsP is the most difficult and expensive material to process and has the poorest thermal stability. However, InGaAsP can be used to produce active devices.

InGaAsP multilayer structures can be grown on InP by a gas source using molecular beam epitaxy (MBE) or metal–organic MBE (MOMBE). The multi-layered structures facilitate quantum layers, strain-matching layers for hetero-structures, and vertical integration of various optical functions. Whereas InGaAsP has no native oxide, SiO_2 and Si_3N_4 can be deposited using low-energy, low-temperature CVD. Buried stop-etching layers and reflectance layers are accom-modated via hetero-MBE.

InGaAsP facilitates various active optical devices including EO switches, LEDs, and laser diodes. These can be coupled to echelle grating for add-drop multiplexers. Currently, processing area is limited to about 5-cm O.D. PBG devices have been fabricated on InP and on GaAs.

11.1.3 $LiNbO_3$

Lithium niobate ($LiNbO_3$) provides planar optical elements and waveguides. It is widely used for Mach–Zehnder modulators, AO tunable filters, spectrum ana-lyzers, and wavelength-division-multiplexing add-drop nodes. Optical switches based on EO and AO effects can also be fabricated from this material.

Hybrid combinations of AO and EO devices can be employed for more complex structures. $LiNbO_3$ on Si technology will most likely replace the use of bulk $LiNbO_3$ substrates for integrated optics as it can provide the additional capabilities of Si for MOEMS microactuators.

11.1.4 Sol-Gels

The sol-gel technique is based on hydrolysis of liquid precursors. The precursors are usually organosilicates yielding silicate sol-gel materials. However, the method is not restricted to the silicon compounds. For example, compounds of zirconium, vanadium, etc., can be used as precursors leading to materials pos-sessing different physical and chemical properties [11.2].

As sol-gels occur in the liquid phase, it is possible to add any substance (as solutions or suspensions) at the early stage of the growth process. Simple mixing provides uniform distribution of the dopant within the liquid host phase. After gelation, the guest molecules become physically entrapped within the now solid host matrix. Furthermore, hydrolysis, doping, and gelation occur usually at ambi-ent temperatures, allowing entrapment of very delicate molecules without their decomposition. The doped matrices usually possess good optical characteristics that are vital for the production of microphotonic devices.

The main advantage of this process stems from the fact that it offers an alternative approach to conventional production of glasses, glass-like materials, and ceramics of various properties and applications. Sol-gel technology enables production of doped glassy materials either as porous dry gels (xerogels) or densified materials.

Another attractive feature of this technology is the fact that sol-gel materials can be obtained as bulks, thin films, and nanopowders. Such matrices, activated by doping, impregnation or covalent bonding, yield materials that can be used as, among other possibilities, optical sensors, smart materials, and amplifiers.

Recently, a single-step sol-gel process has been developed that dramatically cuts the cost of making optical components such as arrayed-waveguide gratings and variable optical attenuators and couplers. In this process the silica layers may be doped to give a refractive index in the range of 1.45 to 1.50. This enables the creation of the so-called superhigh delta waveguides, in which the difference in the index between the core and cladding can be as high as 3%. This will allow manufacturing smaller components and, therefore, lead to a higher density of planar circuit components on a single wafer [11.2].

11.2   FABRICATION

11.2.1   MULTIPLE-CYCLE SURFACE MICROPROCESSING

This technique allows fabrication of multicomponent integrated structures that would be difficult to produce with bulk micromachining.

Multiple-cycle surface microprocessing steps are identical to those used in integrated circuit (IC) fabrication processes. There are several ways by which microprocessing can be realized. The type of material selected determines the processing sequence.

The main steps involved consist of several processes such as the following:

- Fabrication of wafer substrates
- Film growth of substrates
- Doping
- Lithographic patterning
- Deposition of dielectric, semiconductor, and metallic layers
- Deposition of stop-etch and sacrificial layers
- High-aspect reactive ion etching (RIE)
- Anisotropic etching of V grooves
- Dicing
- Wafer-to-wafer bonding for hybrid integration
- Packaging

Devices are usually fabricated on Si substrates, which are grown as bulk materials that are then sliced into wafers.

Active components are built by growing thin films on wafer substrates. The following materials are used as thin films:

- Epitaxial Si
- SiO_2
- Silicon nitride (Si_3N_4)
- Polysilicon (polycrystalline Si)

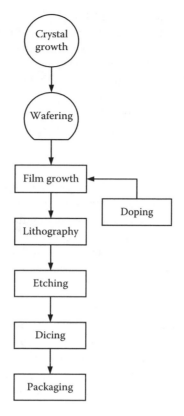

FIGURE 11.3 Fabrication steps.

The films are often doped to modify their mechanical, optical, or electrical properties. The process of thermal diffusion or ion implantation is used for the introduction of dopants.

A desired pattern is imposed on a film by using lithography. A pattern is transferred to a film via a mask. The film is selectively etched away, leaving the required pattern in the film. The wafers are then diced into chips and packaged. A schematic illustration of the fabrication steps is shown in Figure 11.3.

11.2.1.1 Film Growth

Epitaxy is defined as growth of a single-crystal film upon a single-crystal substrate. If the composition of the film is the same as that of the substrate, the process is called *homoepitaxy*. However, if the film composition differs from that of the substrate, the process is called *heteroepitaxy*. Many compound semiconductors, such as gallium arsenide (GaAs) and silicon carbide (SiC), can be grown heteroepitaxially on Si, whereas doped Si layers are homoepitaxial.

SiO_2 is grown by the thermal oxidation of Si. The rate of SiO_2 growth is dependent on the oxygen pressure, temperature, and crystal orientation of the substrate.

Other films can be deposited using CVD. This process is also used for batch coating. In addition to SiO_2 films, metals, polysilicon, and S_3N_4 can be deposited by CVD.

Metal films that are used for interconnections and contacts can also be deposited by vacuum evaporation, sputtering, and plating.

11.2.1.2 Doping

Introducing impurities into the thin-film layer will modify its properties. This step is called *doping*. The doping process is realized by thermal diffusion or ion implantation. Thermal diffusion is based on heating the wafers at a high temperature and passing a gas carrying dopants across them.

In the ion implantation process, the wafer surface is exposed to a beam carrying dopants; this is followed by subjecting the crystal lattice to annealing.

11.2.1.3 Lithography

The technique by which the pattern on a mask is transferred to a film or substrate is known as *lithography*. The pattern transfer is achieved by means of a radiation-sensitive material. The most common form of radiation used for this purpose is electron beam, UV, or x-ray. Lithography consists of two key steps: (1) mask design and mask manufacturing and (2) pattern deposition.

11.2.1.4 Etching

Following pattern deposition, the desired pattern is imprinted on the film by a process known as etching. *Etching* is defined as the selective removal of unwanted regions of a film or substrate.

There are two main types of etching — wet chemical etching and dry etching. Wet etching requires the use of liquid reactants. The most common technique for dry etching is based on the use of reactant gas plasmas.

SiO_2 and Si_3N_4 provide the masks for typical anisotropic etchants. The masks protect areas of Si from etching and define the geometry of the region to be etched.

As it has been mentioned earlier, the Si crystal planes (100) and (110) etch significantly faster than the (111) planes (see Figure 11.1 and Figure 11.2). Therefore, the so-called anisotropic etching may be used to produce holes of different sizes or cavities [11.3]. This is schematically illustrated in Figure 11.4.

For example, V-shaped grooves are produced by etching. V grooves are used for the precise positioning of optical fibers.

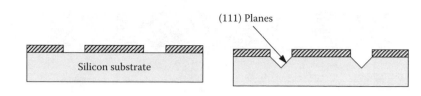

FIGURE 11.4 Anisotropic etching.

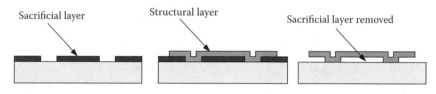

FIGURE 11.5 Surface micromachining.

11.2.1.5 Surface Micromachining

In this technique, various structures are manufactured by using the substrate wafer as a mechanical support on which multiple layers are deposited and patterned. Usually this is a four-step process. The process is initiated by depositing a sacrificial layer, followed by making openings in the underlying substrate, depositing a structural layer and patterning it into the desired geometry. The process is completed by removing the underlying sacrificial material. This process is schematically illustrated in Figure 11.5.

Polysilicon is usually used as the structural material, and SiO_2 is used as the sacrificial material. Si_3N_4 is very often used as an insulator if electrical isolation of the substrate or the other structural components are required. A number of other material systems have also been investigated such as aluminum/polyimide, Si_3N_4/polysilicon, and Si_3N_4/SiO_2 [11.3]. The polysilicon layer can also be incorporated into a wide variety of sensors and actuators such as electrostatic comb drives.

The cycle of deposition, patterning, and etching of each material can be repeated several times to build up multilayered structures.

11.2.1.6 LIGA

LIGA is a widely known process that is being used for fabrication of three-dimensional microstructures in a wide variety of materials, such as metals, polymers, ceramics, and glasses. LIGA is a German acronym for *lithographie, galvanoformung, und abformung*, which means lithography, electroplating, and molding.

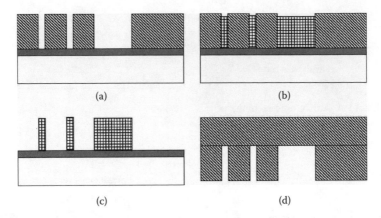

FIGURE 11.6 LIGA process: (a) patterning, (b) electroplating, (c) resist removal, (d) molding.

Molding refers to a fabrication process that uses molds to allow the deposition of the structural layer. After the structural layer is deposited, the fabricated components are obtained when the mold is dissolved in a chemical etchant that does not affect the structural material. In other words, molding is an additive process, i.e., the structural material is deposited only in those areas that are part of the device structure. On the other hand, surface micromachining is an example of a subtractive process, in which the final structure is realized by the material removal.

A schematic presentation of LIGA process is shown in Figure 11.6. Usually x-rays are used to for lithography, and polymethylmethacrylate (PMMA) is used as the resist [11.3]. The openings in the patterned resist are usually plated with metal, allowing for a highly accurate replica of the original resist pattern. The mold is then dissolved.

Synchrotron radiation is used to expose the resist. The short x-ray wavelength allows deep (up to 1 mm) resist layers to be exposed without significant diffraction effects. This enables fabrication of high-aspect-ratio structures.

Less expensive alternative techniques use UV mask aligners to expose the resist and to produce features several-hundred-micrometers thick with aspect ratios of approximately 15. Parts are usually electroplated in nickel. After removal of the resist, they can be replicated in other materials.

Using the LIGA process, it is possible to produce a high-aspect-ratio metallic microstructure. In addition, the larger thickness of high-aspect-ratio structures allows for greater stiffness as well as increased torque in electrostatic actuators [11.3].

11.2.2 FABRICATION OF WAVEGUIDES

The fabrication of waveguide structures is usually done with GaAs/AlGaAs, InGaAsP, SiO_2/Si_3N_4 on SOI, and Si/SiGe/Si on Si or SOI. Considerable expertise

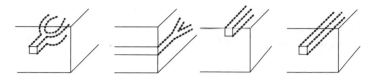

FIGURE 11.7 Examples of waveguide structures.

has been developed for Si/SiO$_2$/Si waveguide technologies and InGaAsP. Si/SiGe/Si waveguides can be prepared by heteroepitaxy.

In the case of glass, wet and dry ion-exchange techniques are commonly used for fabricating most passive components such as splitters and combiners. Dip coating, on the other hand, forms polymer waveguides on glass and other substrates. The most popular material for active devices, LiNbO$_3$, can be processed either using metal in-diffusion technique (usually titanium) or by proton exchange in weak acids.

The waveguide structures may be planar, coupled, channel, or branching. For example, Figure 11.7 illustrates some examples of waveguides structures such as (1) a three-dimensional Y junction, (2) a two-dimensional Y junction, (3) a surface channel, and (4) an embedded channel.

The basic requirements of a waveguide material are (1) their transparency in the wavelength of interest and (2) higher refractive index of the medium in which it is embedded. In channel optical waveguides, light is confined not only across its thickness but also across its lateral direction. In this design, the waveguide in its cross section is surrounded in all directions by a medium having lower indices of refraction. For the raised- or buried-type channel waveguide, transverse confinement in both directions is possible because there exists a high-refractive-index core region that is surrounded by regions of lower refractive indices.

It is expected that in the near future hybrid integration of planar waveguide structures with PBG structures will enable further size reduction and increase of component density through the use of sharper waveguide bends and closer waveguide spacing.

Manufacturing of waveguides typically starts with a lower cladding layer being deposited on the surface of a wafer of the appropriate material, e.g., silicon. The core layer is then deposited on top of the lower cladding layer. The core layer has a slightly higher refractive index, which is required to keep the light confined to the core material. A mask is then applied using lithographic methods to the surface of the core layer. The mask contains the pattern for the waveguide structure. The next step is to etch away the material not protected by the mask, leaving only the core waveguide pattern. A final layer of cladding with refractive index lower than the core is then added to cover the core.

Table 11.1 summarizes some of the most common waveguide materials, fabrication processes, and their properties.

TABLE 11.1
Materials and Processes for Waveguides

Substrate	Waveguiding Layer	Fabrication Process	Properties
Glass or fused quartz	Various glasses Nb_2O_5 Ta_2O_5 Polymers	Sputtering, E-beam evaporation, solution deposition	Lossy, not applicable to AO, EO, and NLO devices
Glass	Mixed metal oxide layers	Ion migration and ion exchange, chemical etching	Low cost and easy fabrication
$LiNbO_3$ or $LiTaO_3$	Metal oxide layers	Ti in-diffusion, proton exchange	Low loss and excellent applicability to AO, EO, and NLO devices
GaAs or InP	$Ga_{1-x}Al_xAs$ $Ga_{1-x}In_xAs_{1-y}P_y$	LPE, VPE, MBE, and MOCVD	Applicable to optoelectronic integration
Quartz crystal	Metal oxides	Ion implantation	Susceptible to defects

11.2.3 FABRICATION OF PBG STRUCTURES

Electron beam (E-beam) lithography is employed to fabricate PBG structures. The lithographic process is used for printing images onto a layer of photoresist that is subsequently used as a protective mask against etching. The pattern is defined using PMMA E-beam sensitive resist. This is usually deposited over an etch mask layer such as SiO_2 or a metallic mask such as Cr. E-beam lithography is employed to form a pattern in the PMMA. This pattern is subsequently transferred onto the mask layer by RIE. The mask layer is subsequently employed as the RIE mask for formation of the dielectric holes comprising the PBG structure. E-beam lithography has a finite continuous write area of about 1 mm². Adjacent patterns have to be stitched together, which may result in subsequent alignment errors. These may be unacceptable errors for larger microphotonic circuits. For example, phase masks for fiber Bragg gratings, involving submicron dimensions, were originally produced mainly by E-beam lithography. However, as the dimensional tolerance requirements on the masks increased for 50- and 25-GHz WDM channel spacings, the phase masks had to be made almost exclusively using holographic UV interference techniques (photolithography). This technique is capable of larger area patterning, without stitching errors.

Conventional photolithographic equipment usually uses an excimer laser that generates UV light at 248 or 193 nm. A series of lenses are used to reduce the original image. The beams are passed through the mask on which the patterns are traced. The image is then exposed on the wafer. Another chemical treatment is used to etch away either the exposed or unexposed areas of the image. Using extreme ultraviolet (XUV), one can embed even smaller features on silicon,

TABLE 11.2
Typical Etching Rates

Etched Material	Gas	Etch Rate (nm /min)
Al, Al-Si, Al-Cu	$BCl_3 + Cl_2$	50
Polysilicon	Cl_2	50–80
SiO_2	$CF_4 + H_2$	50
PSG	$CF_4 + H_2$	80
GaAs	$CCl_4 + O_2$	600
Si	$SF_6 + Cl_2$	100–450

beginning with chips at the 70-nm level. Most materials absorb UV light, and there is a need to come up with materials that reflect this wavelength. Microprocessors that have recently reached the market feature 70-nm elements. It is expected that there will be a transition to manufacture chips with 10-nm elements.

Recently, a new technique called dip-pen nanolithography (DPN) has become an increasingly popular for creating nanoscale patterns on surfaces. It uses an atomic force microscope (AFM) tip coated with a molecular ink to deposit molecules onto a surface. It is expected that features as small as 10 nm could be achieved with this technique.

The figure of merit for etching a PBG structure is the aspect ratio, i.e., the smaller the holes and the deeper they are etched, the better the PBG performance. Aspect ratios of 15 to 20 can be realized with the RIE method. RIE typically employs metal masks such as Cr. Etch depths exceeding 50 μm are possible in Si. Typical etch rates for the selected processes are summarized in Table 11.2.

11.2.4 Fabrication of EDWA

An EDWA consists of waveguides embedded in an amorphous glass substrate. The erbium atoms provide the optical gain in the 1.5-μm transmission window.

The most critical issues to optimize an Er^{3+}-activated waveguide in order to function as an optical amplifier are as follows [11.5]:

- Reduction of optical propagation losses
- Proper concentration of erbium ions (generally, at low Er^{3+} concentration, the lifetime of the metastable level is longer and quantum efficiency is higher, but the total intensity of stimulated emission is lower; at higher concentrations, fluorescence quenching may occur because of ion clustering or ion-to-ion interaction)
- Optimizing the pump power

The process of preparing erbium-doped waveguides consists of two main steps [11.4]. The first step is to prepare the doped glass material. A glass block is prepared by using standard melting techniques, with erbium added in the form of erbium oxide powder. Two types of oxide glasses are used to produce commercially available waveguide amplifiers: aluminosilicate and aluminophosphate. The main characteristic of these glasses is that they accept a very high doping level of erbium, i.e., at a level higher than 10^{26} atom/m^3. This acceptance level is much higher than that possible with, for example, silica glass.

The second step is to make a waveguide structure. There are several methods of making the waveguide structures. The fabrication techniques of EDWAs have been adopted from standard waveguide manufacturing processes. These technologies include: ion exchange, RF magnetron sputtering, sol-gel, and E-beam vapor deposition. Also, techniques such as flame hydrolysis deposition (FHD) and plasma-enhanced chemical vapor deposition (PECVD) have been successfully employed.

In order to fabricate an Er-doped planar waveguide amplifier on Si, one needs to use a material that is transparent to (1) the wavelength used to excite Er and (2) the signal wavelength.

A number of waveguide structures can be used to realize a planar waveguide amplifier. In a waveguide amplifier, light is guided in a high-refractive-index material (the waveguide core) that is doped with Er. Light is confined in the waveguide by total internal reflection. The waveguide core is surrounded by a lower-index material (the cladding).

Some approaches of EDWA fabrication involve the combination of two steps together, i.e., ion exchange to fabricate waveguide and ion implantation to dope the glass with erbium, or FHD and aerosol doping. Thus, ion-exchange and sol-gel techniques have been combined in order to produce channel waveguides. Less conventional waveguide configurations have also been suggested, such as a composite guiding structure consisting of an ion-exchange waveguide in a soda-lime glass and an overlapped Er-doped glass [11.5].

Figure 11.8 schematically shows cross sections of different waveguide amplifier structures. Indicated are (a) a ridge waveguide, (b) a strip-loaded waveguide, and (c) a diffused waveguide. These structures are similar to those used for waveguides. Some of the characteristics of various EDWA materials are summarized in Table 11.3. Table 11.4 presents some examples of EDWAs and their characteristics.

11.2.5 FABRICATION OF MOEMS

The availability of high-performance MOEMS actuators has led to the creation of a new class of devices for microphotonic systems. Many of the large arrays of optical switches have been built by the use of surface micromachining. For example, sensors and actuators are made from patterning thin films, such as polysilicon, on the surface of a silicon substrate and sacrificially etching another material, such as silicon dioxide, to free the mechanical structure.

However, the biggest advantage of MOEMS devices is that they can be designed and manufactured by using multiple-cycle surface microprocessing.

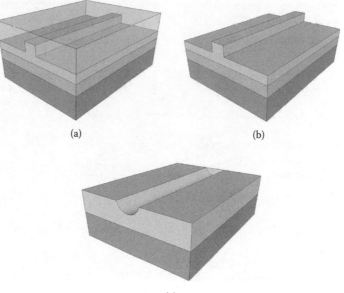

(a)   (b)

(c)

FIGURE 11.8 EDWA structures.

TABLE 11.3
Characteristics of Several Active Optical Waveguides

Material	Dopant	Fabrication Technology	Loss (at 633 nm) dB/cm
Na-modified BK7	2.0 wt% Er_2O_3	K^+-ion exchange	0.8
Soda-lime glass	0.3 mol% Er_2O_3	Diluted Ag^+ ion exchange	—
	3.0 wt% Er_2O_3	Diluted Ag^+ ion exchange	0.7
	0.42 atomic% Er	Diluted Ag^+ ion exchange	0.4
	0.54 atomic% Yb		
	0.2 atomic% Er	K^+ ion exchange	1
Na-Ca-silicate	1.0 atomic% Er	RF magnetron sputtering	1
Aluminosilicate	8.0×10^3 ppm Er	Sol-gel	—
Aluminosilicate	0.5 atomic% Er	Sol-gel	0.5
Silica-titania	1.0 atomic% Er	Sol-gel	—
Germania-silica	0.2 mol% Er	Sol-gel	2–3
Germania-silica	600 ppm Er	E-beam vapor deposition	0.11
P-doped silica	0.5 wt% Er	FHD plus aerosol doping	0.5

TABLE 11.4
Characteristics of EDWAs

Glass	Fabrication Method	Dopant Er	Yb	Maximum Gain (dB)	Length (mm)	Pump Power (mW)
P-doped silica	Flame hydrolysis deposition (FHD)	0.5 wt%	—	13.7	19.4	640
Soda-lime silicate	Radio frequency sputtering (RFS)	14,600 ppm	—	15.0	45.0	280
Soda-lime	RFS	4.1×10^{20} at/cm^3	—	19.0	45.0	300
	Ion exchange (IE)	2 wt% Er$_2$O$_3$	5 wt% Yb$_2$O$_3$	4.3	28.0	90
Borosilicate	IE	3 wt% Er$_2$O$_3$	5 wt% Yb$_2$O$_3$	9.0	39.0	130
Phosphosilicate	plasma-enhanced chemical vapor deposition (PECVD)	0.48 wt%	—	5.0	75.0	420
Silicate	RFS	3.3 wt% Er$_2$O$_3$	—	7.2	17.0	80

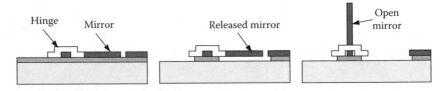

FIGURE 11.9 Fabrication of three-dimensional MOEMS device.

Figure 11.9 illustrates schematically an example of a three-dimensional assembly of an MOEMS device.

This technique may be also used as a platform for the hybrid integration of microphotonic subsystems containing active components such as lasers and detectors. For example, a hybrid integrated transreceiver chip that contains waveguide circuitry, an embedded dielectric filter for demultiplexing, a photodiode receiver, and a laser diode source with monitor photodiode is available commercially in a 1.3- or 1.5-μm package.

However, actuators made by this technique are poorly suited for translating or rotating relatively large and externally fabricated elements such as mirrors or lenses. Therefore, higher force actuators are preferably made by deep reactive ion etching (DRIE) of a silicon substrate.

The actuator fabrication process is relatively simple. First, alignment marks are formed on the front and back surface of an oxidized carrier wafer. Shallow cavities are then plasma-etched into the front surface of the silicon. These cavities define which portions of the actuators is free to move and which will be attached to the carrier. A second wafer is fusion-bonded to the front surface and then polished to the desired thickness. The alignment marks from the bottom surface are then transferred to the top surface to allow alignment with the cavities in the device. After oxidation, contact-hole etch, metallization deposition, patterning, and DRIE are performed. Those portions of the device that are above the cavities are then suspended by narrow flexural elements.

Electrostatic actuation is still the most common technique used, although thermal actuation and magnetic actuation are also implemented in some devices. Devices based on electrostatic actuators can be accurately positioned and easily driven with up to 150 V at very low current. Electrical connections to the moving elements are made through conductive silicon flexures. Electrical insulation is provided by a combination of the oxide layer between the moving or fixed parts of the device and the carrier wafer and by etched trenches that surround device features [11.6].

Mirrors that are part of MOEMS devices introduce some optical losses due to scattering. These losses can be reduced if polysilicon mirrors are replaced with single-crystal mirrors. Single-crystal mirrors are thicker and flatter, whereas polysilicon mirrors are typically 2- to 10-μm thick and they have tendency to bend and curve. Most remaining losses come from array collimators. For example, commercially available 8 × 8 MOEMS switches deliver losses in the range of 1 dB. The losses for 16 × 16 switches are limited to 1.5 dB [11.6].

Packaging of MOEMS is a major concern. MOEMS devices can be easily affected by atmospheric contaminants such as moisture and particles. Particles may lock up MOEMS mechanisms and therefore immobilize its performance. The main packaging problem is related to stiction. The surface-area-to-mass ratio of MOEMS is so large that it magnifies surface effects. Stiction is a friction-like effect that locks together adjacent surfaces. The stiction-breaking force can be several orders of magnitude higher than the force that MOEMS drivers can generate.

There are several ways that this problem is being solved. Firstly, designers minimize surface contact area to eliminate the stiction problems. Secondly, coating of surfaces with Teflon-like materials is used to minimize the stiction. Thirdly, as stiction is closely related to the presence of water vapor, some desiccants are used to keep water vapor at a minimum. Some desiccants are also applied to trap other microscopic particles.

Recently, a new approach to the manufacturing of MOEMS has been developed. It is called *grayscale lithography*. The technique of grayscale lithography is used to create lenses, prisms, and various other structures.

In microphotonic systems, three-dimensional geometries with greater depths are becoming necessary for applications with MOEMS, in which RIE or ion milling are no longer viable. MOEMS fabrication technologies originated directly from IC fabrication. IC devices require only two-dimensional or planar structures because mechanical parts are required. Therefore, structures fabricated for MOEMS devices have been traditionally designed with vertically anisotropically

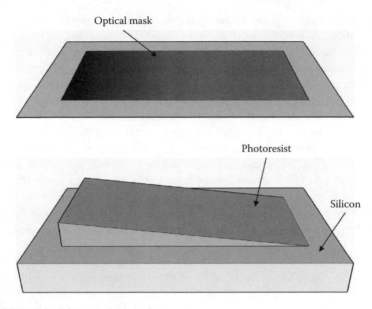

Optical mask

Photoresist

Silicon

FIGURE 11.10 Grayscale lithography.

or isotropically etched sidewalls. However, these technologies are restricted to a limited range of shapes.

Grayscale lithography allows manufacturing of arbitrary-shaped microstructures in various materials. This technique can produce three-dimensional structures in a photoresist layer, which can then be transferred in silicon by use of dry etching.

There are three key steps of grayscale lithography: (1) mask design, (2) grayscale lithography, and (3) dry anisotropic etching. Steps (1) and (2) allow for a precisely designed three-dimensional profile in a photoresist-masking layer by modulating the beam intensity that is irradiating the photoresist surface. Step (3) allows the three-dimensional profile in the photoresist to be transferred into the underlying silicon substrate by dry anisotropic etching. This process is schematically illustrated in Figure 11.10.

11.3 INTEGRATION APPROACHES

Although the advantages of microphotonic devices are obvious, there are still some challenges that have to be overcome before the benefits offered by this technology can be fully realized. One of these challenges is related to integration of various components, devices, and subsystems. Unlike the semiconductor industry, in which electronic integration is a small fraction of the product cost, integration of microphotonic components represents a large portion of the costs.

Microphotonic devices offer significant improvements in terms of manufacturing costs, size, weight, processing speed, and power consumption over distributed bulk-optics systems. However, coupling to external optical I/O channels is a major source of loss of optical signal. Mechanical reliability is the other concern related to integrated microphotonic devices.

There are several innovative approaches to the integration of microphotonic systems that have been developed in the recent years. Some of them are summarized in the following subsections.

11.3.1 MICRO-PICS

The objective of micro-PICs is to integrate the various photonic subcomponents in a single integrated chip. Micro-PICs can be considered as a hybrid suite of technologies that enable the miniaturization and integration of basic functional blocks such as switches, modulators, wavelength-division multiplexers, spectrum analyzers, true optical time-delay lines, beam splitters, etc. Achieving a high functionality on a single micro-PIC entails minimizing the size of the individual generic functional blocks that form the basic subsystems. It is believed that using a micro-PIC approach can minimize the cost of customizing a microphotonic device for specific applications.

Photonic components such as tunable WDM filters, photodiodes, optical waveguides, and electronic components such as laser drivers and transimpedance amplifier (TIA) circuits are all based on different materials and technologies, making them difficult for monolithic integration.

The integration of several functions on a single chip is critically dependent on the architecture and, hence, the size of the individual circuits. In this respect, $1 \times N$ channel structures are inherently smaller than cascades of 2×2 channel devices. Echelle grating (1 cm^2) and superprism (< 2 mm^2) wavelength-division multiplexers are significantly smaller than AWG structures (> 16 cm^2). However, the $1 \times N$ devices entail more complex planar beam shaping and coupling. Optical 2×2 switches based on index-matching microbubbles or tunable resonant cavities can be considerably smaller than traditional EO (about 1 cm^2) and MOEMS optical-switching nodes. The high optical confinement provided by high-index waveguides such as Si (n = 3.45) facilitates new structures for passive devices such as beam splitters. These beam splitters can be several orders of magnitude smaller than traditional LiNbO$_3$ and silica waveguide passive devices.

The technological basis of micro-PIC approach is provided by (1) integrated optics, (2) PBG structures, (3) MOEMS, and (4) thin-film smart materials, all of which can reconfigure their optical characteristics in response to external control signals (see Figure 11.11). Thin-film smart materials offer new possibilities for active devices on substrates such as SOI that traditionally offer limited capabilities for active optical components.

It should also be pointed out that in the near future the emerging technology of quantum photonics will become an important element of micro-PICs technology.

By definition, standardized micro-PICs should provide fundamental building blocks that could be manufactured and incorporated into systems with near plug-and-play ease, compacting the functionality of an entire rack of equipment into

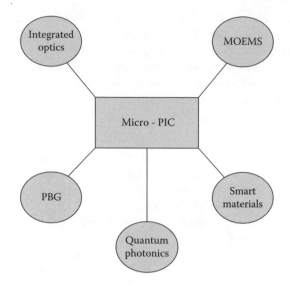

FIGURE 11.11 Basic technologies for micro-PICs.

a space no larger than a matchbox. The benefits include reduced design time, decreased manufacturing costs and, ultimately, rapid deployment [11.7].

The objective of micro-PICs is to integrate the various optical subcomponents in a single photonic integrated chip. The main technical requirements are as follows:

- Fiber waveguide integration
- Suitable substrate material for micro-PICs
- Ability to process microphotonic materials to provide desired optical functions
- Ability to integrate several functions on a single substrate
- Hybrid integration of microphotonics with other technologies

There are a number of key challenges for the integration of micro-PICs systems. First, several different material systems are used in the photonic industry to generate, modulate, multiplex, amplify, and detect the optical signal. These material systems are based on lithium niobate, silicon, silica, and III-V compound semiconductors (e.g., InGaAsP). Each of these materials entails somewhat different manufacturing tools and processes, as well as design rules.

The integration of micro-PICs starts with passive devices built on SOS planar waveguides. Examples include devices such as splitters, taps, couplers, and AWGs that are integrated on a single substrate. The next step is adding switching capabilities to the aforementioned passive devices. These include integration of switches and AWGs for optical add-drop multiplexers (OADM).

The two most promising material systems for micro-PICs are InGaAsP (InP) and SOI. Both SOI and InGaAsP facilitate a high-refractive-index contrast that is desirable for the waveguide structures.

InGaAsP (InP) has the advantage that it can provide active devices such as laser diodes and semiconductor optical amplifiers. Therefore, InP can provide device integration that includes signal generation, detection, amplification, high-speed modulation, combining, routing, and switching, as well as passive splitting. InP yields compact passive devices with small bending radii, enabling more devices to be integrated on the same substrate. Some manufacturers believe that with InP they will be able to integrate different passive and active functions to create an integrated monolithic chipset [11.7].

However, processing InP devices involves significantly higher costs than SOI because of costlier substrates, processing equipment, and handling procedures. Processing InP involves toxic materials and special procedures to avoid surface decomposition of As or P. This requires special safety precautions, toxic gas containment, and monitoring, which entail delays and more expenses.

Recent developments in the area of SOI have gained a lot of attention from the industry. The growth of single-crystal $LiNbO_3$ and GaAs on SOI has been recently demonstrated, and it will enable the future fabrication of various EO devices. SOI can provide most of the required functionalities for micro-PICs, including the following:

- PBG channel waveguides
- Y-junction interferometers
- Resonant coupled cavities
- Tapered input couplers
- SiGe quantum-well structures
- MOEMS, thermo-optical or bubble 2×2 optical switches
- Standard ridge waveguides with right-angle bends
- One-dimensional optical amplifiers and Bragg gratings using active glasses such as Er-doped silica
- Echelle grating and WDM structures
- Superprism planar optics

It is expected that silicon-based components will enable on-chip or chip-to-chip commercial integration. The attractiveness of silicon-based optical devices is the potential integration with CMOS ICs for high-volume manufacturing. The first prototype of such silicon-based optical components includes thermally tunable WDM Bragg filters, high-speed optical Mach–Zehnder interferometer (MZI) modulators, and a Si/Ge high-speed photodiode [11.8].

11.3.2 Optical Interconnects

Minimization of external optical coupling between multifunction components can substantially improve the net system optical throughput and reliability. One of the possible approaches is to implement optical interconnects. High-speed chip-to-chip optical interconnects have low loss and a large transmission bandwidth. Another key advantage is their immunity to electromagnetic interference (EMI).

Optical interconnects have been divided into four categories: (1) box-to-box, (2) board-to-board, (3) chip-to-chip, and (4) on-chip interconnects.

Components that were traditionally housed in a stand-alone box have been implemented into a board-level form factor, which is called a blade. They are plugged into a common chassis. Optical interconnects most often used by the industry are the box-to-box types.

One type of board-to-board optical interconnects is called *backplanes*. These are point-to-point or point-to-multipoint high-speed interconnects with typical lengths of under 1 m [11.8]. Many of these optical interconnects have been demonstrated, including polymer waveguides integrated on Si, planar lightwave circuit interconnects, and fiber ribbon arrays integrated with vertical-cavity surface-emitting lasers (VCSELs) and photodiodes. However, to be widely adopted, optical interconnects must advance toward a smaller form factor with lower power consumption and lower cost.

In all the aforementioned optical interconnects, the major problem is losses due to surface reflections. These losses can be reduced in two ways: (1) zero-gap design and (2) index-matching gels. Zero-gap design requires precision mechanical mating of optical parts, typically achieved with a fusion splice or a mechanical connector.

With the index-matching approach, the gap is filled with an optical gel that serves as a light bridge across the connect area. The gel reduces the need for stringent mechanical tolerances on cleaving and polishing, expensive fusion equipment, and extensive technician training.

Figure 11.12 illustrates the basic geometry of a fiber-optic interconnect. An incident beam enters a splice or connector and encounters the end of the input fiber, often fused silica glass, which has a refractive index of 1.46. At this point the index of refraction changes from 1.46 to the value of 1.0 of the air in the gap. As the beam moves from the gap into the output beam guide, the index changes again. Reflections are introduced at each interface.

For fused silica glass commonly used in fiber-optic links, the typical reflection from an unintended large air gap is about 7%. Industry standards require much lower levels of reflection on the commercial interconnects.

The refractive index of an optical gel is selected to match the refractive index of the fiber. In this way it is possible to eliminate the large differential optical impedance between air in the gap and the optical waveguides.

11.3.3 Optical Coupling

One of the most difficult challenges facing microphotonic systems is efficiently coupling of light into and out of the chip. Particularly difficult is the coupling of light from a standard optical fiber or an external light source to a silicon waveguide. Overcoming these challenges requires the development of processes and structures in addition to the basic devices.

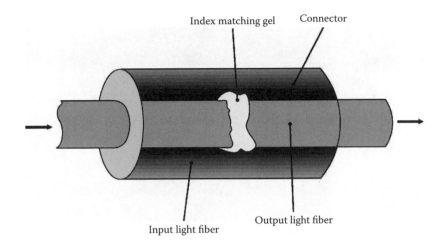

Index matching gel   Connector

Input light fiber   Output light fiber

FIGURE 11.12 Index-matching optical interconnect.

Current fiber attachment techniques rely on the closed-loop optimization of fiber position to ensure low-loss coupling. This technique is time consuming and costly. Passive alignment techniques for fiber attachment remove the need for closed-loop optimization by creating highly precise lithographically defined structures on the silicon surface in order to align the fiber to the waveguide aperture.

Active alignment techniques are typically capable of placement tolerances better than 1 μm. The accuracy required of a passive alignment technique will depend on the overlap of the fiber and waveguide modes, which can be controlled by waveguide tapers.

Other common techniques use fiber tapers. There are two methods of taper coupling — (1) pseudovertical tapering and (2) gradual horizontal and vertical modal tapering [11.9]. A single-mode fiber core with the index of refraction n = 1.5 usually has a diameter of 8 μm. A silicon waveguide, with the index of refraction n = 3.45, is typically only a few micrometers in width. A schematic illustration of these two designs is shown in Figure 11.13.

The key processing parameters essential to the success of either mode-transfer device are lithography and etching processes. For pseudovertical tapering, a horizontally tapered waveguide is patterned on top of another waveguide, and the optical mode is gradually squeezed from the top taper to the smaller, lower waveguide. The most important parameters for this transition are the length of the taper (the longer the length, the more slowly one can transform the mode, which results in lower loss) and the taper tip width (see Figure 11.13a). In order to reduce optical losses associated with the finite size of the tip width, the tip should be designed such that the minimum width is substantially smaller than the wavelength of light transmitted in the waveguide.

Creating taper devices usually relies on significant wet chemical etching of silicon material in order to produce the final taper design. However, it is often difficult to control the shape of sidewalls, particularly when more specific profiles

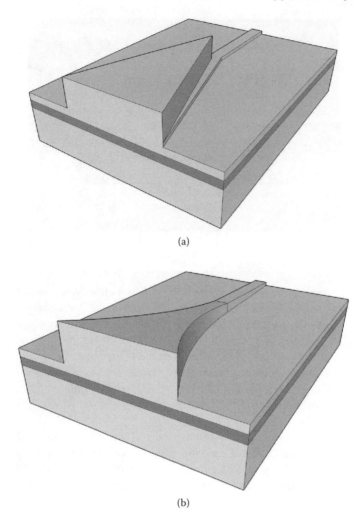

(a)

(b)

FIGURE 11.13 (a) Pseudovertical tapering and (b) three-dimensional taper.

such as a parabolic design are required. Thus, the most common approach is RIE, which has its drawbacks of producing surface roughness on the exposed sidewall. As optical losses due to surface roughness are strongly dependent on waveguide size, the losses due to roughness should be minimal at the beginning of the taper when the taper dimension is large.

Grayscale lithography has also been used to produce tapers. Tapers based on gray scale should result in much lower loss. However, grayscale-based lithography adds additional complexity and cost. A grayscale mask is several times more expensive than a standard lithography mask, and this process is more complicated than standard processing.

11.3.4 Wafer Bonding

Grayscale lithography has certain limitations due to the fact that the microstructure geometry is defined by the internal crystalline structure of the substrate. Therefore, fabricating multiple and interconnected structures is often difficult or impossible. Additional processing technologies have been developed in order to overcome these limitations.

Combining bonding techniques with anisotropic etching allows for manufacturing of complex three-dimensional microstructures. One such bonding technique is the process known as *wafer bonding*. Wafer bonding is used to bond a pair of silicon wafers together directly, face to face.

A method of bonding silicon to glass that appears to be gaining in popularity is electrostatic bonding. The silicon wafer and glass substrate are brought together and heated to a high temperature. A large electric field is applied across the junction, which causes an extremely strong bond to form between the two materials. Figure 11.14 shows a glass plate bonded over a channel etched into a silicon wafer.

In silicon direct bonding (SDB), or fusion bonding, the polished sides of two silicon wafers are contacted face to face and the wafer pair is annealed at high temperatures. During annealing, a bond is formed between the wafers that can be as strong as that in bulk silicon.

SDB is a low-cost technology for silicon and silicon compounds such as SiO_2 or Si_3N_4. In comparison to anodic bonding (silicon–glass), SDB has two main advantages:

* No thermal mismatch between the bond materials
* Both bond partners can be easily structured

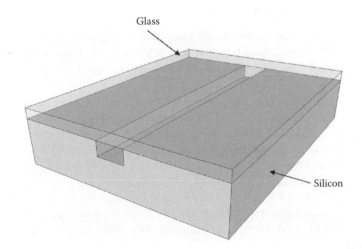

Glass

Silicon

FIGURE 11.14 Silicon bonding to glass.

The maximum bond strength increases with the applied process temperature. At temperatures between 800 and 1200°C, the intrinsic strength of bulk silicon is reached. Thermally sensitive devices can be bonded with sufficient strength for dicing at temperatures between 200 and 400°C by using chemical surface activation methods.

The first step of low-temperature SDB of two structured wafers is the cleaning and hydrophilization in an acid mixture (H_2SO_4 and H_2O_2). After rinsing and spin-drying, one wafer is wetted with silicate solutions such as sodium silicate (NaSi) or tetraethylorthosilicate (TEOS). Next, the wafers are rinsed and dried. The activated wafers are contacted in a bond aligner with an accuracy of better than 10 μm at room temperature. The joined wafer pair is annealed in a furnace at atmospheric pressure in air in a temperature range of 200 to 400°C.

There are a number of different methods available for bonding micromachined silicon wafers together or to other substrates in order to form larger more complex devices. Electrostatic bonding and SDB can form very strong joints. However, these techniques require joined surfaces be extremely flat and clean. Other bonding methods include using an adhesive layer, such as a glass, or photoresist.

Wafer-bonding techniques are used to assemble individually micromachined layers to form a complete structure. Usually entire wafers or individual dies are bonded together. A primary application of wafer bonding is in the fabrication of SOI devices. Wafer bonding and thinning is used to produce a bulk-quality layer of single-crystal silicon over a thin film of thermally grown silicon dioxide. Wafer bonding allows the fabrication of structures that are thicker than a single wafer. This allows the fabrication, for example, of MOEMS actuators, capable of higher forces and displacements.

11.3.5 HYBRID SILICA/MOEMS

A new and promising approach to integration is based on the use of hybrid silica/MOEMS technology. Hybrid silica/MOEMS technology produces optical devices that provide scalable integration using standard volume production techniques.

An example of a hybrid silica/MOEMS device is a switch in which a MOEMS actuator is flip-chip bonded to a SOS waveguide (see Figure 11.15).

The cost is and will be the major limiting factor of integration. It is estimated that, for example, dense wavelength-division-multiplexing (DWDM) systems will not be acceptable for metro link applications until the cost per single node drops to $100,000. Some industry observers believe that only when costs reach $1000 per node will DWDM be acceptable for the residential applications [11.7].

It is believed that hybrid silica/MOEMS technology will speed up the time frame, making photonic DWDM nodes at the metro level possible within the next few years.

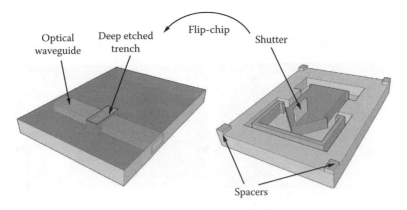

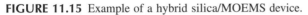

FIGURE 11.15 Example of a hybrid silica/MOEMS device.

11.3.6 HYBRID PBG

A further limiting factor in passive signal distribution within traditional waveguide devices is the bend radius of the channel waveguides. The minimum bend radius depends on the relative difference in the refractive index of the waveguide core and the cladding. For materials such as $LiNbO_3$, this difference is small (0.05) and the optical signal is very weakly confined to the core. As a result, a large bend radius must be employed to distribute the optical signal on the substrate, requiring considerable substrate area. Increasing the refractive index of the waveguide core by using Si in SOI structures and InP facilitates strong optical confinement and relatively sharp bends, up to 90°. This facilitates much more compact integrated optical structures.

In view of the maturity level of PBG technology, a consideration is to develop a hybrid integration of PBG structures with traditional integrated optical structures to afford a net reduction of the device size and improve device performance. This includes the use of PBG structures to improve the optical confinement and reduce the bend radius of channel waveguides. PBG concepts could also be employed to improve the efficiency of waveguide optical amplifiers and laser diodes through the reduction of spontaneous emission and improved optical confinement.

11.4 FABRICATION OF SMART COATINGS

Laser ablation deposition (LAD) is the common technique for the production of thin-film smart coatings. *Laser ablation* is defined as the removal of material by an intense optical beam. Laser ablation ejects target atoms that can combine with reactive gases such as oxygen to form thin-film material molecules that are deposited onto the substrate.

In most metals, glasses, and crystals, laser ablation is carried out by vaporization of the material by heat. The laser beam produces a plume, i.e., a plasma-like substance consisting of molecular fragments, neutral particles, free electrons

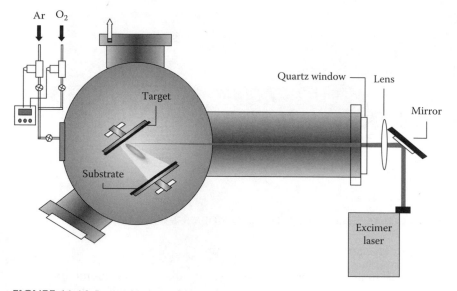

FIGURE 11.16 Laser ablation system.

and ions, and chemical reaction products. Control of the plume is used to realize and control the deposition of a given material on a specially prepared substrate.

The deposition process is controlled by the selection of the laser beam fluence, i.e., a parameter that is defined as laser energy per unit area on the work material:

$$\text{fluence (joule/cm}^2) = \text{laser pulse energy (joule)/focal spot area (cm}^2)$$

Important deposition parameters include (1) laser beam wavelength, (2) pulse duration, (3) pulse repetition, (4) beam quality, (5) assist gases (Ar/O_2) ratio, (6) deposition pressure, and (7) substrate temperature.

A schematic illustration of a LAD system is shown in Figure 11.16. As shown in Figure 11.16, the excimer laser beam is reflected by 90°-angle mirror and focused on the target. Typically, lasers used for thin-film deposition are pulsed excimer lasers, which have a relatively low duty cycle. A photograph of a KrF laser-based LAD system is shown in Figure 11.17.

In order to ensure a homogeneous and uniform thin-film coating, the following three dynamic processes act simultaneously:

- The substrate holder rotates around its central symmetric axis.
- The target rotates around its central symmetric axis.
- A reflecting mirror periodically sweeps the laser spot along a diameter of the target.

Coating uniformity is very critical for the properties of the deposited thin films. However, the thickness distribution of the deposited material by a single

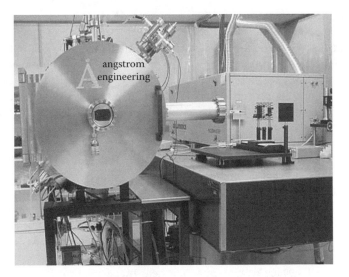

FIGURE 11.17 KrF-based laser deposition system.

laser pulse follows a Gaussian shape. Hence, the deposited thin film may not be geometrically uniform and its properties may vary spatially over the substrate. By sweeping the laser beam and rotating the substrate holder and the target rotation, the complete area of the substrate is homogeneously sputtered by laser ablation.

Multiple targets can be used as they enable the deposition of several different layers *in situ* during a single pump-down. This minimizes interface contamination that could degrade the deposited material.

A typical LAD system includes the following features:

- Precise total pressure and flow rate process control
- *In situ* crystal thickness monitor
- *In situ* target temperature sensors
- *In situ* spectrometer to monitor the material composition

In the case of VO_2 smart materials, the deposition chamber is pumped down to about 5×10^{-6} Torr. This vacuum is necessary to minimize contamination with hydrocarbons and other impurities. The deposition is usually performed in a controlled gas background such as O_2/Ar. It has been observed that the O_2 concentration in Ar gas and the total deposition pressure are critical in stabilizing the single VO_2 phase.

The required film thickness may vary from about 0.05 to about 0.60 μm. Therefore, the manufacturing reproducibility is enhanced by using *in situ* diagnostics that includes a thickness monitor.

The substrate temperature is an important parameter as it affects the grain size of the deposited film. The substrate temperature controls the surface mobility of adsorbed radicals during the film formation. Higher deposition temperatures generally facilitate better film crystallinity and a sharper variation in the film characteristics between the metallic and insulating states. On the other hand, a higher temperature, for example, near the substrate melting point may affect the film quality. Therefore, there is an optimal deposition temperature that maximizes the thin-film coating properties. For example, a substrate temperature of 520°C or more influences negatively the grain size of the single VO_2 phase. If the substrate's temperature decreases, then the metal–insulator transition will be weak.

Substrate surface cleaning is also an important factor for achieving a good adhesion with the coating and can also affect coating morphology and microstructure by influencing nucleation and growth patterns.

Thin-film smart coatings may be deposited on quartz, sapphire, and Si substrates. The coatings may be deposited not only on crystalline substrates such as single-crystal silicon, but also on materials such as Kapton and Al. This is very advantageous for many applications as smart coatings can be directly deposited on the desired surface to maximize the heat transfer and mechanical integrity while minimizing the overall device weight.

Another critical parameter for smart-coating fabrication is the control of stoichiometry. For example, at higher concentrations of O_2 in the gas mixture and higher pressures, a mixed-phase structure containing VO_2 and V_2O_5 could be observed. As the total pressure is reduced, and the concentration of O_2 in the mixture is reduced too, then the deposited films are mainly VO_2. X-ray diffraction (XRD) may be used to study the structure and stoichiometry of the resulting coatings. An example of XRD measurements that were used to verify the quality of VO_2 coatings is shown in Figure 11.18. The peak in Figure 11.18 at 27 degrees

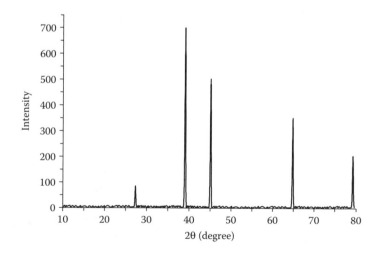

FIGURE 11.18 X-ray diffraction spectra of VO_2 deposited on aluminum.

corresponds to the presence of VO_2. The other four peaks at higher angles characterize the substrate (i.e., aluminum). The presence of V_2O_5 would be indicated by a diffraction peak at 21 degrees. Therefore the diffraction spectra in Figure 11.18 confirm that the deposited coating is mainly VO_2.

The substrate texture also has a strong influence on the film growth and the crystallinity of the coatings. Therefore, the substrate may be polished and then patterned in order to increase the materials' performance characteristics.

There are three approaches for the fabrication of nanoengineering of VO_2 smart coatings:

- Nanoclusters of VO_2 that are formed on the substrate using specific deposition conditions
- Nanocrystallites of VO_2 formed on periodic microstructured substrates
- Postdeposition patterning of the VO_2 using lithography and dry etching

The metal–insulator transition temperature of VO_2 can be controlled by incorporating a suitable donor-like dopant such as W, or acceptor-like dopants such as Ti. In this way the transition temperature can be shifted over a wide range, from above 70°C to below 20°C, as may be required for a given application.

REFERENCES

1. Joshkin, V.A., Oktyabrsky, V.A., Moran, P., Saulys, D., Kuech, T.F., and McCaughan, L., Growth of oriented lithium niobate on silicon by alternating gas flow chemical beam epitaxy with metalorganic precursors, *Appl. Phys. Lett.*, 76(15), 2125–2127, 2000.
2. Eldada, L., Optical Networking Components, http://photonics.dupont.com/downloads/OpticalNetworkingComponents.pdf (September 2, 2005).
3. Mehregany, M. and Roy, S., Introduction to MEMS, *Microengineering Aerospace Systems*, Helvajian, H. (Ed.), The Aerospace Corporation, Los Angeles, 2004, http://www.aero.org/publications/helvajian/index.html. (Accessed September 5, 2005.)
4. Barbier, D., Erbium-doped waveguide amplifiers promote optical-networking evolution, *Lightwave*, 144–146, November 2000.
5. Righini, G.C., Pelli, S., Ferrari, M., and Brenzi, M., Erbium-activated silicate waveguides and amplifiers, *8th Microoptics Conference*, Osaka, Japan, Paper K6, October 24–26, 2001.
6. Berger, J.D. and Anthon, D., Tunable MEMS devices for optical networks, *Optics and Photonics News*, 43–49, March 2003.
7. Bjorklund, G., Integrating the metro, *SPIE's OE Magazine*, 28–29, August 2001.
8. Herve, P. and Ovadia, S., Optical technologies for enterprise networks, *Intel Technol. J.*, 8, 2004.
9. Salib, M., Liao, L., Jones, R., Morse, M., Liu, A., Samara-Rubio, D., Alduino, D., and Paniccia, M., Silicon photonics, *Intel Technol. J.*, 8, 144–160, 2004.

12 Advanced Microphotonic Devices

There are a number of advanced microphotonic devices that have reached their commercial maturity. It is interesting to point out that all these devices employ a set of generic functional blocks or subsystems that have been introduced in the previous chapters, such as:

- Fiber or waveguide couplers
- SOI channel waveguides and $1 \times N$ branches
- MOEMS switches
- Waveguide amplifiers
- Waveguide interferometers
- Waveguide Bragg gratings

Typically, each generic functional block has substantial market value on its own. They can be integrated to enable various more complex multifunctional devices. As the component library expands, more sophisticated devices can be realized at minimal additional development cost. Microphotonic device design methodology parallels the highly successful development methodology that was employed for the evolution of integrated electronic circuits.

12.1 PHOTONIC COMPUTER

Over the past decade the need for high-performance computers has significantly increased. As has been described in Chapter 2, many performance improvements in conventional computers have been achieved by miniaturizing electronic components so that electrons only travel short distances within a very short time. Increased speed of computers was achieved by miniaturizing electronic components to a micron scale and by increasing the density of interconnections necessary to link the electronic gates on microchips. This approach has been based on the steadily shrinking size of the elements that are drawn on microchips. This has led to the development of very large scale integration (VLSI) technology. Currently, the smallest dimensions of VLSI are in the range of 80 μm. However, there is growing concern that even these technologies may not be capable of solving the computing limitation problems.

One of the most promising approaches to the computing limitation problems is optical computing. Optical computing is based on (1) optical calculation of transforms, (2) optical pattern matching, and (3) optical storage of data.

A photonic computer is a device that, instead of electronic currents, uses optical beams to perform digital computations. An electric current flows at only about 10% of the speed of light. This limits the rate at which data can be exchanged over long distances, and is one of the factors that led to the evolution of optical computation. It is believed that a photonic computer may be developed that could perform operations 10 or more times faster than a conventional electronic computer. It is expected that the photonic computer will revolutionize computing in much the same way the semiconductor chip revolutionized electronics.

Optical beams, unlike electric currents, pass through each other without interacting. Many laser beams can be used in such a way that their paths intersect, but there is no interference among the beams even when they are confined essentially to two dimensions. Electric currents must be guided around each other, and this makes three-dimensional wiring necessary. Thus, an optical computer, besides being much faster than an electronic one, could also be smaller.

12.1.1 SEEDs

The first computer chips containing multiple optical switches or transistors known as self-electro-optical-effect devices (SEEDs) have been developed, and they will be the basic components of optical computers. SEEDs act as optical modulators that operate by applying a voltage across several quantum wells.

The SEED consists of a semiconductor PIN diode structure with multiple quantum wells, as shown in Figure 12.1. An optical beam enters the SEED through the n-doped semiconductor and passes through the quantum wells and the barriers. The n and p semiconductors are used to apply bias voltage to the quantum wells and to modulate the beam by using the quantum-confined stark effect (QCSE) (see Chapter 3). Bragg gratings are used to reflect the optical beam such that it passes through the quantum wells for a second time where its modulation is further enhanced.

There are two types of two-dimensional SEED arrays: (1) transmissive and (2) reflective.

In the transmissive device the optical beams enter the SEED in a direction perpendicular to the array and then exits at the opposite end as a modulated signal. In the reflective mode, the modulated optical beam is reflected back from the array. Reflective SEEDs have lower insertion loss and greater modulation depth. They are more commonly used as they are easier to implement.

12.2 OPTICAL MEMORY STORAGE DEVICES

In optical computing two types of memory are being considered: (1) memory based on arrays of one-bit-store elements and (2) memory that uses mass storage based on optical disks or holographic storage systems. The main advantage offered by holographic optical data storage over the currently implemented storage technologies includes (1) significantly higher storage capacities and (2) faster readout rates.

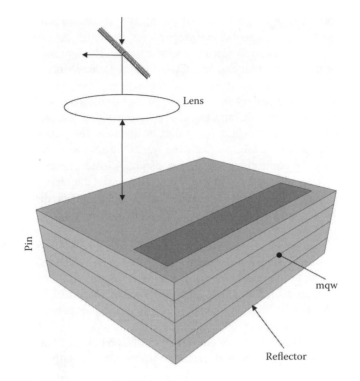

FIGURE 12.1 Self-electro-optical-effect device.

It is expected that the current development of compact, high-capacity, rapid and random access data storage devices will form the basis of future massive capacity and fast-access data archives.

The spatial light modulators (SLMs) described in Chapter 5 are used in these new optical data storage devices. The goal is to expand the capacity through stacking layers of optical material. For example, a minidisk has been developed that uses special compression to shrink a standard CD's 640-MB storage capacity onto the smaller polymer substrate. The minidisk may use one of two methods to write information onto an optical disk. Along with minidisk development, standard magneto-optical CD technology has expanded the capacity of the 3.5-in.-diameter disk from 640 MB to commercially available 1 GB storage media [12.1].

In addition to these existing systems, there are several other systems under development. These systems use multiple layers of optical material on a single disk. For example, the magnetic super-resolution (MSR) system uses two layers of optical material. In this approach the data is written by using a magnetically modulated laser. The data is then copied from the lower to the upper layer. In this way, data can be stored much closer together on the bottom layer. By using this method, capacities of up to 20 GB on a 5.25-in. disk have been obtained [12.1].

Advanced storage magneto-optical (ASMO) systems are built by using the MSR approach. These systems have much larger capacities than standard optical disks. This larger capacity is realized by much more efficient use of disk space. This may be compared with classical CDs [12.1]. CDs have mechanical grooves along with areas that are not modified between them. These grooves are used as guides for the writing and reading lasers. However, standard systems only record data in the grooves, i.e., the spaces between grooves are not used. The ASMO systems record data on the entire space, i.e., on and between grooves. The system can eliminate the cross talk between the tracks by choosing adequate groove depth. By closely controlling the groove depth, ASMO systems eliminate the problem of cross talk and at the same time maximize the signal-to-noise ratio (SNR) [12.1]. In this way it is expected to produce optical disk drives with capacities between 6 and 20 GB on a 12-cm optical disk, which is the same size as a standard CD with capacity of 640 MB.

A further twofold increase in storage capacity of the ASMO systems has been achieved by using multilayered polymer disks. Therefore, when the data is copied from the bottom to the upper layer, it is expanded in size, amplifying the signal. These systems are called magnetic amplifying magneto-optical systems (MAM-MOS). MAMMOS technology could help bridge the gap between optical disk drives and holographic memories [12.1].

Other techniques under development are based on the use of active-molecule-doped polymers to store optical data holographically. Usually a thin polymer layer of PMMA doped with phenanthrenequinone (PQ) is used in this technique. In this approach, the interference patterns are generated when the PMMA material is illuminated by two coherent beams. The interference patterns are created as a result of PQ molecules being bonded to the host matrix, i.e., constructive interference patterns correspond to brighter areas, whereas destructive patterns correspond to darker areas. In this way, a hologram may be written into the polymer material [12.1].

Conventional systems use memory addresses to track and retrieve data at a given location. Currently there are several systems under development that aim at implementation of associative data access capabilities of holographic memories. Unlike the standard magneto-optical storage devices, the associative data access approach enables a parallel search of the entire memory space for the presence of a keyword or search argument. A number of applications can benefit from this mode of operation, including management of large multimedia databases, video indexing, image recognition, and data mining [12.1]. These applications may be realized because different types of data, such as formatted and unformatted text, grayscale and binary images, video frames, alphanumeric data tables, and time signals, can be interleaved in the same medium.

The associative memory search system may consist of an SLM and holographic memory cubes (see Figure 12.2). An SLM is used to interrogate holographic cubes for simultaneous search of the entire memory storage. By changing the angle of the reference beam, more data can be written into the cube. To search the data, a binary or analog pattern that represents the search argument is loaded

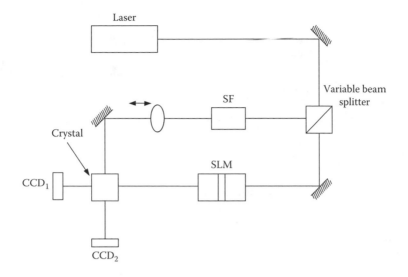

Laser

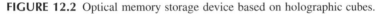

FIGURE 12.2 Optical memory storage device based on holographic cubes.

into an SLM which then modulates the laser beam. The light diffracted by the holographic cube on a charge-coupled device (CCD) generates a signal that indicates the pages that match the sought data [12.1].

12.3 PHOTONIC-BAND-GAP SENSORS

High-order photonic-band-gap (PBG) structures that have recently been developed could be used to facilitate various types of sensors that would be useful for both scientific and commercial applications (see Chapter 10). Figure 12.3 shows a sensor consisting of a linear PBG structure that is chemically treated to react with a specific biochemical agent.

The presence of the target chemical results in a reaction that shifts the spectral characteristics of the sensor. As the area of each sensor is very small, many sensors could be used in arrays to facilitate detection of a broad range of biomarkers, for example, as a miniature laboratory on a chip. The potential applications are vast and almost limitless.

12.4 CASCADE LASERS

A certain type of semiconductor heterostructure has been developed that allows for construction of a number of microphotonic devices. The quantum cascade (QC) laser is an example of such a device. The QC laser operates in a fundamentally different manner than a standard semiconductor laser. In conventional semiconductor lasers, a photon of light is emitted when a negative charge (an electron) jumps from a semiconductor's conduction band to a positive charge (or "hole")

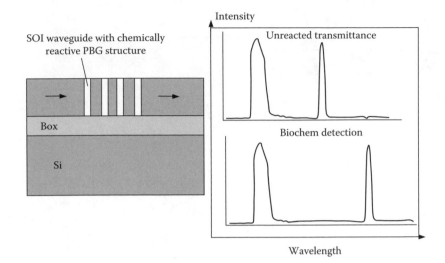

FIGURE 12.3 Conceptual design of a PBG biochemical sensor.

in the valence band. Once an electron has been neutralized by a hole, it can emit no more photons.

Cascade lasers, first developed in 1994 by Federico Capasso and Jerome Faist at Bell Labs, operate in the midinfrared wavelength region (4 to 12 μm). QC lasers are unique in that the output wavelength of light is determined by the geometry of structure.

QC lasers are based on subband transitions (see Figure 12.4). The QC laser contains a series of electron traps, or quantum wells. They are very often compared to an electronic waterfall. When an electric current flows through a QC laser, electrons cascade down energy levels, emitting a photon at each step (see Figure 12.5). The semiconductor material in the laser is arranged in such a way as to sandwich an electron in two dimensions as it passes through, coaxing the electron into a quantum well. As it exits a well, it emits a photon and loses energy. When the lower-energy electron leaves the first well, it enters a region of the material where it is collected and sent to the next well. Active wells are arranged in a QC laser in such a way that each is at a slightly lower energy level than the preceding one. In this way a cascade effect can be generated.

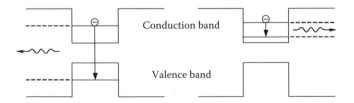

FIGURE 12.4 Subband transition in QC laser.

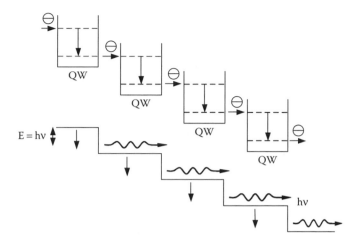

FIGURE 12.5 Cascading process.

The QC laser structure is manufactured by a layer of atoms at a time by molecular beam epitaxy (MBE). By simply changing the thickness of the semiconductor layers, the laser's wavelength can be changed as well. Therefore, selecting the design of the layered structure controls the wavelength. Fine-tuning of the wavelength may be realized by varying the operating temperature of the laser. In practice it is possible to shift the wavelength of the emitted light by 15 wavenumbers.

A typical structure is made up of periodic repetitions of two sections. One of the sections is used as an injector collector region, the other section acts as an active region (see Figure 12.6). The electrons are injected from the injector region into the upper laser energy level of the active section in which laser generation takes place. After that, the electrons decay by a nonradiative transition and are tunneled to the next structural period.

Recently, PBG structures have been used to construct QC microlasers [12.2]. In this design a PBG microcavity is employed as the source of optical feedback within a QC laser structure. PBG techniques allow for a greatly scaled-down version of QC lasers, enabling miniaturization and on-chip integration, with potential applications such as multiwavelength two-dimensional laser arrays spectroscopy.

QC lasers are usually operated in pulsed mode at room temperature and have average emission powers in the tens of milliwatts over a small spectral window. The spectral density of these lasers is therefore several orders of magnitude higher than a glow bar that is used as a standard light source in state-of-the-art Fourier transform infrared (FT-IR) spectrometers.

The QC laser's unique properties, i.e., their output power, tuning range, and ability to work in a pulsed mode at room temperature, make it an ideal tool for applications in the remote sensing. Gases and vapors have characteristic chemical absorption spectra that are unique to their chemical structures.

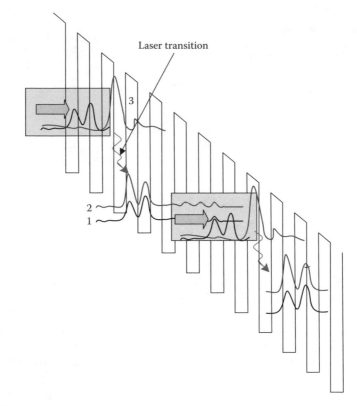

FIGURE 12.6 Periodical structure of a typical quantum cascade laser.

The laser's wavelength can be tuned to match a specific spectrum of gases of interest, and in this way it can then be used to determine their presence. Cascade lasers have been successfully used in remote sensing of environmental gases and pollutants.

Future applications may include cruise controls, radars, industrial process control, and medical diagnostics.

12.5 MINIATURIZED IR SPECTROMETERS

Optical spectroscopy can provide quantitative information about the chemical composition and bonding configuration of molecular structures. Most molecules that comprise solid, liquid, or gaseous samples have characteristic vibrational modes that are associated with their chemical bonds and can interact with photons. This results in optical absorption bands, mainly in the infrared spectral range, that are indicative not only of the chemical bond but also of the actual bonding configuration and the local chemical environment. Noncontact infrared (IR) spectral reflectance or transmittance measurements probe these characteristic molecular vibrational modes of a sample to provide valuable information about the chemical bonding and composition in solids, liquids, and gases.

IR spectroscopy, therefore, can yield more information than other methods that merely provide data about the relative content in a sample. This technique has been widely employed in laboratories using relatively large, bulk-optics dispersive near-infrared (NIR) spectrometers and FT-IR spectrometers as a diagnostic tool for the characterization of suitable samples and to assist process development for the synthesis and modification of materials.

Tabletop dispersive spectrometers generally consist of a free-space arrangement of input optics, an input slit, mirrors to fold the optical path, a reflecting replicated grating, output optics, and an output slit or detector array. For spectrographs using a single detector, additional scan motors are employed to vary the position of the detector and thus vary the output wavelength impinging on the photodetector. FT-IR instruments employ an optical interferometer with a moving mirror in one arm of the interferometer to provide a spatial transform of the desired signal spectrum. This requires precise positioning of the movable mirror and an additional reference laser. As the spectral information is not measured directly by the FT-IR instrument, substantial additional data processing is required to estimate the desired spectral information using mathematical transform techniques. Free-space IR spectrometers are sensitive to effects caused by the ambient medium, such as optical absorption by water vapor. As a result, FT-IR and IR dispersive spectrometers are generally purged using a supply of dry N_2. The alignment of the free-space optics is sensitive to temperature fluctuations and mechanical vibrations. This often requires additional costly environmental controls (i.e., temperature control and vibration damping) that add to the net system size and power consumption to assure the spectrometer performance.

There are myriad applications for IR spectroscopy in nonideal conditions outside the laboratory that require performance characteristics comparable to a tabletop FT-IR but in a more robust environment and with an instrument that has a reduced size and mass. Potential applications include, but are not limited to, the detection of biohazards in the working environment, blood analysis, analysis of soil samples and mineralogy by diffuse reflectance spectroscopy, monitoring of water and air quality, and the analysis of pharmaceutical powders and liquids. Many of these applications, including diffuse reflection measurements and biohazard detection, require a substantial system SNR to facilitate satisfactory resolution of the desired signal. For biohazard detection, detectivities in the range of a particle per million (ppm) to a particle per billion (ppb) are desirable, requiring very high SNR for a spectrometer system to be viable. For general diffuse reflection measurements, the return signal from an optically scattering target can be less than 5% of the incident illumination signal, yet a minimum SNR exceeding 1000 is desired for the spectral characterization of the diffusely reflected signal. Moreover, very intense illumination sources are undesirable as they can damage or alter the sample under investigation. Therefore, it is desirable to have a portable, lightweight infrared spectrometer capable of performance comparable to tabletop dispersive or FT-IR instruments but with improved tolerance to environmental perturbations.

Therefore, some attempts have been made to remedy the problems of size, weight, and complexity relative to typical laboratory-grade IR spectrographs and FT-IR instruments. However, these attempts do not provide a solution to the additional requirements of the spectrometer performance and signal detectivity with regard to the laboratory-grade instruments. This is especially critical for miniature spectrometers in order to provide a viable solution in the IR spectral range (spanning from 1.5 to beyond 12 μm) as the detectivity of typical uncooled infrared detectors such as PbSe or HgCdTe is several orders of magnitude less than that of Si (visible detector) and InGaAs detectors. In many of the applications in which a compact, lightweight, portable spectrometer system is highly desirable, such as biohazard detection, geological surveys of minerals in rock samples, analysis of contaminants in soil, water and air samples, and hyperspectral planetary surveys for space exploration, the detectivity and SNR of the spectrometer system is of prime concern.

Recently, various waveguide-based miniaturized NIR spectrometers have been devised. In general, an optical slab waveguide consists of a sandwich structure of three basic layers: an upper cladding, an intermediate core layer, and a lower cladding. The main requirements for waveguides are that the core and cladding layers be transmissive to the desired optical signal and that the refractive index of the core layer be larger than that of the upper and lower cladding layers.

Optical slab waveguides offer an important cost advantage over the use of bulk IR-transmissive materials for the realization of integrated IR spectrometers as the vertical optical confinement provided by the waveguide structure substantially reduces the amount of typically expensive IR material that is required to fabricate the spectrometer.

Figure 12.7 shows a schematic of a miniaturized spectrometer that employs an IR waveguide structure in which thin-film upper and lower cladding layers are deposited on an IR-transmissive core layer by using standard vacuum deposition techniques [12.3]. Typically, waveguide structures are prepared by depositing core and cladding layers onto a substrate. Core thicknesses employed in integrated optics are about 10 μm for operation at 1550 nm. However, such an approach has significant disadvantages in the realization of waveguide spectrometers as a large core thickness of 1 mm or more for the waveguide is desirable to provide efficient coupling to the typical optical sources employed for spectroscopy. Depositing such a core thickness on a substrate is technically challenging and costly. The design shown in Figure 12.7 uses upper and lower cladding layers that are deposited onto the thick core layer using standard thin-film vacuum deposition techniques. These form an integral part of the slab waveguide structure. The thickness of the core layer can be selected to provide optimum performance for a selected, or a range of selected, applications. For a silicon core, either metallic cladding layers that function as internal mirrors or a $Si_{1-x}C_x$ alloy with a refractive index smaller than that of Si are two potential structures. The upper and lower cladding layers control the numerical aperture (NA) of the waveguide

1. Waveguide structure

2. Diffraction grating

3. Input face

4. Output face

5. Input aperture

6. Programmable shutter/chopper

7. Linear detector array

8. Pins

9. Sockets

10. Detector mechanical support

11. Heat sink

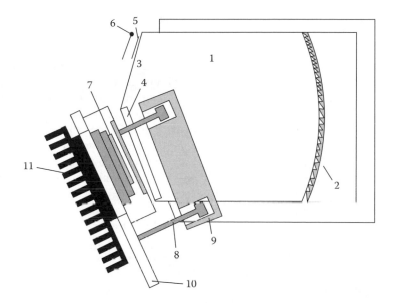

FIGURE 12.7 A miniaturized NIR spectrometer.

and reduce the effect of scattered light on the desired output optical signal. Slab waveguide spectrometers usually employ a bare Si slab waveguide core with the surrounding or ambient medium acting as the upper and lower cladding. But the choice of cladding layers can prevent the surrounding medium from affecting the desired optical signal within the waveguide.

The optical signal is typically coupled into a waveguide by focusing the optical signal onto the core of the waveguide, as shown schematically in Figure 12.8. This can be performed using either refractive or reflective optics. The focal length of the focusing optics is usually matched to the acceptance NA of the spectrometer by the following formula:

$$NA = f/2d = n_{wg} \sin(\theta_{wg}) = \sin(\theta_{air})$$

where f is the focal length, d is the diameter of the desired input optical signal, n_{wg} is the refractive index of the waveguide core, θ_{wg} is the maximum angle of

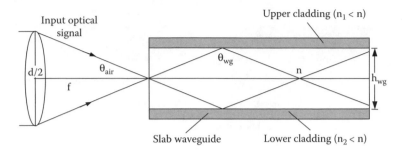

FIGURE 12.8 Optical coupling into slab waveguide structure.

propagation of the optical signal within the waveguide, and θ_{air} is the corresponding maximum divergence angle of the output signal in air. For typical spectrometers, the NA is in the range of 0.20 to 0.25. For good coupling efficiency, the thickness of the waveguide core should be comparable to the diameter of the attainable focal spot. This is typically much larger for the nonideal extended broadband optical sources used in spectroscopy as compared to the laser illumination sources employed in optical telecommunications. Typical spot diameters can vary from about 0.5 mm for optical arc sources such as Xe lamps to more than 5 mm for the collection and use of ambient illumination, as is employed for remote-sensing applications. Therefore, a methodology for the fabrication of high-performance waveguide spectrometers that can provide a core thickness, h_{wg}, of several millimeters is desirable for good input optical coupling.

As the optical signal emerges from the output portion of a slab waveguide into air, it diverges in the vertical direction according to Snell's law of refraction. The amount of divergence depends on the NA of the input optical signal. If the waveguide is nonideal such that there is substantial scattering of the optical signal within the waveguide, the effective NA can increase relative to the original input NA. Precise control of the output focal length in the horizontal direction is also required. Using a conventional slab waveguide structure, therefore, results in considerable loss of the output optical signal if the detector array is located some distance from the output face of the waveguide. This produces a considerable challenge when coupling the output of slab waveguide spectrometers to detector arrays. Typically these detector arrays are cooled, and thus the cooled detector must be vacuum or hermetically sealed, resulting in additional separation required between the output portion of the slab waveguide and the detector array (to accommodate an additional window and provide thermal isolation).

The diffraction grating can be fabricated directly on the waveguide structure by micromachining techniques. However, this is costly and results in some surface roughness and imperfections that can degrade the grating performance. The preferred method is to manufacture the grating elements separately in a thin crystalline material such as Si using standard Si-processing technology. This can yield very-high-quality gratings with measured peak diffraction efficiencies approaching theoretical limits (85 to 90%) and low scattering of the optical signal.

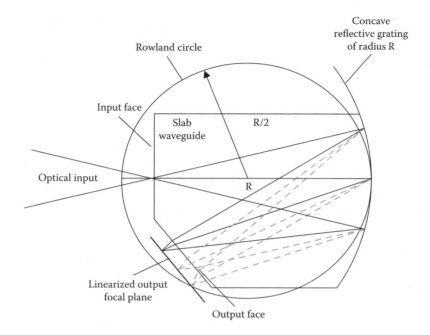

FIGURE 12.9 Spectrometer optical geometry.

Several gratings can be manufactured on a single Si wafer using batch processing, substantially reducing the cost of the individual gratings. The processed wafers are diced to provide individual gratings. The gratings are subsequently coated with gold and affixed to the concave end of the waveguide structure. This methodology provides diffraction gratings in batch production at a low cost but with performance that approaches theoretical limits in diffraction efficiency and has no physical limitations on the height of the grating. The grating does not require any special chirp, resulting in substantially relaxed tolerances for manufacturing.

The input and output faces of the waveguide structure can be used as optical elements in directing and tailoring the characteristics of the desired optical signal. This approach uses difference in the refractive index between the waveguide and the surrounding medium to assist the shaping of the desired optical signal. For example, the output focal plane lies along a curve as shown in Figure 12.9. This provides a number of significant advantages [12.3]. First, by adjusting the spectrometer length relative to the grating radius of curvature R, the output focal length of the dispersed optical signal in the horizontal plane of the waveguide can be tailored to accommodate the spacing required by various cooled detector arrays.

The output face of the waveguide structure serves as an integrated optical element to linearize the focal plane outside the slab waveguide. Using standard optical design methods and the additional capabilities afforded by the waveguide structure, linear output focal planes over relatively broad spectral spans can be achieved.

Using optical polishing techniques, a focusing curve in the vertical direction of the output plane is possible. This curvature provides focusing of the desired optical signal in the vertical direction at the output plane of the integrated spectrometer. This enables additional windows and filters to be positioned between the slab waveguide structure and the detector array, providing efficient optical coupling to cooled detector arrays.

An input aperture is required at the input face of the slab waveguide. The desired input signal is focused onto the aperture. The desired optical signal can also be introduced into the spectrometer by aligning and coupling a suitable input fiber or array of fibers with respect to the input aperture. The input aperture can be directly applied to the input face by vacuum-evaporating a suitable optically opaque material such as Al and masking the desired portion of the input face to provide a window for the desired optical signal. Alternatively, the input aperture can be mechanically fixed to the slab waveguide structure to enable the mechanical or electronic interchange of input apertures of different sizes. The purpose of the input aperture is to define the optical input and to control the spectral resolution of the miniature spectrometer. The width of the input aperture in the horizontal direction affects the operating resolution of the spectrometer system. Reducing the input slit width provides higher spectral resolution, up to the intrinsic spectral resolution capabilities of the waveguide spectrometer itself.

A programmable optical shutter or chopper can be placed between the spectrometer input face and the desired optical signal. This requires additional control electronics and data processing software. The shutter can be remotely opened and closed using a suitable control signal to either pass or interrupt the desired optical signal. This enables the differentiation of the desired optical signal from unwanted noise in the spectrometer system. This can significantly remedy many of the deficiencies, drift, and nonidealities exhibited by typical IR detector arrays. IR detectors, particularly materials such as PbSe, can exhibit significant instabilities and the corresponding undesirable drift in the detector signal, even with temperature stabilization of the detector array. Moreover, linear detector arrays employ additional signal multiplexing that can substantially increase their noise relative to unmultiplexed detectors.

The detector's detectivity, or its ability to differentiate a desired optical signal from background noise, is measured in terms of the specific detectivity $D*(T)$. This is related to the noise equivalent power (NEP) of the detector by

$$\text{NEP} = \text{SQRT}(A_{det}\, BW)/D*(T)$$

where $A_{det} = w_{det}\, h_{det}$, w_{det} is the width of the detector element, h_{det} is the corresponding height of the detector element, and BW is the bandwidth of the signal. NEP represents the optical power required by an individual element of the detector array for a SNR of 1. This excludes the additional noise contribution of any electronics in the processing chain between the detector element and the data acquisition system. It is desirable to have A_{det} and BW as small as possible and $D*(T)$ as large as possible in order to minimize the NEP and, hence, to maximize

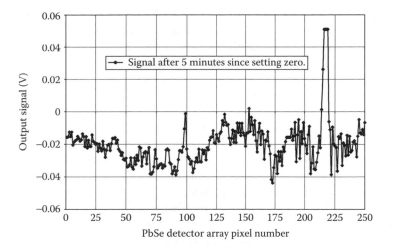

FIGURE 12.10 Measured drift in the background signal of a 256-channel PbSe array.

the signal detectivity of the system. Typically, the height of the detector elements, h_{det}, is constrained to be approximately equal to the thickness of the slab waveguide to ensure that the entire optical signal at the output of the slab waveguide is transferred to the detector elements.

The attainable detectivity of an IR detector array can be significantly less than that indicated by its NEP rating because of intrinsic and extrinsic nonidealities. Even with cooling and nominal temperature stabilization, IR detector arrays, such as PbSe, operating at longer optical wavelengths, can exhibit some unwanted signal drift and instability that are comparable to the desired signal in magnitude for lower illumination levels. Figure 12.10 shows the measured drift in the signal of a commercially available 256-channel PbSe detector array in a 5-min span after a 60-min warm-up, with the detector array temperature-stabilized at 260 K using a PI temperature controller rated for ± 0.002 K. The detector signal from all the pixels was offset to an initial value of 0 V at t = 0 min.

Miniature IR spectrometers employing multiplexed linear detectors arrays are at a significant disadvantage compared to large tabletop instruments in terms of the noise performance of their detector elements. Experience with commercially available PbSe detector arrays indicates that their attainable noise performance can be an order of magnitude greater than the theoretical NEP of an individual element.

A spectrometer using a programmable shutter is operated in synchronization with the detector array. The detector array measures alternatively the illuminated (V_{ill}) and the dark signal (V_{dark}) as obtained by opening and closing the programmable shutter. The user can adjust the relative rate of measuring the dark and illuminated signals. This is synchronized with the acquisition of the data from the detector array. The measured detector signal V_{meas} actually consists of the desired optical signal as converted to electronic form by the detector array (V_{des})

and the unwanted background signal V_{dark} that includes the detector dark signal, detector intrinsic noise, and electronic noise inherent in the signal path between the detector and the data acquisition system (multiplexer nonidealities and amplifier noise):

$$V_{meas}(i) = V_{des}(i) + V_{dark}(i)$$

It should be noted that the unwanted signal, V_{dark}, contains both statistically random, varying noise components and nonrandom pixel signal drift. Usually such systems would employ linear detectors that tend to average the measured signal $V_{meas.}$ However, this is influenced by the nonidealities and dark signal drift exhibited by typical IR detectors. Thus, even with averaging, one does not obtain an accurate representation of V_{des} under small-signal conditions. An example of this is shown in Figure 12.11.

The data was obtained using a commercially available 256-channel PbSe array. The detector signal was measured and summed for each individual pixel for 100 sequential scans. The average shown in the Figure 12.11 as $V_{meas}(\text{avg})$ is the traditional summed value for each pixel divided by the total number of scans, 100. This average value exhibits peak excursions of about ± 10 mV. Averaging for longer times does not further improve this estimate of V_{des} due to nonrandom variations in the background signal of the PbSe detector array.

The performance of miniature spectrometers may be further improved by integrating them with additional microphotonic devices such as Fabry–Perot (FP) filters and programmable multislit shutters. These devices are briefly described in the following sections.

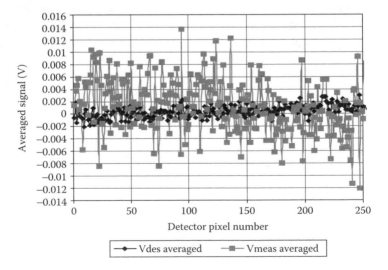

FIGURE 12.11 Comparison of typical PbSe detector pixel noise.

12.6 MINIATURE FP FILTERS

Recently, significant advances have been made in achieving both high-spectral-resolution and broadband operation in very compact spectrometers by innovatively combining narrowband FP filters (see Chapter 3) with miniature broadband waveguides.

For example, an FP filter can provide a resolution on the order of 0.01 nm or better. The trade-off is a narrow spectral range of operation of about 100 to 200 nm because of the limited free spectral range (FSR) and sequential spectral measurements.

The small FSR of typical FP filters is usually a hindrance to their application for spectroscopy. Figure 12.12 and Figure 12.13 show an innovative approach in which a single tunable FP filter operates in series with the waveguide spectrometer. In this case, the concept benefits from the periodic transmittance characteristics of the FP filter. The FSR of the FP filter is designed to match the dispersion characteristics and spectral spacing of the output detector pixels. As a result, the FSR required is only about 4 to 8 nm. For a given tuning of the FP cavity air gap spacing, the spectrometer dispersion demultiplexes the narrowband transmission peaks provided by the FP filter such that each neighboring peak falls on separate detector pixels. This requires matching the FP transmittance periodicity with the detector pixel spacing. As the detector pixel positions are determined by standard lithography, they can be tailored for operation with the FP filter characteristics.

The use a single FP filter at the optical input to the spectrometer to effectively digitize the optical signal will result in a serially multiplexed stream of narrowband signals into the spectrometer. The waveguide spectrometer then demultiplexes the digitized spectral signal onto the 256 detector pixels. By tuning the

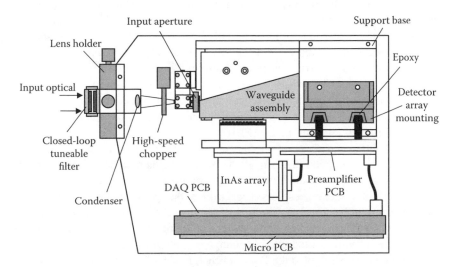

FIGURE 12.12 High-resolution FP infrared spectrometer.

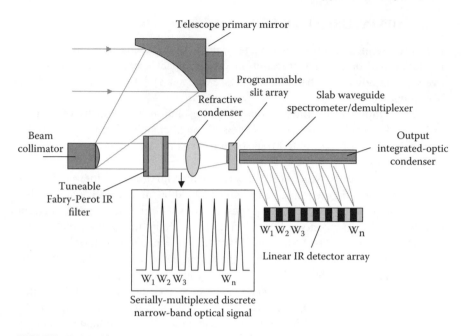

FIGURE 12.13 Operation of the high-resolution FP-MEMS spectrometer.

FP cavity spacing in N segments over the FSR of the filter, a total of $N \times 256$ spectral measurement points are feasible.

Using dielectric mirrors can provide a peak transmittance of about 90% at the FP cavity resonance if high-quality dielectric layers are used. Broadband operation from 1.15 to 2.45 μm is feasible by chirping the thickness of the dielectric layers. The required FP mirrors can be constructed on a quartz or sapphire substrate using a few alternating layers of SiO_2 ($n = 1.45$) and amorphous silicon (a-Si) with an index of refraction of about 3.45. This provides the spectral operation between 1.4 and 2.4 μm. Figure 12.14 shows a schematic of a simple three-layer dielectric mirror design on quartz. If a high-contrast SiO_2/a-Si dielectric coating for the FP mirrors is used, then only about three to five layers per mirror are required, significantly reducing the fabrication cost.

Some of the transmittance characteristics, as simulated by using Zeemax, are shown in Figure 12.15 and Figure 12.16.

Broadband spectral operation can be realized by selecting the radius $R(\lambda)$ characteristic of the two mirrors in such a way that it is matched to the operating spectral range of the wavelength-division demultiplexer.

For current moderate-resolution spectrometers, operation with a tunable laser diode source can be set up for the m = 1 and m = 2 diffraction orders. Although individual detector pixels have a bandwidth of about 7 nm due to their width, the spectrometer produces a measurable change in the output detector signal with an input wavelength shift of only 1 nm, even for m = 1, as the output diffracted signal shifts in position relative to the adjacent detector pixels. This implies that

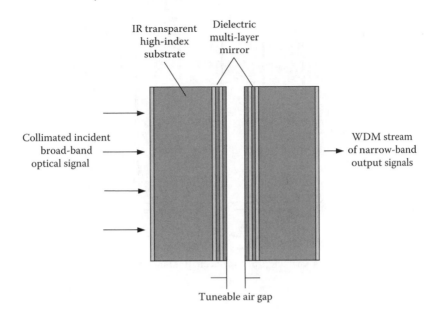

FIGURE 12.14 Tunable FP broadband filter.

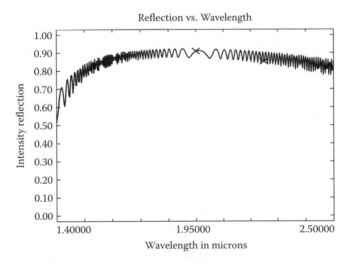

FIGURE 12.15 Zeemax simulations of a-Si/SiO$_2$ dielectric mirror on quartz.

the spectrometer itself is capable of intrinsically tracking or resolving small changes in wavelength at the output on the order of 1 nm or better. Using the m = 2 diffraction order effectively doubled the spectral resolution of the input spectral signal, with input signals 4-nm apart exhibiting distinct diffraction peaks on adjacent pixels (see Figure 12.17).

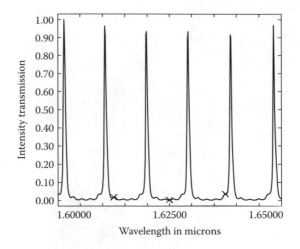

FIGURE 12.16 FP filter characteristics between 1600 and 1650 nm.

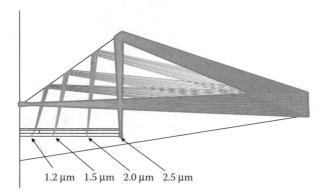

FIGURE 12.17 Zeemax simulations of a FP spectrometer output resolution for a range of 1.2 to 2.5 μm.

By reducing the detector pixel width and increasing the spectrometer output spectral dispersion, a 1-nm spectral resolution is feasible by just the use of the basic waveguide design. The output dispersion efficiency for m = 2 can be increased by blazing the diffraction grating. This will allow full coverage from 1.2 to 2.4 μm. The spectrometer output dispersion can be further enhanced by increasing the grating radius of curvature, R_g, from 8.7 to 12 cm. Table 12.1 lists some of the basic design parameters for miniature spectrometers with FP filters.

Using a ZnSe core in place of the c-Si core increases the spectral dispersion by a factor of n (ZnSe)/n (c-Si), which is equal to 2.4/3.45. Using ZnS cladding on ZnSe ensures good high-temperature stability, as this waveguide structure is well-matched thermally.

TABLE 12.1
Selection of Basic Parameters of a Waveguide Spectrometer Design

Spectrometer Core	Blazing	R_g (cm)	Spectral Range (μm)	Output Dispersion (nm/μm)	Input Slit Width	Pixel Bandwidth (nm)
c-Si	m = 1	8.7	1.2–5.0	0.15	52	7.8
c-Si	m = 1	12.0	1.2–5.0	0.11	52	5.7
c-Si	m = 2	12.0	1.2–2.4	0.05	52	2.9
c-Si	m = 2	12.0	1.2–2.4	0.05	20	1.1
c-Si	m = 3	12.0	1.5–2.2	0.04	30	1.1
ZnSe	m = 2	12.0	1.2–2.4	0.04	52	2.0
ZnSe	m = 2	12.0	1.2–2.4	0.04	30	1.1
ZnSe	m = 3	12.0	1.5–2.2	0.02	40	1.0

For the aforementioned MEMS-based FP filters, the mirror motion required is only several microns as the required wavelength tuning range per channel is typically less than 8 nm. This approach is very promising because of the potential for operation over a broad spectral range. The required mirror motion is achievable using a fairly simple collinear planar parallel-plate mechanical structure with electrostatic actuation, which should offer minimal friction and good long-term reliability.

There are several other techniques that are considered as the basic design for FP filters. For example, the electro-optical effect exhibited by liquid crystals may also be used to provide tunable FP filters. Their advantage is that they do not need moving mechanical parts. In this case, the tunability of the peak transmittance wavelength is achieved using voltage variation of the refractive index of the liquid crystal. The useful optical transmittance range is mainly in the visible and NIR spectral ranges, i.e., up to about 1.8 μm. Electro-optical techniques generally require substantially high voltages (10^3 V) to obtain a significant change in the refractive index. Therefore, the cavity length is limited by the required electric field and by the optical transmittance loss of the liquid crystal. Generally, a relatively thin cavity of a few microns is employed.

However, owing to the small FSR required of 8 to 10 nm, they could be viable option if a suitable IR-transmissive material with a substantial electro-optical coefficient is identified.

12.7 MINIATURE SHUTTER ARRAYS

In dispersive optical instruments, such as miniature waveguide spectrometers, higher resolution is obtained by using narrower input slit widths. However, this also limits the input optical collection efficiency, reducing the attainable SNR for a given collected signal.

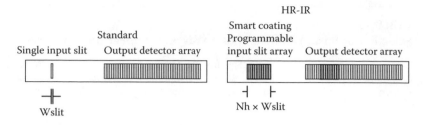

FIGURE 12.18 Comparison between single-slit and multislits spectrometers.

Optical input coding, such as the well-known Hadamard transform spectroscopy (HTS), may be used to significantly improve the performance of diffractive spectrometers by orders of magnitude. This requires the traditional single slit of a diffractive spectrometer to be replaced by an array of N programmable shutters to maximize the input light coupling of the instrument. The programmable input slit array increases the input luminosity by a factor of $N/2$ relative to single-slit spectrometers. This remedies an important deficiency exhibited by miniature spectrometers — very limited input luminosity due to the restrictions associated with a single miniature input slit. A shutter array may be controlled individually to either pass or block the desired optical signal. This shutter array is employed to code the optical input to the signal analyzer [12.2]. This is schematically illustrated in Figure 12.18.

Each individual slit provides a distinct spectrum that is multiplexed together by the programmable shutter array. HTS provides the same performance benefits as are inherent in typical Fourier transform spectroscopy (FTS).

In the case of HTS the effective input aperture and luminosity, for a given spectral resolution, is increased by a factor of $N/2$. The coded optical input is deconvolved using a relatively simple Hadamard inverse transform. Using Hadamard processing for a N_i input shutter or slit array and binary weighting (i.e., +1 or 0), the mean square error in the spectral estimate can be reduced by a factor of $\sqrt{(Ni)/2}$. Each shutter or slit is of width w, corresponding to the width of the detector pixels.

Figure 12.19 shows an experimental setup that integrates a mechanically selectable 16-element binary code array. For the photomask used, each slit was 60-μm wide, when open, with a period of 80 μm.

A 16 slit code was employed based on the Wallis transform matrix (see Table 12.2). Here, 1 corresponds to a slit in the open position, and 0 corresponds to the slit in the closed position.

The binary values of the code are given by matrix S(j, k). The row index j corresponds to the slit position. The column index k corresponds to the code sequence for each measurement. The measured multiplexed data is a 256×16 matrix (256 rows by 16 columns) corresponding to each sequence k of the binary code, as given by:

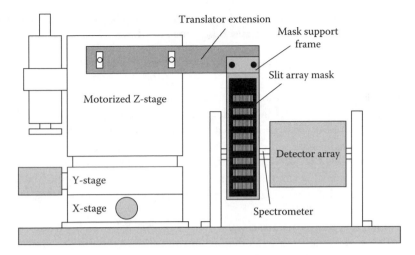

FIGURE 12.19 HTS experimental test setup and mechanically selectable 16 array input mask.

TABLE 12.2
Coding Matrix S(j,k)

S(j, k)	1	2	3	4	5	6	7	8	9	10	11	12	13	14	15	16
1	1	0	0	1	0	1	1	0	1	1	0	1	0	0	0	1
2	0	1	0	1	1	0	1	0	0	1	1	0	1	0	0	1
3	0	0	1	1	1	1	0	0	0	0	1	1	0	1	0	1
4	1	1	1	1	0	0	0	0	0	0	0	1	1	0	1	1
5	0	1	1	0	0	1	1	0	1	0	0	0	1	1	0	1
6	1	0	1	0	1	0	1	0	0	1	0	0	0	1	1	1
7	1	1	0	0	1	1	0	0	1	0	1	0	0	0	1	1
8	0	0	0	0	0	0	0	0	1	1	1	1	1	1	1	1
9	0	1	1	0	1	0	0	1	1	1	0	1	0	0	0	1
10	1	0	1	0	0	1	0	1	0	1	1	0	1	0	0	1
11	1	1	0	0	0	0	1	1	0	0	1	1	0	1	0	1
12	0	0	0	0	1	1	1	1	0	0	0	1	1	0	1	1
13	1	0	0	1	1	0	0	1	1	0	0	0	1	1	0	1
14	0	1	0	1	0	1	0	1	0	1	0	0	0	1	1	1
15	0	0	1	1	0	0	1	1	1	0	1	0	0	0	1	1
16	1	1	1	1	1	1	1	1	1	1	1	1	1	1	1	1

$$\text{Out }(i, k) = \Sigma \text{ In}(i, j) \times S(j, k), \text{ for } j = 1 \text{ to } 16$$

where i is the row index from pixel 1 to pixel 256, and k is the column number of the binary matrix S(j, k), corresponding to code sequence k. Here, $In(i,j)$ is the spectrum corresponding to single slit j, for $j = 1$ to 16. The multiplexed data can be deconvolved into the 16 spectra estimated using:

$$In(i, j) = \Sigma \ Out(i, k) \times S^{-1}(k, j), \text{ for } k = 1 \text{ to } 16$$

where $S^1(k, j)$ is the inverse of the coding matrix.

This signal inversion is independent of the exact characteristics of each slit and of the uniformity of their illumination. It mainly depends on the presence of an inverse transformation for the binary optical signal coding. However, the alignment accuracy of each row of slits is important and will affect the accuracy of the inverse transformation. In this respect, fixed prealigned programmable slit arrays are crucial to obtaining the maximum SNR with the input binary coding technique.

For the preceding code matrix, the inverse matrix, $S^{-1}(k, j)$, is given by Table 12.3. The inverse transformation mainly consists of the addition and subtraction of the multiplexed values and one division per final data element.

Figure 12.20 shows a comparison of a typical single-channel dispersive spectrometer scan and the corresponding HTS one for the same measurement time. This clearly illustrates the dramatic improvement in SNR in the HTS scan.

There are several technologies that have been used to facilitate the realization of arrays of miniature shutters or mirrors to provide the required Hadamard input binary coding.

TABLE 12.3
Inverse Transform Matrix $S^{-1}(k, j)$

S1(k, j)	1	2	3	4	5	6	7	8	9	10	11	12	13	14	15	16
1	1	-1	-1	1	-1	1	1	-1	-1	1	1	-1	1	-1	-1	1
2	-1	1	-1	1	1	-1	1	-1	1	-1	1	-1	-1	1	-1	1
3	-1	-1	1	1	1	1	-1	-1	1	1	-1	-1	-1	-1	1	1
4	1	1	1	1	-1	-1	-1	-1	-1	-1	-1	-1	1	1	1	1
5	-1	1	1	-1	-1	1	1	-1	1	-1	-1	1	1	-1	-1	1
6	1	-1	1	-1	1	-1	1	-1	-1	1	-1	1	-1	1	-1	1
7	1	1	-1	-1	1	1	-1	-1	-1	-1	1	1	-1	-1	1	1
8	-1	-1	-1	-1	-1	-1	-1	-1	1	1	1	1	1	1	1	1
9	1	-1	-1	-1	1	-1	1	1	1	-1	-1	-1	1	-1	1	1
10	1	1	-1	-1	-1	1	-1	1	1	1	-1	-1	-1	1	-1	1
11	-1	1	1	-1	-1	-1	1	1	-1	1	1	-1	-1	-1	1	1
12	1	-1	1	1	-1	-1	-1	1	1	-1	1	1	-1	-1	-1	1
13	-1	1	-1	1	1	-1	-1	1	-1	1	-1	1	1	-1	-1	1
14	-1	-1	1	-1	1	1	-1	1	-1	-1	1	-1	1	1	-1	1
15	-1	-1	-1	1	-1	1	1	1	-1	-1	-1	1	-1	1	1	1
16	1	1	1	1	1	1	1	1	1	1	1	1	1	1	1	-7

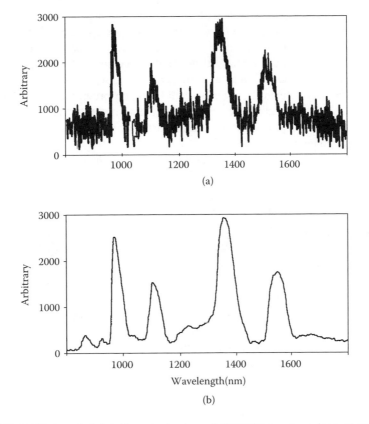

(a)

(b)

Wavelength(nm)

FIGURE 12.20 A typical data for a single-channel dispersive spectrometer scan and HTS scan.

Early Hadamard transform spectrometers, primarily operating in the visible spectral range, were first realized in the 1970s. They typically employed arrays of slits on a rotary wheel that were mechanically positioned to obtain the Hadamard optical code. The early versions of the visible HTS spectrometers exhibited some problems associated with the mechanical positioning of the Hadamard mask array (stickage, mechanical misalignment, etc.).

Another approach was to use liquid crystals as stationary voltage-controlled optical shutters [12.4]. Opaque and transparent states of liquid crystals were generated by controlling the rotation of polarized illumination through two parallel polarizers. However, some problems were encountered. First, the spectral operating range was largely limited to the visible and short NIR spectral ranges because of optical absorption by the liquid crystal. It was also found that the transition between on states and off states was less than ideal. Moreover, the response time of the liquid crystal shutters was relatively slow, i.e., in the range of 10 msec.

More recent work employed MEMS-based digital micromirror arrays (DMA) to provide optical multiplexing between the output plane of a bulk-optics diffraction grating and a single detector. A Hadamard transform spectrometer operating in the 1- to 1.6-μm spectral range using an 800-element DMA was developed to multiplex the diffracted output signal from a bulk-optics grating spectrometer to a single InGaAs detector. The DMA could also be programmed to simulate a classical dispersive spectrometer with a single output slit. In this case, the DMA was programmed to illuminate the InGaAs detector sequentially, a single output wavelength channel at a time.

A brief review of the latest developments in the area of miniature shutter arrays is summarized in the following sections.

12.7.1 VO$_2$-BASED SHUTTER ARRAYS

One potential solution to the realization of a programmable array of optical shutters without moving components is the use of smart coatings that exhibit a metal–insulator transition (see Chapter 3). This has significant benefits in terms of the mechanical simplicity, reliability, and the achievable packing density.

In this approach, the input aperture window contains a thin-film electrochromic or thermochromic shutter, such as an alloy based on VO$_2$. This provides an integrated thin-film programmable shutter (window) that can be switched over a broad spectral range, from transparent to opaque, using a control signal, such as an applied voltage or temperature. The added benefit is that there are no moving mechanical components (see Chapter 10). The electrochromic or thermochromic material can be deposited onto the input slit using vacuum deposition techniques. Optically active alloys, based on VO$_2$ in particular, can provide optical chopping or modulation over a broad spectral range.

The long (2 to 5 mm), narrow (50 to 100 μm) shutters required for the input coding are readily achievable using established lithography techniques to pattern the smart coating.

Various thin-film alloys based on the VO$_2$ material system have been developed for use as programmable optical mirrors and shutters. VO$_n$ exhibits one of the largest observed variations in electrical and optical characteristics due to the metal–insulator transition. The transition temperature increases with the oxygen content, varying from 126 K for VO, to 140 K for V$_2$O$_3$, and to 341 K for VO$_2$.

In the insulating state, VO$_2$ exhibits high transmittance, typically exceeding 75% that extends from the NIR (approximately 1 μm) to beyond 12 μm in the far infrared. Optical absorption by VO$_2$ is very low in this insulating state, indicating a relatively clean band gap, free of defect states. In the metallic state, the optical transmittance of VO$_2$/Si is very low, typically below 0.5% even for 1000-Å-thick films. This implies that a high-quality smart coating can provide optical switching or modulation over a very broad spectral bandwidth spanning from the NIR to the far infrared.

VO$_2$ in the insulating state is a semiconductor that can be doped with suitable donors and acceptors to tailor its characteristics. This is a major advantage of the

TABLE 12.4
Effect of W and Ti Doping

	Dopant Composition			
W doping	0	1.5%	3%	3%
Ti doping	0	0%	0%	1%
Transition temperature	68°C	30°C	20°C	0°C

VO_2-based material system and facilitates a strong spin-off potential for other micro- and nanotechnology applications. In this case, impurities such as W and Mo provide an extra d-orbital electron as compared to vanadium and act as donors. Impurities such as Ti, with one fewer d-orbital electron than V, act as classical acceptors. The effects of doping VO_2 with dopants such as W and/or Ti are summarized in Table 12.4. As indicated, the transition temperature can be shifted from 68°C to below −20°C, as may be required for a given application.

VO_n thin-film shutters can be heated by directly passing a current through the film. It has been estimated that, for a shutter area of $50 \times 1000 \ \mu m$, the heating time is about 1 msec for an applied power of about 1 mW. This is more than adequate for most shutter array requirements. Heat loss to the substrate will increase the heater power required for a given off time. Summary of estimated VO_2 thermally activated on/off switch times is presented in Table 12.5.

The VO_2-based optical switch is usable over a broad spectral range from about 1.5 μm to above 12 μm. It is very attractive due to the lack of any mechanical motion and the high-potential operational reliability (see Table 12.6).

Moreover, VO_2 slit elements can be spaced closely together and have no limitations on the achievable slit height. However, this approach does introduce a coating into the beam path such that the on-state optical transmittance has wavelength dependence. Shutter operation between 1.5 and 3.5 μm can be based on microthermal heating. However, VO_2 has a finite thermal emittance in the insulating and metallic states, with a nominal value of about 0.2 to 0.3, depending on the thickness of the VO_2 layer. VO_2 optical switch operation at longer wavelengths will require cooling and some thermal-control circuitry to minimize the self-emittance from the VO_2 windows.

TABLE 12.5
VO_2 Thermally Activated On/Off Switch Times

Transition	Power	Slit Area	Time (msec)
Heat	1 mW	$50 \times 1000 \ \mu m$	1
Cool	Ambient	$50 \times 1000 \ \mu m$	5

TABLE 12.6
Performance Data for VO$_2$ Shutter

Parameter	Description	Experimental Value
V$_{closed}$	Voltage required to switch from insulating to metallic state	50 to 60 V (for 50-μm-wide slits)
I$_{closed}$	Current required to maintain metallic state	50 μA
T$_{switch}$	VO$_2$ switching temperature	on at 35°C off at 50°C (for 1.5% W-doped VO$_2$)
t$_{close}$	VO$_2$ shutter close time (insulator to metallic transition)	< 20 msec
t$_{open}$	VO$_2$ shutter open time (metallic to insulator transition)	< 20 msec
Tr (on)	Optical transmittance between 1.5 and 3.2 μm in the on state	> 55% at 1550 nm
Tr (off)	Optical transmittance between 1.5 and 3.2 μm in the off state	< 2% (dependent on VO$_2$ thickness)

12.7.2 MEMS SHUTTER ARRAYS

12.7.2.1 Shutter Arrays Based on Comb Electrodes

A MEMS shutter array is shown in Figure 12.21a. Each shutter consists of a 60-μm-wide Al/Si slider, about 1-mm long, that covers a 50×900 μm window micromachined in the underlying substrate (see Figure 12.21b). The required actuator displacement is about 50 μm. Each slider (shutter) is supported by two Si flexure springs, which provide the restoring force. Top and bottom interdigitated "comb" electrodes provide activation. The application of a control voltage pulls the movable slider towards the fixed electrodes, opening the window. Si flexure springs provide the return restoring force. Power is only required during the "open" transient of the shutter in order to provide the displacement current to charge the electrostatic comb capacitor. By using a 25- to 30-μm-thick doped c-Si layer for the actuators, the actuation voltage can be reduced below 75 V. The actuator can be fully suspended from the substrate to eliminate friction.

In the open position of the shutter, the optical signal is allowed to pass through the SOI device window without any obstructions. This provides an ideal on transmittance for each shutter, Tr (on) = 100%, that is independent of the wavelength of the optical signal. In the closed position of the shutter, the window is totally obscured by the slider. An Al coating applied to the Si device layer allows the shutter to be totally opaque to provide an ideal off transmittance for each shutter, Tr (off) = 0%, that is independent of the wavelength.

Figure 12.22 shows the SEM micrograph of several comb electrode elements comprising the linear-drive mechanism.

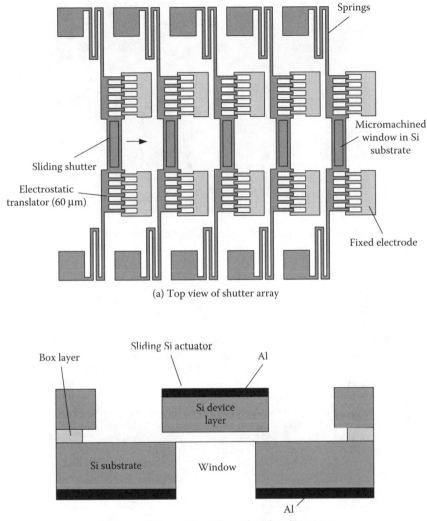

(a) Top view of shutter array

(b) Cross-section of the slider in the "closed" position

FIGURE 12.21 MEMS shutter array based on comb electrodes.

12.7.2.2 Shutter Arrays Based on Sweeping Blades

In this design, selective electroplating technology is used combined with thick photoresist masks to build sweeping blades [12.5]. One blade is fabricated for each slit. The displacement of the blades is parallel to the substrate plane. Two electrodes located on both sides of each blade are used to either close or open a particular slit (see Figure 12.23). The fabrication of these electrodes is based on selective electroplating and thick photoresist masking. This electrode scheme allows for low actuation voltages. It has been estimated that the actuation voltage

FIGURE 12.22 SEM micrograph of electrostatic comb drive with 50-μm translation.

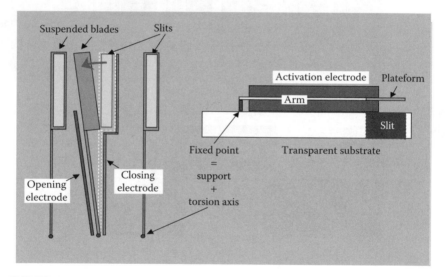

FIGURE 12.23 MEMS shutter array based on sweeping blades.

in this case could be as low as 20 V and that the blade's response could be in the range of a few milliseconds.

12.7.2.3 Shutter Arrays Based on Zipping Actuators

This approach is based on a zipping actuator to close and open the slits. In this case, a stress gradient is introduced in the structural layer used to define the actuator [12.5]. This stress gradient is obtained by varying the fabrication parameters during deposition of the actuator layer on top of a sacrificial layer. Once

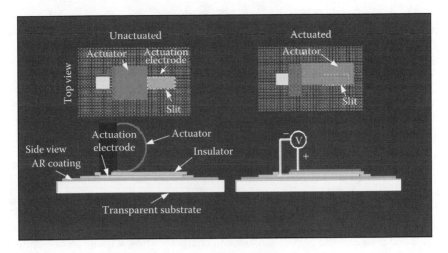

FIGURE 12.24 MEMS shutter array based on zipping actuator.

the sacrificial layer is removed, the stress in the material relaxes, causing the actuator to bend out of the substrate plane (see Figure 12.24). Electrodes are positioned on the substrate and surround the slit. When a sufficiently high voltage is applied between these electrodes and the actuator, the generated electrostatic force pulls the actuator down to the substrate, which closes the slit. It has been estimated that an actuation voltage is as low as 30 V with the response time on the order of a few milliseconds for the slit closure. Opening of the slit relies primarily on the restoring force in the actuator.

A brief comparison between MEMS and VO_2 shutter arrays is summarized in Table 12.7.

TABLE 12.7
Comparison of Technologies for Shutter Arrays

Parameter	MEMS Shutter Arrays	VOn Shutter Arrays
Transmittance	Wavelength independent	Dependent on spectral transmittance characteristics of VO_n and substrates
Tr (on)	> 90%	55% to 90%
Tr (off)	0%	< 2%
T_{open}	~ 1 msec	< 20 msec
T_{close}	~ 1 msec	< 20 msec
Power	Transient charge current during switching	1 mW/slit
Switching voltage	30–75 V	50–60 V
Reliability	Stiction and wear issues	No moving components (estimated to be greater than 10^8 cycles)
Complexity	Electrostatic actuation	Critical deposition conditions for VO_2

In terms of optical performance, MEMS shutters can offer the best optical performance with wavelength independent characteristics that can offer an ideal Tr (on) of 100% and Tr (off) of 0% and switching speeds in the range of 1 msec.

However, the fabrication of the electrostatic actuators for the MEMS approach still requires some further development. The main issues that need to be resolved are the actuator stiction, deflection angle, and switching speeds.

12.8 SUPERPRISM

As it has been mentioned in the preceding text, traditional beam-deflecting devices — such as MEMS — have limitations for the deflection angle and switching speed. PBG structures offer another solution in the manufacturing of deflection devices. These PBG-based devices operate using a principle that allows a very wide deflection angle. This principle is called the *superprism effect* and is one of the peculiar dispersion characteristics, which can be applied to many devices such as wavelength-division multiplexers or demultiplexers and dispersion compensators.

According to this principle, a beam incident on a PBG can be widely deflected inside the PBG structure by a slight change of wavelength or the incident angle of light [12.6]. This is schematically illustrated in Figure 12.25.

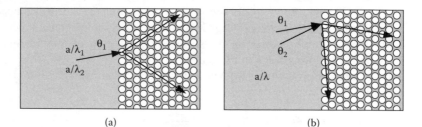

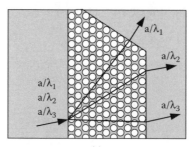

FIGURE 12.25 Superprism effects.

A wide-angle deflection may be realized when the output end of the PBG structure is tilted onto the input end. Analysis of the dispersion surfaces indicates a deflection angle of ± 50° in a two-dimensional PC composed of triangular lattice of air holes by changing the incident angle by ±2° or the wavelength by 2% [12.6].

The beam deflectors may be used in microphotonic devices that perform such functions as deflection, refraction, focusing, and birefringence.

REFERENCES

1. Goswami, D., Optical computing, *Resonance*, 56–71, 2003.
2. Srinivasan, K., Painter, O., Colombelli, R., Gmachl, C., Tennant, D.M., Sergent, A.M., Sivco, D.L., Cho, A.Y., Troccoli, M., and Capasso, F, Lasing mode pattern of a quantum cascade photonic crystal surface-emitting microcavity laser, *Appl. Phys. Lett.*, 84(21), 4164–4166, 2004.
3. Kruzelecky, R.V., High Performance Miniature Spectrometer, Patent pending, 2005.
4. Hammaker, R.M., DeVerse, R.A., Asunskis, D.N., and Fateley, W.G., Hadamard transform near-infrared spectrometers, *Handbook of Vibrational Spectroscopy*, Vol. 1, John Chalmers and Peter R. Griffiths (Eds.), John Wiley & Sons, New York, 2002, pp. 453–460.
5. Le Noc, L. and Picard, F., Programmable microslit array for miniature spectrometer, National Optics Institute (unpublished document), 2005.
6. Baba, T. and Nakamura, M., Photonic crystal light deflection devices using the superprism effect, *IEEE J. Quantum Electron.*, 38, 909–914, 2002.

13 Quantum Photonic Systems

Recent developments in quantum photonics have provided a new way to manipulate and encode information that has no classical counterpart (see Chapter 3). This has led to design of a range of new algorithms, cryptographic protocols, and higher-precision measuring techniques. Many designs have been devised to build devices that will be able to harness the quantum world and turn the theoretical advantage of quantum photonics into a practical implementation.

Quantum photonics has demonstrated a high control of the quantum particles because photons interact weakly with each other and thus have long decoherence times. Demonstration of quantum cryptography have even led to prototypes that are presently reaching the market. Several quantum photonic techniques and devices have been patented in the U.S. and worldwide.

However, these devices still have severe limitations with regard to the distance over which they can be used. To go beyond these first prototypes by extending the maximum distance for which they can be deployed or to increase their reliability, better control of quantum systems must be achieved in order to be able to implement quantum repeaters and error correction schemes. Such schemes will require quantum gates and other quantum functional operators that are able to control the interaction between photons.

Microphotonic devices will play a crucial role in the future quantum optics devices because they will provide the required compactness, efficiency, and design flexibility.

13.1 QUANTUM COMMUNICATIONS

The growing interest in quantum communication applications generates the necessity to extend the distances over which quantum information can be distributed. A principle of quantum communication may be illustrated with the example of a single-stage entanglement link that has been described in Chapter 3. Another scheme for distributing quantum information over arbitrary distances is based on quantum teleportation of entanglement photons.

The possibility to distribute quantum entanglement over arbitrary distances may lead to the establishment of communication networks that may operate on a global scale. Presently, the possibilities of transmitting quantum information directly either via optical fibers or through optical free-space links are rather limited in distance because of absorption and environmental influences. Experiments based on fiber technology have demonstrated that entangled photon pairs

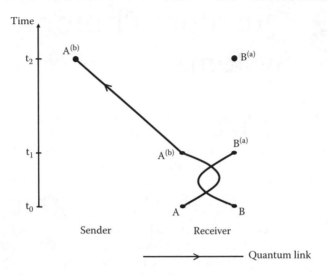

FIGURE 13.1 Formation of a quantum communication link.

can be separated by distances ranging from several hundreds of meters up to 10 km.

13.1.1 QUANTUM DENSE CODING

Entangled states permit a completely new way of encoding information. This new way of encoding was suggested by Charles Bennett of the IBM Research Division in Yorktown Heights, NY, and Stephen Wiesner of Brookline, MA, in 1992.

Selecting two photons, A and B, and then entangling them is the first step in the establishing the communication link (see Figure 13.1). This operation is initiated at the receiver end of the link (step t_0 in Figure 13.1). Storing one entangled photon ($B^{(a)}$) at the receiver end and sending the other photon ($A^{(b)}$) through a quantum channel to the data transmitter completes the formation of the communication link (step t_2 in Figure 13.1).

As mentioned earlier, each photon must be kept isolated from its surroundings to ensure that the entanglement is maintained.

The data transmission process is triggered at the sender end by performing one of the predetermined operations on the stored photon ($A^{(b)}$). The operations have to be unitary to maintain the quantum mechanical coherence of the particle. For example, a single photon can have only two distinguishable polarization states, e.g., "vertical" and "horizontal." These operations may be performed by using a quantum gate (e.g., U-type gate) that is an equivalent of a quantum modulator (see Figure 13.2). These operations can include (1) doing nothing or (2) changing the photon polarization.

It is sufficient to manipulate only one of the two qubits locally in order to switch the two-qubit state from any one of the four Bell states to all others (the Bell states have been defined in Chapter 3). These operations, although performed

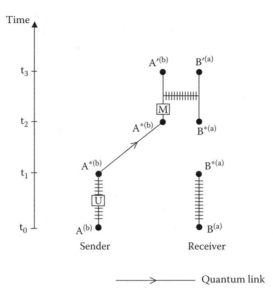

FIGURE 13.2 Quantum data transmission.

only on one photon (i.e., $A^{(b)} \Rightarrow A^{*(b)}$) at the sender end, affect the entangled quantum state of the second photon (i.e., $B^{(a)} \Rightarrow B^{*(a)}$) that is stored at the receiver end (step t_1 in Figure 13.2),

The result of these operations cannot be verified by measurements on these two photons separately. They can be determined only by measuring both of them together (i.e., $A^{*(b)}$ and $B^{*(a)}$). Therefore, the sender sends the "modulated" photon to the receiver, who by measuring them jointly can determine which of the operations have been performed (step t_2 in Figure 13.2). The receiver needs to use another type of quantum gate (e.g., M-type gate) to perform the Bell-state measurements (BSM) that allow for decoding the message (step t_3 in Figure 13.2). This type of photonic gate works as a quantum decoder. As there are four possible outcomes of this measurement, the sender has sent twice as much information as can be sent classically with a two-state particle.

This example illustrates a novel way to encode the information that leads to the possibility of transmission of two bits of information by sending only one qubit. This application of quantum communication is known as *quantum dense coding*. Quantum dense coding was the first experimental demonstration of the basic concepts of quantum communication.

13.1.2 TELEPORTING SYSTEM

Quantum teleportation allows for the transmission of quantum information to a distant location despite the fact that it is impossible to directly measure this information.

Quantum teleportation is defined as the transfer of key properties from one particle to another without the use of a physical link. It is a kind of "remote copying" of an unknown quantum state that could be compared to a simultaneous exact measurement of all observables of the system, including noncommuting ones. However, it has to be performed in such a manner that it does not violate the uncertainty principle.

The basis for the teleportation is the same as for quantum dense coding. The first step is to form a quantum communication link (Figure 13.1). The two entangled photons $A^{(b)}$ and $B^{(a)}$ are then placed at the sender and receiver location, respectively (the step t_0 in Figure 13.3).

The next step is to select a photon C — it is this photon state that is to be teleported. The photon C is then entangled with the photon $A^{(b)}$ at the sender location. This is illustrated in Figure 13.3 as the step t_1. At the end of this process there are three mutually entangled photons — $A^{(b,c)}$, $C^{(a,b)}$ at the sender location, and $B^{(a,c)}$ at the receiver location.

Now the system is ready to perform the process of teleportation of the photon C. At the specified time t_2 photons $A^{(b,c)}$ and $C^{(a,b)}$ are measured jointly at the

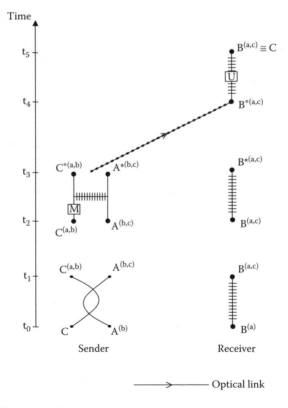

FIGURE 13.3 Quantum teleportation.

sender location by using a photonic gate that is similar to the quantum decoder (i.e., BSM). This quantum decoder allows determining the quantum state of the two-photon system.

This measurement projects the state of the particles $A^{(b,c)}$ and $C^{(a,b)}$ (step t_3 in Figure 13.3) into one of the four Bell states. At the time of the measurement, the particle $B^{(a,c)}$ is transformed into a new quantum state $B^{*(a,c)}$.

The result of this measurement is a piece of information that includes two classical bits that the sender passes through a conventional link to the receiver. According to the measurement result, the receiver performs one of the specified unitary operations on particle $B^{*(a,c)}$ using a quantum modulator. The purpose of this modulation is to reverse the changes that were introduced by the decoder at the sender end, i.e., bring $B^{*(a,c)}$ back to its previous state $B^{(a,c)}$. This can be explained in the following way: As the result of the original entanglement, both photons $A^{(b)}$ and $B^{(a)}$ are mutually orthogonal. A mutually orthogonal system of three photons is created by entangling the photon $A^{(b)}$ with the photon C. If $A^{(b)} \perp B^{(a)}$ and $A^{(b)} \perp C$, then $B^{(a)} \equiv C$ for a system in which only two directions of polarization are allowed.

The end effect of this operation is that photon $B^{(a,c)}$ is exactly in the same state as the original photon C was at the sender location. The photon B has been put into an exact reproduction of the former state of C. The photon C itself is no longer in that state, having been thoroughly disrupted by the measurement.

This means that through the phenomenon of quantum teleportation it is possible to deliver exactly that part of the information in an object that is too delicate to be measured out and delivered by conventional methods. Entanglement of photons over distances as long as 10 km has been demonstrated at the University of Geneva. It is expected that it is quite feasible that the teleportation-type communication will be implemented over long distances.

Quantum teleportation is theoretically perfect, yielding an output state that equals the input with a fidelity F = 1. It is important to note that the fidelity is an average over all input states. Therefore, it is a measure of the ability to transfer an arbitrary, unknown superposition from the sender to the receiver.

By contrast, a sender and receiver who share only a classical communication channel cannot hope to transfer an arbitrary quantum state with such a high fidelity. In practice, however, fidelities less than one are realized because of imperfections in the entangled pair of photonic qubits, the Bell measurement, and the receiver's unitary transformation.

13.2 BUILDING BLOCKS

One of the main challenges in the photonic approach to quantum information processing is the requirement for very efficient logic devices. It has been demonstrated that in addition to nonlinear optical devices it is possible to devise a number of building blocks that are based on linear optical elements. These devices have succeeded in producing the desired logical output with a probability that can approach unity. These linear optical building blocks may form the basis for a scalable approach to quantum photonic processing systems. Among these quantum devices are photon entanglers, modulators, decoders, and repeaters.

13.2.1 Entanglers

There are various ways in which qubits can be realized using single photons. Every degree of freedom that is available can, in principle, be exploited. The available properties are the photons' polarization, spatial mode, emission time, or their optical frequency.

Two types of photonic qubits have been investigated experimentally in communication schemes. These are (1) time-bin qubits and (2) polarization qubits.

13.2.1.1 Time-Bin Qubit Entanglers

A time-bin qubit is generated by a coherent superposition of photon at two time intervals. In practice, passing a single photon through an interferometer can generate a time-bin qubit. Such a qubit may be presented by the following expression:

$$\left|\Psi\right\rangle = a\left|0\right\rangle + be^{i\theta}\left|1\right\rangle$$

The preceding equation is a more general form of the Equation 3.50 (Chapter 3), where θ is the phase between the states $\left|0\right\rangle$ and $\left|1\right\rangle$. In the case of time-bin qubits, the states $\left|0\right\rangle$ and $\left|1\right\rangle$ represent photon at two different times. Passing the photons through an interferometer (different path length) may induce the time difference. This is schematically illustrated in Figure 13.4, where an entangler is made from fiber-optic components: a splitter, a combiner, a phase shifter, and a switch.

It has been demonstrated that this type of quantum information encoding is applicable for long-distance transmission in optical fibers because it is insensitive to polarization fluctuations and polarization mode dispersion [13.1].

13.2.1.2 Polarization Qubit Entanglers

The nonlinear three-wave mixing effect known as the parametric down-conversion (PDC) is most often used to produce entangled photons (see Section 3.10 in

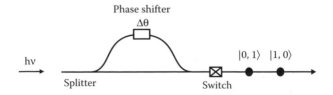

FIGURE 13.4 Time-bin qubit entangler.

Chapter 3). There are several nonlinear optical materials that can be used as the media for the spontaneous or optically induced PDC. In the spontaneous case, only one of the three interacting photons is initially excited. There exists a quantum mechanical probability that the initial photon will decay into two photons. In other words, the nonlinear crystal converts a single ultraviolet photon into a pair of infrared photons with entangled polarizations. The rate of this process is proportional to the pump intensity and the magnitude of the nonlinear coefficient of the nonlinear medium. Among the most commonly used nonlinear crystals are KD_2PO_4 (DKDP), $LiNbO_3$, and BaB_2O_4 (BBO).

In a standard teleportation setup, a laser is directed at a crystal with nonlinear optical properties.

There are two major schemes that are used in PDC systems: type I and type II. These two types refer to various geometries of the nonlinear crystals that are required for optimization of phase matching (i.e., to comply with the momentum conservation law) between pumping and down-converting photons. The practical difference is that a type-I scheme generates momentum or energy–time entanglement, whereas a type-II scheme additionally generates polarization entanglement. It turns out that the arrangements generating polarization-entangled states are much more efficient than the energy–time-entangled states. Therefore, a type-II has proven to be a more practical one because it generates directly the polarization-entangled photons. Type-I arrangements require additional optical processing to realize the polarization entanglement.

Figure 13.5 illustrates type-I PDC experimental setup. When a laser beam passes through an optically nonlinear material there is a probability that one of the photons will spontaneously decay into a pair of photons with longer wavelengths. The photons are emitted in two cones and propagate in directions symmetric to the direction of the original photon.

This type of the down-conversion may be used to produce polarization entanglement if the polarization of one of the photons is rotated by 90° (e.g., by using a half-wave plate) before it is superimposed with the other photon (see Figure 13.6). A beam combiner may be used to superimpose the two photons.

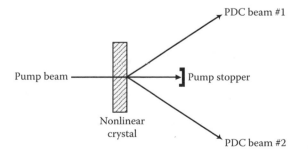

FIGURE 13.5 Type-I parametric down-converter.

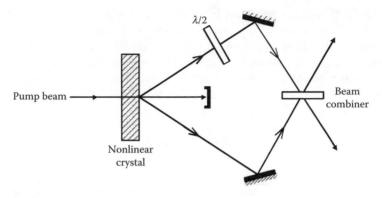

FIGURE 13.6 Polarization entangler source.

The state of the created polarization-entangled photons may be described by the following expression:

$$\left|\Psi_{AB}^{+}\right\rangle = \frac{1}{\sqrt{2}}\left[\left|H_A\right\rangle\left|V_B\right\rangle + \left|V_A\right\rangle\left|H_B\right\rangle\right]$$

The preceding expression indicates that this entanglement state can be transformed into one of the Bell states.

The type-II PDC scheme is shown in Figure 13.7. In this case, one of the photons is polarized horizontally and the other, vertically. It is possible to arrange the experiment such that the cones will partially overlap (see Figure 13.8).

As shown in Figure 13.8, the horizontally polarized photons (*H*) are generated on the upper cone; the vertically polarized photons (*V*) are generated on the lower cone.

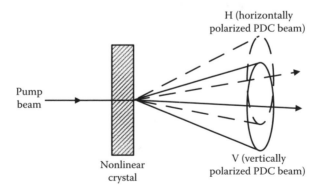

FIGURE 13.7 Type-II parametric down-converter.

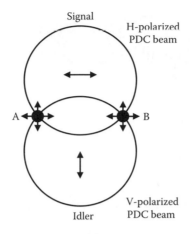

FIGURE 13.8 Cross section of polarization-entangled beam.

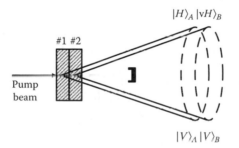

FIGURE 13.9 Double-crystal PDC entangler source.

There is another way to realize the polarization entanglement — by using two thin type-I nonlinear crystals next to each other to generate PDC. The nonlinear crystals are oriented in such a way that their optical axes are at 90° to each other (see Figure 13.9).

The incident (pumping) beam and optical axis of the first crystal define the vertical plane of polarization; the incident beam and optical axis of the second crystal define the horizontal plane. If the incident beam is vertically polarized then the down-conversion occurs only in the first crystal. If the incident beam is horizontally polarized, then the down-conversion occurs only in the second crystal. For the case in which the incident beam is linearly polarized and oriented at 45° with respect to the optical axes of the nonlinear crystals, it is equally probable that the down-conversion occurs in either crystal.

The created polarization entanglement may be described by the following expression:

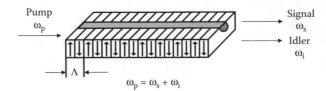

FIGURE 13.10 PPLN-based entangler source.

$$\left|\Phi_{AB}^{+}\right\rangle = \frac{1}{\sqrt{2}}\left[\left|H_{A}\right\rangle\left|V_{B}\right\rangle + \left|V_{A}\right\rangle\left|H_{B}\right\rangle\right]$$

The intersection of the cones will lead to the realization of the polarization entanglement.

13.2.1.3 PPLN-Based Entanglers

A new kind of photon entangler has been proposed that is based on the so-called periodically poled lithium niobate (PPLN) waveguides. PPLN waveguides are optically active waveguides that are periodically poled in the direction of the highest nonlinear coefficients (see Figure 13.10). The periodically poled structure allows for an efficient phase matching because of extended interaction length between the pumping and down-converted photons generated. This phase-matching structure leads to an improvement of the entanglement efficiency by several orders of magnitude as compared to the entanglers that are based on bulk-optics elements.

In Figure 13.10, the poling period is indicated by λ (in the range of 12 μm). Frequencies ω_p, ω_s, and ω_i correspond to the pump, signal, and idler beams.

These types of entanglers allow constructing of long-distance communication links. One advantage is that they can be pumped by using conventional semiconductor laser diodes. They can also be integrated with the other quantum photonic fiber-optic and waveguide devices.

13.2.2 Modulators

Most of the quantum communication protocols require a means to encode the quantum state of a two-particle system. Such experimental determination is usually implemented through the unitary operations with the help of sets of polarizers and wave plates.

A pair of entangled photons has four orthogonal basis states, i.e., the Bell states. These states can be connected by operations on a single photon. The Bell states offer the possibility of switching from any one of the four states to any other by simply performing an operation on one of the two qubits. For example, it is possible to switch from $\left|\Psi^{+}\right\rangle$ to $\left|\Psi^{-}\right\rangle$ by simply applying a phase shift, i.e., $\left|H\right\rangle \rightarrow \left|-H\right\rangle$, $\left|V\right\rangle \rightarrow \left|V\right\rangle$. The state $\left|\Psi^{+}\right\rangle$ can be obtained by "flipping" the second qubit, i.e., $\left|\Psi^{-}\right\rangle$. The state $\left|\Phi^{-}\right\rangle$ can be obtained by the combination of a phase shift and flipping [13.1].

The following expressions describe the possible manipulations on the Bell states by single-photon operations:

$$\left|\Psi^{+}\right\rangle = \frac{1}{\sqrt{2}}\left(\left|H\right\rangle\left|V\right\rangle+\left|V\right\rangle\left|H\right\rangle\right) \rightarrow \left|\Psi^{-}\right\rangle = \frac{1}{\sqrt{2}}\left(\left|H\right\rangle\left|V\right\rangle-\left|V\right\rangle\left|H\right\rangle\right) \text{ (phase flip)}$$

$$\left|\Phi^{+}\right\rangle = \frac{1}{\sqrt{2}}\left(\left|H\right\rangle\left|H\right\rangle+\left|V\right\rangle\left|V\right\rangle\right) \rightarrow \left|\Phi^{-}\right\rangle = \frac{1}{\sqrt{2}}\left(\left|H\right\rangle\left|H\right\rangle-\left|V\right\rangle\left|V\right\rangle\right) \text{ (phase flip)}$$

$$\left|\Psi^{+}\right\rangle = \frac{1}{\sqrt{2}}\left(\left|H\right\rangle\left|V\right\rangle+\left|V\right\rangle\left|H\right\rangle\right) \rightarrow \left|\Phi^{+}\right\rangle = \frac{1}{\sqrt{2}}\left(\left|H\right\rangle\left|H\right\rangle+\left|V\right\rangle\left|V\right\rangle\right)$$

(polarization flip)

$$\left|\Psi^{-}\right\rangle = \frac{1}{\sqrt{2}}\left(\left|H\right\rangle\left|V\right\rangle-\left|V\right\rangle\left|H\right\rangle\right) \rightarrow \left|\Phi^{-}\right\rangle = \frac{1}{\sqrt{2}}\left(\left|H\right\rangle\left|H\right\rangle-\left|V\right\rangle\left|V\right\rangle\right)$$

(polarization flip)

All of the aforementioned operations are unitary and they do not change the total probability of finding the system in the states $\left|0\right\rangle$ and $\left|1\right\rangle$ (see Table 13.1). In working with the Bell states it is common to refer to them as: (1) the phase shift, (2) the bit flip, (3) the combined phase shift and bit flip, and (4) the identity operator, which does not change the state on which it operates. All four operations are relatively easy to perform experimentally with photons.

Usually quarter- and half wave polarization plates (plates that shift the phase between the two polarization states of a photon by $\lambda/4$ and $\lambda/2$, respectively) have been used to make the unitary transformations between the Bell states.

TABLE 13.1
Bell-States Modulation

Modulator Setting		Phase Shift	Bell State	
$\lambda/2$	$\lambda/4$	(φ)	Initial	Final
$0°$	$0°$	$0°$	$\left|\Psi^{+}\right\rangle$	$\left|\Psi^{+}\right\rangle$
$0°$	$45°$	$\pi/4$	$\left|\Psi^{+}\right\rangle$	$\left|\Psi^{-}\right\rangle$
$90°$	$0°$	$\pi/2$	$\left|\Psi^{+}\right\rangle$	$\left|\Phi^{+}\right\rangle$
$90°$	$45°$	$3\pi/4$	$\left|\Psi^{+}\right\rangle$	$\left|\Phi^{-}\right\rangle$

Schematic illustrations of several configurations of quantum modulators are presented in Figure 13.11. Figure 13.11a shows a simple bulk-optics arrangement that consists of two of wave plates (i.e., $\lambda/2$ and $\lambda/4$). Figure 13.11b illustrates a fiber-optic-based arrangement in which the phase shift $\Delta \varphi$ is introduced by a controlled optical delay. A waveguide-based quantum modulator is shown in Figure 13.11c. In this case, a phase shift can be electrically induced and controlled by using, for example, a microheater. Similar control may be implemented in a PBG-based modulator that is shown in Figure 13.11d.

13.2.3 DECODERS

One of the most critical elements of future quantum communication and computation systems will be an efficient quantum decoder. A decoder task is to perform a BSM on two particles that were independently generated.

Multiqubit operations on photons can be performed with linear optical elements such as wave plates and beam splitters. This approach is exploited in the so-called linear optical quantum computation (LOQC; see Subsection 13.3.2) to generate multiphoton-entangled states.

A basic concept of such processing is illustrated in Figure 13.12. Figure 13.12 shows two photons that are directed from opposite sides on a semitransparent beam splitter.

In such a case, there are four possible outcomes: two photons may leave the beam splitter together in one or the other output port, both photons may be reflected into opposite ports, or both photons may be transmitted. The probability amplitudes for two photons of the last two outcomes have opposite phase, and therefore they cancel each other completely if the beam splitter reflects or transmits with equal probability. If two photons are incident on a 50/50 beam splitter then at the output side of the splitter they will emerge in the same beam.

However, in the case of entangled photons, the output will be more sophisticated. The photons that are in $\left| \Psi^- \right\rangle$ Bell state will emerge in different beams, whereas the photons in the other three Bell states will emerge in the same beam. This is the so-called *bunching* effect. Therefore, if there are two detectors inserted into each of the possible output paths, then they can be used to decode the state of the entangled photons. For example, if the two detectors register photons at the same time, then it will be known that the photons are in a $\left| \Psi^- \right\rangle$ state. However, there is no way of knowing which way the photons reached the detectors. The beam splitter could have transmitted them both, or both could have been reflected.

Figure 13.13 illustrates a two-stage decoder consisting of linear optical elements that can be used for the Bell measurement on polarization qubits. Hong, Ou, and Mandel first demonstrated this type of decoder in 1987 [13.2], and since then it is known as the *HOM interferometer*. HOM interferometers have since been utilized in quantum information protocols, and they are used to construct quantum optical logic gates.

Two photons A and B shown in Figure 13.13 are projected onto a beam splitter (BS). After passing through the beam splitter, each photon is subjected

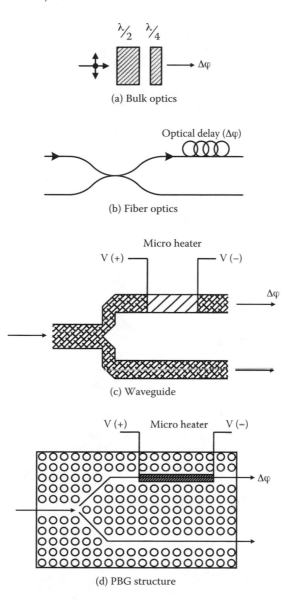

(a) Bulk optics

(b) Fiber optics

(c) Waveguide

(d) PBG structure

FIGURE 13.11 Quantum modulators.

to a single qubit projection on the horizontal and vertical axis using a set of two polarizing beam splitters (BS_1 and BS_2). Their particular state is recorded by a coincidence between a set of four detectors. The detectors incidences interpretation is summarized in Table 13.2.

It has to be pointed out that such a two-stage decoder can distinguish between two different Bell states ($\left| \Psi^{\pm} \right\rangle$), i.e., it will recognize only two out of the four

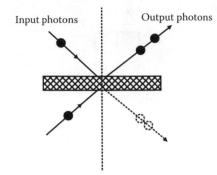

FIGURE 13.12 Splitter for entangled photons.

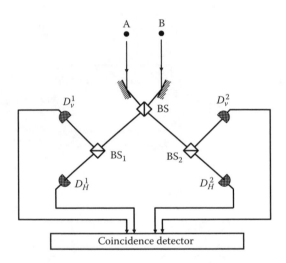

FIGURE 13.13 Two-stage decoder.

Bell states. For example, if both photons exit the splitter in the same beam but are still detected in orthogonal states, then the two-photon state is projected on to the $\left|\Psi^+\right\rangle$ state. This state will be recorded by a coincidence between D_V^1 and D_H^1 (or D_V^2 and D_H^2). If their particular state is recorded by a coincidence between detectors D_V^1 and D_H^2 (or between D_V^2 and D_H^1), then the two-photon state is projected on the $\left|\Psi^-\right\rangle$ state.

There is no experimental possibility to differentiate between all of the Bell states, at least by using linear optics only. However, by being able to discriminate between three different possibilities $\left|\Psi^+\right\rangle$, $\left|\Psi^-\right\rangle$, and $\left|\Psi^\pm\right\rangle$, it would be possible to send one "trit" of information with just one photon, even though the photon had only two distinguishable polarization states.

TABLE 13.2
Bell Measurements Decoding Matrix

Modulator Setting		Encoded Bell State	Decoder Reading (Coincidence Logic)
$\lambda/2$	$\lambda/4$		
$0°$	$0°$	$\left\vert \Psi^{+} \right\rangle$	D_V^1 and D_H^1
			or
			D_V^2 and D_H^2
$0°$	$45°$	$\left\vert \Psi^{-} \right\rangle$	D_V^1 and D_H^2
			or
			D_V^2 and D_H^1
$90°$	$0°$	$\left\vert \Phi^{+} \right\rangle$	2 photons in either
			D_H^1, D_V^1, D_H^2, or D_V^2
$90°$	$45°$	$\left\vert \Phi^{-} \right\rangle$	2 photons in either
			D_H^1, D_V^1, D_H^2, or D_V^2

By using a set of polarizers, beam splitters, mirrors, and detectors, it is possible to identify photons in the $\left\vert \Psi^{+} \right\rangle$ state. However, more advanced nonlinear quantum gates will be needed to be able to identify the $\left\vert \Phi^{+} \right\rangle$ and $\left\vert \Phi^{-} \right\rangle$ states.

The aforementioned procedure is an example of a postselected multiqubit operation because it generates correlations and entanglement without optical nonlinearities. Therefore, it may be used to produce entangled states for Bell measurement tests and to make probabilistic BSMs for quantum teleportation and entanglement swapping.

However, a nonlinear quantum gate will be needed to distinguish between the four Bell states. Such a gate would depend on a nonlinear interaction between the two photons; various theoretical and experimental groups are working on this challenge.

13.2.4 SINGLE-PHOTON DETECTORS

Quantum photonic applications require single-photon states be detected with efficiency approaching unity. Therefore, a great effort has been made in recent years toward developing high-efficiency, low-noise, and high-count-rate detectors.

There are a number of approaches that have been implemented. The most promising solutions have been provided by avalanche photodetectors (APDs). Some encouraging progress has also made with the use of detection techniques based on nonlinear up-conversions.

13.2.4.1 APDs

APDs for detection of single photons are used because they are efficient and they do not require cryogenic cooling. They are based on the generation of avalanche gain when one or more photons are absorbed.

Silicon-based APDs have the best performance in the visible range of the electromagnetic spectrum. They demonstrate very low-noise avalanche gains. Commercially available single-photon counting modules have quantum efficiencies of 50 to 70% and counting speeds up to 10 to 15 MHz.

Other types of the counters are modified Si-based devices. There are two types: visible light photon counters (VLPC) and solid-state photomultiplier (SSPM). Their quantum efficiency is in the range of 95%.

APDs based on InGaAs are used in the infrared range of the electromagnetic spectrum, more specifically in the 1.0- to 1.6-μm range. Their performance is better than those based on Ge.

13.2.4.2 Frequency Up-Converters

Up-conversion effects are used as the basis of some detectors. The idea here is to convert weak IR photons into visible signals, where single-photon detection is more efficient and more convenient. Frequency up-conversion uses wave-mixing generation in nonlinear optical crystals (see Chapter 3). If a weak input signal at ω_{in} is mixed with a strong pump at ω_p frequency, then a higher-frequency output field at $\omega_{out} = \omega_{in} + \omega_p$ is generated. With sufficient pump power this up-conversion can occur with near-unity efficiency even at the single-photon level.

It is possible to detect photons at 1.55 μm with this type of detector. The efficiency of this detection could be as high as 90%. PPLN may be used for the up-conversion. The PPLN crystal may be embedded inside a pump cavity that imposes a spatial mode for the input photons.

An additional advantage of this technique is its applicability to other processes required for the quantum photonic systems. For example, these devices may also be used to up-convert quantum states, i.e., to convert a single photon in an arbitrary quantum polarization state. For example, the photonic qubits can be prepared at wavelengths with the most convenient and efficient methods, and then they can be converted with near-unity efficiency to wavelengths that are optimal for photonic logic gates.

13.2.5 Tomographers

A quantum computation is vulnerable to errors and to environmental decoherence, which may destroy the entanglement. Characterization of quantum operations including errors and decoherence is an important issue for quantum information processing. The technique of quantum process tomography (QPT) has been used to characterize the quality of quantum operations. QPTs have been demonstrated for single qubits and for mixed ensembles of two-qubit systems.

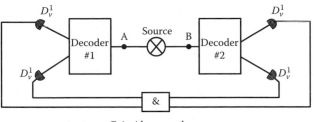

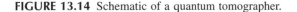

Coincidence analyzer

FIGURE 13.14 Schematic of a quantum tomographer.

The most general approach to measure a quantum state requires the reconstruction of its entire density matrix. Such a procedure is known as *quantum tomography*. A typical tomographer scheme is depicted in Figure 13.14. In contrast to the BSMs described in the previous sections, the density matrix is reconstructed from the statistical outcomes of different joint projection measurements.

In general, there is some freedom in the choice of the measurements that can ultimately be performed. A density matrix for a partially mixed state in a d-dimensional Hilbert space contains $d^2 - 1$ independent parameters. Therefore, one needs at least as many independent measurements to be able to reconstruct such a matrix. For example, for a two-qubit state this would require 15 measurements.

In the setup shown in Figure 13.14, each of the two photons is analyzed by using a qubit decoder. The qubit gates are based on two vectors *1* and *2*, respectively, which specify the bases to be projected on.

A source emits correlated qubits, each of which can be described in a two-dimensional Hilbert space. The qubits then propagate toward two decoders that are used for measurements. The outputs of the measurements are correlated to Bell-states projections. Some examples of typical tomographer outcomes are illustrated in Figure 13.27 and Figure 13.31 (see Section 13.3).

13.2.6 QUANTUM NODE

Several types of quantum nodes have been devised and experimentally demonstrated. These type of nodes are quantum equivalent of the nodes introduced in Chapter 4. It is expected that the future quantum networks will be based on nodes in which qubits will be stored and manipulated. Quantum channels will be used to send qubits. Although the development of the quantum nodes is in its very early phase, remarkable progress has already been achieved. Most of the demonstrated designs have been based on bulk optics. Recently a number of miniaturized building blocks have been successfully tested.

13.2.6.1 Quantum Node for Dense Coding

A conceptual design of a quantum node that may be used for the dense-coding communication link is illustrated in Figure 13.15. This design is an experimental

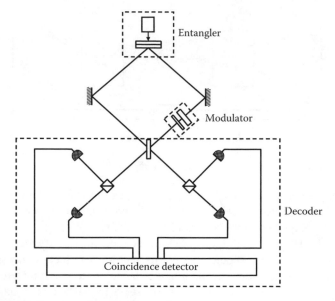

FIGURE 13.15 Quantum dense-coding node.

realization of the concept that has been shown in Figure 13.2. Polarization-entangled photons are used in the dense-coding application.

The node consists of (1) double-crystal PDC entangler, (2) bulk-optics modulator, and (3) bulk-optics decoder. The sender locally modulates the joint state of the entangled photons that are generated by an entangler. Then, the photon is sent to the receiver. The receiver decodes the two-photon state by performing a Bell measurement. In this particular node arrangement, the sender can distinguish between two different Bell states.

13.2.6.2 Teleporting Node

Figure 13.16 illustrates an arrangement of the quantum node that can be used for teleporting of polarization-entangled photonic qubits. This type of the quantum node was built and demonstrated at the University of Innsbruck in 1997 [13.1].

The aim of the node is to teleport the polarization of photon *A* to photon *C*. This is done by making a joint measurement on photons *A* and *B*, which changes the polarization of the latter in such a way that photon *C* — which is entangled with it — always acquires the same polarization as the first photon.

A PDC entangler is used to generate two independent entangled pairs of photons. The first pair of photons (i.e., photons *B* and *C*) is generated during the first passage of the laser beam through the nonlinear element. This pair of photons is in orthogonal states of polarization. The second pair of photons (i.e., *A* and *D*) is produced during its back propagation, i.e., after the pumping beam is reflected from a mirror. The photon *D* is used only to trigger the detectors. Photon *A* is

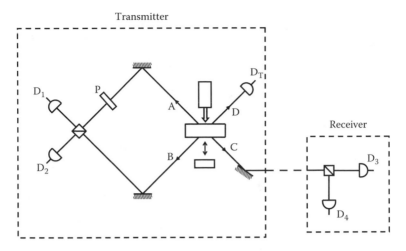

FIGURE 13.16 Teleporting node.

the photon that is being teleported. It is prepared in a known polarization state by using a polarizer P.

Photons A and B are passed trough a beam splitter that acts as an entangler. The measurements of the detectors allow determining the state of the photon A. This information is then transmitted via a classical channel to the receiver. In the case of the node based on linear optics, two out of four Bell states have to be discriminated at the sender site. The outcome of such a BSM has to be used to actively switch the teleported state at the receiver site to finalize the teleportation.

As the result of interaction between photons A and B, the quantum state of the photon C has been transferred into the state that is opposite to the state of the photon B, which is opposite to the state of the photon A. Therefore, the state of the photon C is identical to the state of the photon A. This information is then verified by the measurement by a detector at the receiver's site.

The node allows for using glass fiber-optics channels to distribute polarization-entangled photons between the stations. In order to complete the transmission, there is a requirement for a direct conventional connection to communicate the results of a BSM at the sender's location.

One of the features of the aforementioned node is its limitation to single-qubit teleportation scheme, i.e., the teleported photon cannot be utilized any further for a subsequent teleportation that may be required in order to implement a long-distance link. There is a requirement for a teleporting scheme that would allow for a teleported qubit to be used in the subsequent teleportation stage.

A team of scientists from Vienna [13.1] has proposed another version of the teleporting node that allows for multistage use of transmission. This version of qubit-preserving node is schematically illustrated in Figure 13.17.

The same entangler source as in the previous design may be used. The photons A and B are produced and manipulated by the sender. The photon C is sent to the receiver, and the photon D is used as a detection trigger. A variety of mirrors,

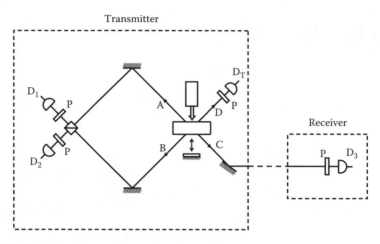

FIGURE 13.17 Qubit-preserving teleporting node.

beam splitters, and polarizers are used to direct the photons to four detectors labeled D_1, D_2, D_3, and D_T (trigger).

Polarizers are used to prepare the teleported photon in a well-defined state. The operation of the node involves preparation of the teleported photon in various different polarization states — horizontal, vertical, +45°, –45°, and right-hand circular.

A polarizer P in front of the trigger detector D_T is used to prepare the photon A nonlocally in the state to be teleported. The photons A and B are directed onto a polarizing beam splitter to perform the BSM. Fiber-coupled single-photon detectors monitor the two outputs of the polarizing beam splitter.

Coincident photons detections at D_1 and D_2 project the photons A and B onto a $|\Psi\rangle$ state in which they have different polarizations. As the sender's ancillary photon B and the receiver's photon C are also entangled, this detection collapses the receiver's photon into an identical replica of the original photon.

The node operation is set up in such a way that that detectors D_1, D_2, and D_T all register photons at the same time when teleportation takes place. This design of the node allows the teleportation of the required polarization state of photon A to photon C without actually having to detect it. In this design of the node, the detection of the teleported quantum state becomes obsolete, thus enabling the subsequent manipulation (or the next teleportation) of the teleported qubit. Such a procedure allows the teleported photons to be used again, e.g., as "quantum repeaters" in long-distance communication links.

13.2.6.3 Time-Bin Qubit Quantum Node

This type of quantum information transmission is robust for long-distance links in optical fibers because it is insensitive to polarization fluctuations and polarization mode dispersion. Such a time-bin qubit quantum node was designed by a team of scientists from Vienna University [13.1]. They used time-bin qubits

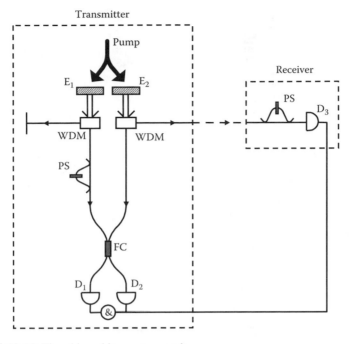

FIGURE 13.18 Time-bin qubit quantum node.

carried by photons at 1310 nm that were teleported onto a photon at the telecom wavelength of 1550 nm. The total length of the link was 6 km.

A schematic design of the node is presented in Figure 13.18. A laser beam is split into two parts that are used to pump two entanglers E_1 and E_2. The entangler E_1 is used to produce the qubit to be teleported, and the entangler E_2 is used to generate the entangled qubit pair.

The entangler E_1 produces a pair of twin photons at telecom wavelengths (1310 and 1550 nm). The twin photons are collimated into an optical fiber and then separated by a wavelength-division multiplexer. The 1550-nm photon is discarded, and the 1310-nm photon is sent to the phase shifter (PS) in which the time-bin qubits to be teleported are created. The entangler E_2 is used to produce nondegenerate entangled time-bin qubits. Then, the 1310-nm photon is sent to the fiber coupler (FC) in which a partial BSM is performed. The 1550-nm photon is sent to the receiving station that is connected with the transmitter via an optical fiber.

The receiver analyses the photon to verify that indeed the state encoded by the sender has been teleported.

13.2.6.4 Microphotonic Quantum Nodes

Figure 13.19 presents a conceptual design of a waveguide-based quantum node. This concept is based on PPLN-waveguide entanglers (E_1 and E_2) and time-bin photonic qubits are used to teleport information.

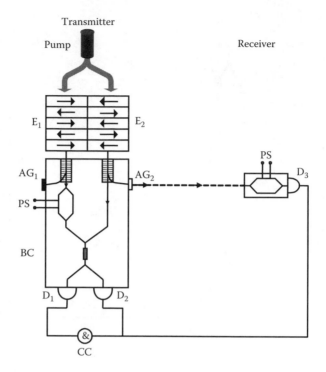

FIGURE 13.19 Waveguide-based quantum node.

A laser diode is used to pump the PPLN entanglers. As in the previously described node, one of the PPLN entanglers is used to produce the qubit to be teleported, whereas the other is used to generate the entangled qubit pair.

The entangler outputs are coupled to two electroacoustic gratings (AG_1 and AG_2) that act as wavelength-division multiplexers. One of the multiplexers (AG_1) separates the twin photons from the entangler E_1. One of the photons is discarded, and the other photon is sent to a waveguide-based phase shifter (PS) in which the time-bin qubits to be teleported are created. The output of the second entangler (E_2) is used to produce nondegenerate entangled time-bin qubits. One of the photons is sent to a waveguide-based beam coupler (BC) in which a partial BSM is performed. The other photon is sent to the receiver that is connected with the transmitter via an optical fiber.

The transmitter outputs and the receiver output are directed to avalanche photodiodes (D_1, D_2, and D_3). The coincidence rate is obtained by using a coincidence counter (CC). The receiver verifies the state that was encoded by the sender to complete the teleportation.

It is expected that any future commercial implementation of quantum photonic systems will have to be based on waveguides or PBG structures. Only these microphotonic approaches will provide the full benefit that quantum photonic technology offers.

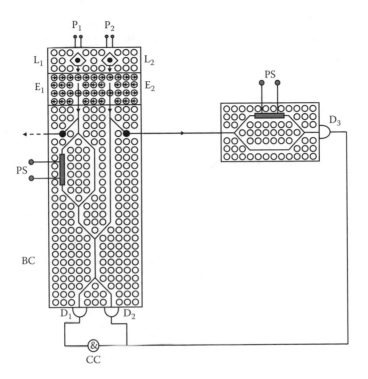

FIGURE 13.20 PBG-based quantum node.

Figure 13.20 illustrates a conceptual design of a quantum node that is based on a PBG structure. This design is a PBG equivalent to the waveguide-based node that is shown in Figure 13.19.

A pair of zero-threshold lasers (L_1 and L_2) is built into a PBG structure (see Chapter 3). These lasers are pumped by external sources (P_1 and P_2).

Similarly, to the previously described node, a couple of PPLN-type entanglers (E_1 and E_2) are implemented into the PBG structure. One of them (E_1) is used to produce the qubit to be teleported, whereas the other (E_2) generates the entangled qubit pair.

The entangler outputs are coupled to PBG waveguiding channels. Each channel is connected to wavelength-selective multiplexers. One of the multiplexers separates the twin photons from the entangler E_1. One of the photons is discarded, and the other photon is sent to a PBG-based phase shifter (*PS*) in which the time-bin qubits to be teleported are created. The output of the second entangler (E_2) is used to produce nondegenerate entangled time-bin qubits. One of the photons is sent to a PBG-based beam coupler (*BC*) in which a partial Bell-state measurement is performed. The other photon is sent to the receiver that is connected with the transmitter via an optical fiber. A PBG-based phase shifter (*PS*) is used to carry out the measurement at the receiver location. The phase shifter output is connected to the single-photon detector D_3.

The transmitter and receiver outputs are directed to a single-photon avalanche photodiodes (D_1, D_2, and D_3). The coincidence rate is obtained by using a coincidence counter (CC).

13.2.7 REPEATERS

There are two main issues that affect distributing of entangled photons over long distance: (1) photon absorption in transmission channels and (2) depolarization of photonic qubits because of decoherence. For optical channels such as optical fibers and waveguides, the probability for both absorption and depolarization of photonic qubits grows exponentially with the length of the channel.

One way of solving these problems is to introduce a quantum repeater. The function of the repeater in the quantum link is to enhance a signal while preserving its quantum nature. This can be achieved by implementing multistage entanglement that enables teleportation of an arbitrary quantum state.

A transmission channel may be divided into several segments ($T - S_1$, $S_2 - S_3$, ... , $S_N - R$) whose lengths are determined by absorption and decoherence losses (see Figure 13.21). T and R in Figure 13.21 indicate the sender and the receiver, respectively. Quantum nodes may be introduced between segments. Making the BSM can subsequently connect the segment pairs. Such a process is known as *entanglement swapping* (see Subsection 13.3.3). A number of entangled pairs of photons are then created between nodes S_1 and S_2, S_3 and S_4, ... , S_N and S_R.

A schematic illustration of experimental setup that can be used to generate entanglement swapping is illustrated in Figure 13.22. Entanglement sources E_1 and E_2 are used to generate pairs of entangled photons. Photons S_1 and S_2 may be entangled together by performing the BSM.

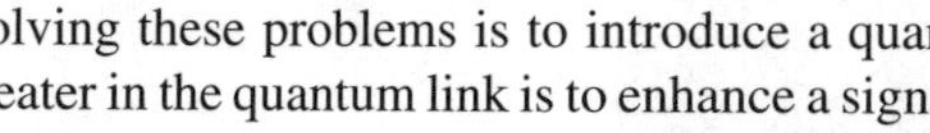

$$T \qquad S_1 \; S_2 \qquad\quad S_3 \qquad\qquad S_N \qquad\quad R$$

FIGURE 13.21 Quantum communication link with repeaters.

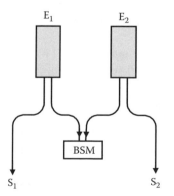

FIGURE 13.22 Entanglement swapper.

13.3 QUANTUM COMPUTERS

Conventional computers work with logic gates. Any arithmetic task can be achieved using a combination of logic gates. The presence of two states is required to provide the basic element of the binary number system on which conventional computers operate. The states on and off voltages provide such a requirement for a conventional computer. For example, an AND gate takes two inputs and the output is 1 if both inputs are 1, otherwise the output is 0. Specifically, any function on classical bits can be computed from the composition of the universal gate, which is a combination of gates NOT and AND (NAND).

On the other hand, a particle — according to quantum mechanics — is effectively in all possible states that are called *superposition* (see Section 13.2.1.1).

It is only when this particle interacts with another particle that its state becomes fixed. The act of measuring the state of a particle is an interaction that fixes its state. However, until the measurement is taken, this particle could be in any one of the possible states. It turns out that quantum particles can be manipulated in ways similar to the logic found in conventional computers. Any transformation on qubits can be done from composition of any two quantum gates.

The nature of a quantum computer is such that it will calculate the results for all possible combinations of input, and this is done in parallel using just one sequence of interactions between particles. Thus, the uncertainty about the state of individual particles is the strength of the quantum computer. As the state of a particle represents a bit of information, the uncertainty about its state also applies to the qubit.

One can think of qubits working their way through the quantum photonic computer, and at the end of the calculation all possible answers are available. The last step is to convert qubits back to bits in order to retrieve the required information. This involves measuring the states of qubits. It is interesting to note that any attempt to do this measurement during the calculation process would destroy the qubits, and the calculation would be disrupted.

It is expected that several classes of so-far unsolvable computing problems could be addressed by quantum computing. One of them is the task of finding the factors of a large number. This task is of great interest for computer security because factors are used for encryption systems. In brief, a binary number consisting of 50 digits has 2^{50} combinations, and it is estimated that the most powerful conventional computer available today would need over a million years to do this. In theory, a quantum computer could solve this task with just a few hundred qubits. A similar efficiency could apply to finding an item in a randomly sorted database, debugging software, etc.

This may be illustrated by an example of two qubits. A system consisting of two horizontally and vertically polarized photonic qubits may be described by four possible states: both vertically polarized (1,1), both horizontally polarized (0,0), and two combinations of one vertical and one horizontal, (0,1) and (1,0). Quantum physics rules allow assigning any probability weight for each combination, as long as the total probability adds up to 100%. In practice, this means

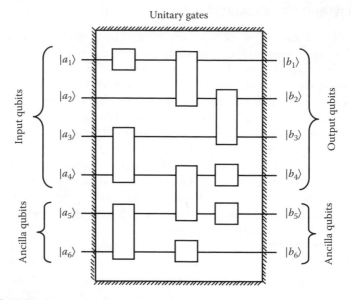

FIGURE 13.23 Schematic of a quantum computer.

that three numbers are needed to completely describe the system, and the fourth number is defined by the 100% constrain. In a similar manner, it may be shown that there are 7 numbers needed to fully describe 3 qubits system, 15 numbers for 4 qubits system, 31 for 5 qubit system, and so on. It is quite clear that the complexity of the quantum system quickly becomes incredibly large. A relatively simple system consisting 100 photonic qubits would require over 10^{31} numbers to fully describe its quantum states. This is many trillion times more then the storage capacity of all computers ever made.

A schematic picture of a quantum computer is shown in Figure 13.23. A set of qubits forms a quantum state input. The input is interfaced with a network of logical gates that perform one- and two-qubit transformations. The network operates on the input qubits as well as on a number of additional (ancilla) qubits to generate an output state. In Figure 13.23, the boxes represent various unitary gates.

13.3.1 COMPUTER BUILDING BLOCKS

It has been demonstrated that nondeterministic quantum logic operations can be performed using schemes that include: (1) linear optical elements, (2) additional photons (ancilla), and (3) postselection based on the output of single-photon detectors [13.3].

This idea of linear optical elements has lead to the concept of LOQC. LOQC has been exploited in a number of schemes that may be used for computation and generation of multiphoton-entangled states. These schemes are probabilistic and employ the so-called postselection techniques, i.e., they use the detection to

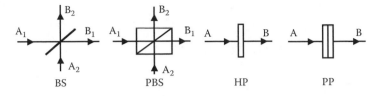

FIGURE 13.24 Basic LOQC elements.

indicate if and when the correct operation has taken place. Also, an extra photon (an ancilla) may be used as a control qubit to perform some unitary transformations on the target qubits. In all systems presented in the following text, it has been assumed that the polarization states of photons are implemented as qubits.

13.3.2 LOQC

A crucial element in the schemes of LOQC is classical feed-forwardability, i.e., it must be possible to detect when the gate has succeeded by performing some appropriate measurements on ancilla photons. This information can then be feed-forwarded for conditional future operations on the photonic qubits.

There is a set of basic elements that are sufficient for practical implementation of LOQC networks. This set includes: (1) a 50/50 beam splitter (*BS*), (2) a polarizing beam splitter (*PBS*), (3) half-wave plates (*HP*), and (4) phase-shifting plates (*PP*). These elements may be used to redirect single photons to the output ports (see Figure 13.24).

The performance of each of these basic elements may be described as follows:

- A 50/50 beam splitter redirects each photon and adds a phase factor i to the reflected beam:
  1. If a photon is directed to port A_1, then

$$a|H\rangle_{A1} + b|V\rangle_{A1} \rightarrow \frac{1}{\sqrt{2}}\left(a|H\rangle_{B1} + b|V\rangle_{B1}\right) + \frac{1}{\sqrt{2}}i\left(a|H\rangle_{B2} + b|V\rangle_{B2}\right)$$

  2. If a photon is directed to port A_2, then

$$a|H\rangle_{A2} + b|V\rangle_{A2} \rightarrow \frac{1}{\sqrt{2}}i\left(a|H\rangle_{B1} + b|V\rangle_{B1}\right) + \frac{1}{\sqrt{2}}\left(a|H\rangle_{B2} + b|V\rangle_{B2}\right)$$

- A polarizing beam splitter reflects the vertically polarized photons without affecting the horizontally polarized photons:
  1. If a photon is directed to port A_1, then:

$$a|H\rangle_{A1} + b|V\rangle_{A1} \rightarrow a|H\rangle_{B1} + b|V\rangle_{B2}$$

2. If a photon is directed to port A_2, then:

$$a\big|H\big\rangle_{A2} + b\big|V\big\rangle_{A2} \rightarrow a\big|H\big\rangle_{B2} + b\big|V\big\rangle_{B1}$$

- A half-wave plate changes polarization:

$$\frac{1}{\sqrt{2}}\left(a\big|H\big\rangle_A + b\big|V\big\rangle_A\right) \equiv \big|+\big\rangle_A \rightarrow \big|V\big\rangle_B$$

$$\frac{1}{\sqrt{2}}\left(a\big|H\big\rangle_A - b\big|V\big\rangle_A\right) \equiv \big|-\big\rangle_A \rightarrow \big|H\big\rangle_B$$

- A phase-shifting plate adds a phase shift:

$$a\big|H\big\rangle_A + b\big|V\big\rangle_A \rightarrow e^{i\theta}\left(a\big|H\big\rangle_B + b\big|V\big\rangle_B\right)$$

13.3.3 HOM Interferometer

A HOM interferometer described in Subsection 13.2.3 plays a crucial role in the LOQC schemes and is itself an example of a postselected multiqubit operation. It may be used to produce entangled states, to make probabilistic BSMs for quantum teleportation and entanglement swapping, and to test indistinguishability of consecutive photons from a single-photon source.

As shown in Subsection 13.2.3, two photons entering a 50/50 beam splitter can leave by different output ports only if they are in some way distinguishable. In the case of photonic qubits, they are usually indistinguishable in wavelength and spatial mode. Therefore, the polarization or their arrival time at the beam splitter is used to distinguish them.

The aforementioned properties of LOQC elements may be used to analyze performance of a HOM interferometer that is shown in Figure 13.25. The interferometer consists of two beam splitters, two mirrors (M), and a detector.

A single photon in $\big|\Psi\big\rangle$ state is injected into port A_1 of the experimental setup shown in Figure 13.25. The port A_2 is empty. It may be shown that, by applying the aforementioned rules, the initial state of the photon, after passing through the first beam splitter and being reflected from two mirrors, will be transformed into

$$\big|\Psi\big\rangle_{A1} \rightarrow \frac{1}{\sqrt{2}} i\left(\big|\Psi\big\rangle_{B1} + ie^{i\theta}\big|\Psi\big\rangle_{B2}\right)$$

After passing through the second beam splitter, the photon will be transformed into

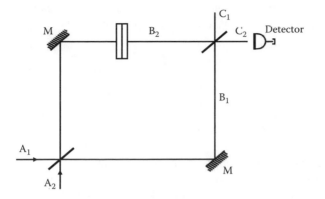

FIGURE 13.25 HOM interferometer.

$$|\Psi\rangle_{A1} \rightarrow \frac{1}{2}i\left(1-e^{i\theta}\right)|\Psi\rangle_{C1} - \frac{1}{2}\left(1+e^{i\theta}\right)|\Psi\rangle_{C2}$$

One may determine the probability P_{C2} of finding a photon in output C_2:

$$P_{C2} = \left\|\frac{1}{2}\left(1+e^{i\theta}\right)|\Psi\rangle_{C2}\right\|^2 = \frac{1}{4}\left|1+e^{i\theta}\right|^2 = \cos^2\left(\frac{1}{2}\theta\right)$$

Similarly, it can be shown that two photons simultaneously entering a beam splitter through two different inputs would always exit the setup through the same output port. This effect of vanishing probability for two photons in different outputs is known as the HOM dip. This is in agreement with the bunching effect that is described in Subsection 13.2.3.

13.3.4 BEAM SPLITTER ENTANGLER

Another example of a function that can be realized with LOQC elements is a splitter-based entangler.

In this application two photons are entering a 50/50 beam splitter, such as that shown in Figure 13.24. If one assumes that one of the photons is prepared in $|H\rangle$ state and the other photon is in $|V\rangle$ state, then it may be shown by using the aforementioned beam splitter transformation characteristics that

$$|H\rangle_{A1}|V\rangle_{A2} \rightarrow \frac{1}{2}\left(|H\rangle_{B1} + i|H\rangle_{B2}\right)\left(|V\rangle_{B2} + i|V\rangle_{B1}\right)$$

$$= \frac{1}{2}\left(|H\rangle_{B1}|V\rangle_{B2} - |H\rangle_{B2}|V\rangle_{B1}\right) + \frac{1}{2}i\left(|H\rangle_{B1}|V\rangle_{B1} + |H\rangle_{B2}|V\rangle_{B2}\right)$$

If only one photon per output port is detected, then the photon state will be projected onto one of the maximally entangled Bell states:

$$\left|H_{A1}\right\rangle\left|V\right\rangle_{A2} \rightarrow \frac{1}{2}\left(\left|H\right\rangle_{B1}\left|V\right\rangle_{B2} - \left|H\right\rangle_{B2}\left|V\right\rangle_{B1}\right)$$

In this particular case, the interferometer acts as a filter for the Bell state $\left|\Psi^{-}\right\rangle$ in which the photons have orthogonal polarizations.

The preceding relationship can be experimentally verified by correlating one of the polarization measurements performed in output ports B_1 and B_2. Experiments show that detecting a photon in the $\left|H\right\rangle$ state in one port is always correlated to the detection of a $\left|V\right\rangle$ photon in the other output port.

13.3.5 QUANTUM PARITY CHECKER

It has been demonstrated that polarization-encoded qubits and postselection could be used to implement a number of quantum logic operations such as a quantum-parity-check and a destructive controlled-NOT (CNOT) gate.

In quantum logic devices there are two inputs and two outputs for the processed photons, i.e., two qubits in the form of two single photons are incident at two inputs.

The function of a quantum parity checker is to transfer the value of the target qubit to the output, provided that its value is the same as that of the control qubit. If these qubits have different values, then the device will produce no output [13.4].

The parity-check operation can be understood from the basic properties of a polarizing beam splitter (PBS). If only one photon is to be detected, both of the incident photons must be transmitted or both must be reflected. In either case, the polarizations and the corresponding values of the qubits must be the same. This ability to compare the polarizations of two photons has been proposed for use in entanglement purification and a variety of other quantum information processing applications.

A polarizing beam splitter is the main component of the quantum parity checker [13.4]. It transmits horizontally polarized photons and reflects vertically polarized photons (see Figure 13.26).

A polarization-sensitive detector in output path B_2 detects any photons in either the horizontal–vertical (H/V) computational basis or another basis rotated by 45°. Two single-photon detectors may be used together with a rotating beam splitter to implement a polarization-sensitive detecting module. The output of the devices is accepted only for those cases in which the detector registers one and only one photon. Table 13.3 presents the truth table for the operation of a parity checker.

Quantum-computing applications require that they be performed without measuring or determining the values of either qubit. For the quantum parity checker illustrated in Figure 13.26, this is accomplished by orienting the

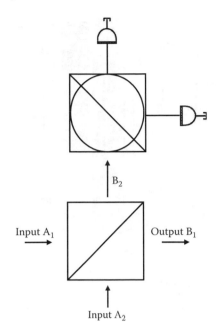

FIGURE 13.26 Quantum parity checker.

TABLE 13.3
Truth Table for Parity Checker

Input A₁ (Control)	Input A₂ (Target)	Output B₁ (Control)	Output B₂ (Target)
$\lvert H \rangle$	$\lvert H \rangle$	1 photon	$\lvert H \rangle$
$\lvert H \rangle$	$\lvert V \rangle$	0 or 2 photons	No defined output
$\lvert V \rangle$	$\lvert H \rangle$	0 or 2 photons	No defined output
$\lvert V \rangle$	$\lvert V \rangle$	1 photon	$\lvert V \rangle$

polarization-sensitive detecting module in a basis rotated by 45°, which essentially converts linearly polarized photons into circularly polarized ones. In this way it erases any information regarding the value of either of the input qubits. This type of postselection technique process provides the nonlinearity that is required for the logic operations as the detection process itself is inherently nonlinear [13.4].

When the inputs consist of an arbitrary superposition of states, this quantum erasure technique combined with the postselection process maintains the required coherence of the probability amplitudes.

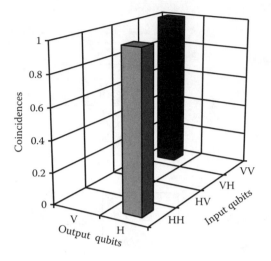

FIGURE 13.27 Performance data for a parity checker.

An example of expected data that characterize the performance of the quantum parity checker is graphically illustrated in Figure 13.27, which shows the coincidence events corresponding to *H* or *V* for all possible values of the input qubits.

13.3.6 CNOT GATE

As described in Chapter 3, a CNOT gate for polarization-entangled qubits flips the target bit if and only if the control bit has the logical value *V* while the control bit remains unaffected.

The CNOT gate along with one-qubit rotations constitutes a universal set of gates in quantum computing. It turns out that any quantum logical circuit can be designed by using one-qubit rotations and a CNOT operation.

The truth table for a CNOT is summarized in Table 13.4 (this table is a modification of Table 3.1).

The design of the CNOT gate was first proposed in 2001 [13.5]. A schematic illustration of a quantum photonic CNOT gate design is shown in Figure 13.28.

The implementation of the destructive CNOT gate shown in Figure 13.28 is similar to the quantum parity checker, except that the polarizing beam splitter is oriented in a basis rotated by 45° and the polarization-sensitive detecting module is oriented in the H/V basis. The goal of the destructive CNOT gate is to flip the logical value of the target qubit (e.g., H ↔ V) if the control qubit is vertically polarized and to transfer it without a change if the control qubit is horizontally polarized. The output of the gate shown in Figure 13.28 is that of a conventional CNOT, except that the polarization-sensitive detector destroys the information contained in the control qubit. Therefore, this design does not allow for feed-forwarding. That is why this design is called a *destructive CNOT gate*.

TABLE 13.4
Truth Table for CNOT Gate

Input A_1 (Control)	Input A_2 (Target)	Output B_1 (Control)	Output B_2 (Target)
$|H\rangle$	$|H\rangle$	$|H\rangle$	$|H\rangle$
$|H\rangle$	$|V\rangle$	$|H\rangle$	$|V\rangle$
$|V\rangle$	$|H\rangle$	$|V\rangle$	$|V\rangle$
$|V\rangle$	$|V\rangle$	$|V\rangle$	$|H\rangle$

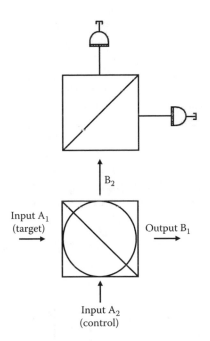

FIGURE 13.28 Destructive CNOT gate.

The operation of the CNOT gate can be understood by projecting the input qubit polarization states into the 45° basis of the polarizing beam splitter and then reprojecting the output states into the H/V basis of the detector.

It can be shown by using the preceding characteristics of the polarizing beam splitter and a phase-shifting plate that an arbitrary target state will be flipped if the control photon in the input 2 is vertically polarized. However, it will be unaltered if the control photon in the input 2 is horizontally polarized. This may be described analytically, e.g., for the control qubit that is vertically polarized ($|V\rangle_2$):

$$|\Psi\rangle_{1'} = a|H\rangle_1 + b|V\rangle_1 \rightarrow a|V\rangle_{1'} + b|H\rangle_{1'}$$

and for the control qubit that is horizontally polarized ($|H\rangle_2$):

$$|\Psi\rangle_{1'} = a|H\rangle_1 + b|V\rangle_1 \rightarrow a|H\rangle_{1'} + b|V\rangle_{1'}$$

Another version of a CNOT gate has been suggested [13.5] that is based on a destructive CNOT gate and the quantum parity checker combined with a pair of entangled photons. This design allows for implementation of a nondestructive (conventional) CNOT gate that satisfies the feed-forwarding criterion.

The experimental implementations of a CNOT gate that operates on two polarization qubits carried by independent photons were reported in 2004 and 2005 [13.6, 13.7]. The experimental scheme illustrated in Figure 13.29 combines a destructive CNOT gate and an encoder. The encoder consists of the polarizing beam splitter PBS_1 and the entangler generating photons A_3 and A_4; its function is to encode the control qubit in the channels A_4 and B_1 [13.6].

As shown in Figure 13.29, a nondestructive CNOT gate for photons A_1 (control qubit) and A_2 (target qubit) is realized by performing a quantum parity check on photons A_1 and A_3 and a destructive CNOT operation on photons A_2 and A_4. The polarizing beam splitter PBS_2 is rotated by 45°.

The output qubits are delivered through spatial modes B_1 (control qubit) and B_2 (target qubit). Consider a particular case when the ancilla photons are in the maximally entangled Bell state $|\Phi^+\rangle$:

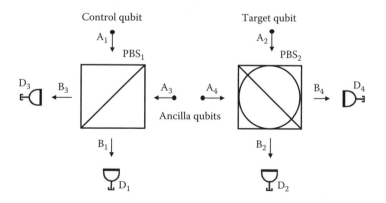

FIGURE 13.29 Nondestructive CNOT gate.

$$\left|\Phi^+_{A_3A_4}\right\rangle = \frac{1}{\sqrt{2}}\left(\left|H\right\rangle_{A_3}\left|H\right\rangle_{A_4} + \left|V\right\rangle_{A_3}\left|V\right\rangle_{A_4}\right)$$

As mentioned previously, this scheme works in those cases in which only one photon is found in each of the channels B_3 and B_4.

The most general form of the input state for the design illustrated in Figure 13.29 may be given by

$$\left|\Psi_{A_1A_2}\right\rangle = \left|H\right\rangle_{A_1}(a_1\left|H\right\rangle_{A_2} + a_2\left|V\right\rangle_{A_2}) + \left|V\right\rangle_{A_1}(a_3\left|H\right\rangle_{A_2} + a_4\left|V\right\rangle_{A_2})$$

For the gate to work properly, the preceding state should be converted into the following output state:

$$\left|\Psi_{A_1A_2}\right\rangle \rightarrow \left|H\right\rangle_{A_1}(a_1\left|H\right\rangle_{A_2} + a_2\left|V\right\rangle_{A_2}) + \left|V\right\rangle_{A_1}(a_3\left|V\right\rangle_{A_2} + a_4\left|H\right\rangle_{A_2})$$

Because the polarizing beam splitter that transmits horizontally polarized photons and reflects vertically polarized photons, the successful detection at the port D_3 of the state $\left|+\right\rangle_{A_1} \equiv \left|H\right\rangle_{A_1} + \left|V\right\rangle_{A_1}$ postselects the following transformation of the arbitrary input channel A_1:

$$a\left|H\right\rangle_{A_1} + b\left|V\right\rangle_{A_1} \rightarrow a\left|HH\right\rangle_{A_4B_1} + b\left|VV\right\rangle_{A_4B_1}$$

The control photon that is horizontally polarized travels through the PBS_1 to the channel B_1. As required from the truth table, this photon will be horizontally polarized at the output B_1. In order for the scheme to work, a photon needs to arrive also at the detector D_3. This additional photon comes from the ancilla pair, because the input control qubit is already in the channel B_1. This photon in the channel B_3 is also H polarized as it has been passed through the polarizing beam splitter PBS_1. The ancilla photons A_3 and A_4 are entangled, and therefore the photon in A_4 is also horizontally polarized. Taking into account the $-45°$ rotation of the polarization in the A_2 and A_4 paths, the input in the PBS_2 is in the state $\left|HH\right\rangle_{A_2A_4}$ for a target qubit horizontally polarized or in the state $\left|VH\right\rangle_{A_2A_4}$ for a target qubit vertically polarized. This input state is transformed by PBS_2, and the subsequent change at the H/V basis of the detector into the state $(\left|HH\right\rangle \pm \left|VV\right\rangle)_{B_2B_4}$ or $(\left|HV\right\rangle \pm \left|VH\right\rangle)_{B_2B_4}$. The detection of a horizontally polarized photon in the detector D_4 will indicate that:

- A horizontally polarized photon is in the channel B_2 if the target qubit A_2 was horizontally polarized.
- A vertically polarized photon is in the channel B_2 if the target qubit A_2 was vertically polarized.

Therefore, the preceding result is in compliance with the requirements for a CNOT gate.

The performance of the system may be summarized as follows: (1) the control qubit is encoded in photons A_4 and in B_1, (2) the photon in channel A_4 is the control input to the destructive CNOT gate, (3) the second photon in channel B_1 is the output control qubit, and (4) the nondestructive CNOT gate for photons A_1 and A_2 is accomplished on the condition of detecting a $|H\rangle$ photon in channel B_3 and a $|H\rangle$ photon in channel B_4.

The preceding system and an analysis of its performance allows for a conceptual design of a truly microphotonic CNOT gate that is based on a PBG structure. A very preliminary concept of such a gate is schematically illustrated in Figure 13.30.

An entangler (in the center) is used to produce both the ancilla pair A_3, A_4 and the two input qubits A_1 (control qubit) and A_2 (target qubit). Phase shifters are placed in the paths of all four photons to control the required polarization and the phase shifts ($\Delta\Theta$). These phase shifters are used to prepare the required input states — they rotate the polarization and compensate the birefringence effects introduced along the traveling paths. The same phase shifters may be also used to adjust the phase between the entangled photons to produce the required Bell state.

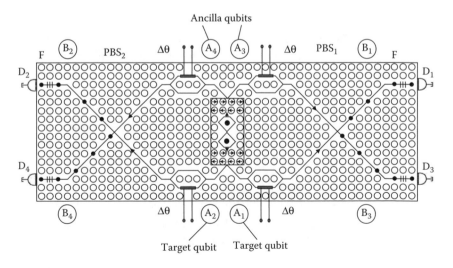

FIGURE 13.30 PBG-based CNOT gate.

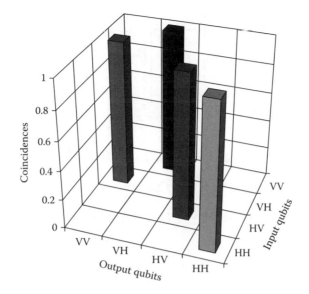

FIGURE 13.31 Experimental verification template data for a CNOT gate.

Beam splitting is implemented by using different defect patterns of the PBG structure. These defect patterns are used to form two polarized beam splitters PBS_1 and PBS_2.

The indistinguishability between the overlapping photons may be further improved by introducing narrow bandwidth spectral PBG filters at the outputs (F). Fiber-coupled detectors (D) monitor the outputs. A fourfold coincidence detection protocol may be used to confirm that photons actually arrive in the output B_1 and B_2.

Figure 13.31 illustrates theoretical results that are expected for the aforementioned CNOT gates. A set of input qubits would have to be prepared in four states HH, HV, VH, and VV to experimentally demonstrate the gate performance. The gate would have to perform according to the data shown in Figure 13.31 to confirm the desired characteristics.

This graph shows coincidences for all the possible (16) combinations of inputs and outputs of V- and H-polarized photonic qubits. It can be seen that when the control qubit is in the logical value H, then the gate works as the identity gate. When the control qubit is in the logical value V, then the gate works as a NOT gate, flipping the second input qubit.

It can also be demonstrated that the gate also works with superimposed states. The special case in which the control input is a 45° polarized photon and the target qubit is an H photon is particularly interesting. In this case, the state $\left| H+V \right\rangle_{A_1} \left| H \right\rangle_{A_2}$ evolves into the maximally entangled state $\left| HH \right\rangle_{B_1 B_2} + \left| VV \right\rangle_{B_1 B_2}$. This shows that CNOT gates can transform separable states into entangled states and vice versa.

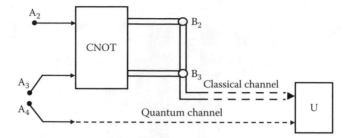

FIGURE 13.32 Quantum photonic teleportation circuit.

13.3.7 Teleportation Photonic Circuit

Another important characteristic of the nondestructive CNOT gate may be applied to teleportation. It has been shown that the gate can be used to simultaneously identify all four Bell states in a quantum teleportation protocol [13.7]. A schematic illustration of a quantum photonic teleportation circuit is shown in Figure 13.32.

A photon A_2 is to be teleported by the sender. This photon is in an unknown polarization state $\left|\Phi\right\rangle_{A2} = a\left|H\right\rangle_{A2} + b\left|V\right\rangle_{A2}$. A pair of photons A_3 and A_4 is generated in an entangled state $\left|\Psi^-\right\rangle_{A3A4}$. In order to implement the teleportation process, it is necessary to discriminate the four Bell states of photons A_2 and A_3, i.e., $\left|\Psi^\pm\right\rangle_{A3A2}$ and $\left|\Phi^\pm\right\rangle_{A3A2}$.

After passing through the CNOT gate, the four Bell states will be transformed into one of the four orthogonal separable states:

$$\left|\Psi^\pm\right\rangle_{A3A2} \rightarrow \left|\pm\right\rangle_{B3}\left|H\right\rangle_{B2}$$

$$\left|\Phi^\pm\right\rangle_{A3A2} \rightarrow \left|\pm\right\rangle_{B3}\left|V\right\rangle_{B2}$$

Therefore, applying a nondestructive CNOT and performing a subsequent polarization analysis on photons A_2 and A_3 can result in the BSMs. Depending on the sender's measurement results, the receiver can then perform a unitary transformation on photon A_4 to convert its state into the initial state of photon A_2.

13.4 QUANTUM CRYPTOGRAPHY

Quantum photonics allows the construction of qualitatively new types of logic gates that may form a basis for a new type of absolutely secure cryptosystems.

Quantum cryptography is an extension of quantum communication as it is based on the same quantum effects. Unlike traditional cryptography, which employs various mathematical techniques to restrict eavesdroppers from intercepting the contents of encrypted messages, quantum cryptography uses such effects as superposition and teleportation. The process of sending and storing information is carried out by photons in optical fibers or waveguides. In this context, eavesdropping can be seen as photonic measurements that disturb the information carrier. Therefore, any attempt of eavesdropping leaves traces that can always be detected.

Two or more parties that exchange sensitive information over an insecure communication channel must use a cryptographic protocol to ensure that communications are secure. The sending party has to use a cryptographic key to encode the information (encryption), and the receiving party has to use a key to decode the information (decryption). If a third party acquires the decoding key, then this may lead to unauthorized interception of sensitive information.

In today's implemented protocols there are two main ways to distribute the cryptographic keys — private and public. The parties have to share a secret key before they send and receive a message in private (symmetrical) key cryptosystems. A cryptographic algorithm that uses a key that is randomly generated every time and used only once is called the *one-time pad* or *Vernam cipher*. It is believed that this is the only currently used cryptosystem today that is truly secure. The usefulness of private key cryptosystems is limited by the need for secure key distribution, either through a trusted communication channel or courier. Security is limited by the fact that classical communication channels are used to exchange a private key. Therefore, an eavesdropping third party may be able to copy the key without being detected.

A public key is accessible by anyone. The security of the asymmetrical system relies on computational complexity that grows exponentially with the number of bits in the key.

The key distribution scheme is based on the intractability of the factorization of large numbers.

For example, the RSA system was the first algorithm known to be suitable for encryption. This algorithm was proposed by Ron Rivest, Adi Shamir, and Len Adleman from MIT in 1977. The letters RSA are the first letters of their surnames. A message encrypted with RSA can be deciphered by factoring the public N-numbers key, which is the product of two prime numbers. Known classical algorithms cannot do this in time, and they quickly become infeasible as the number N is increased.

The RSA system is still widely used in electronic commerce protocols, and it is believed to be secure if sufficiently long N-numbers keys are used. The security of such systems relies on the assumption of limited processing power of the currently available computers.

Quantum key distribution (QKD) takes advantage of the basic quantum principles that the measurement of a quantum state of a system changes its state and that unknown quantum states cannot be copied. When exchanging quantum information, the two communicating parties are able to determine if a third party has compromised the quantum channel before they start the key transmission. They repeat the test process until they find a secure quantum channel over which they can safely exchange the secret key. Thus, the implementation of Vernam ciphers using QKD will make a truly unbreakable practical cryptosystem for transmitting classical messages, combining the unbreakability of the one-time pad with the ease of distribution of public key systems [13.8].

As computing power increases, and new classical computational techniques are developed, the length of time that a message can be considered secure will decrease, and numerical keys will no longer be able to provide acceptable levels of secure communication.

An additional threat to existing public key cryptosystems is the so-called Shor's algorithm. Shor's algorithm relies on the laws of quantum physics to provide enhanced capabilities for a class of problems that includes number factoring. Similar to all quantum computer algorithms, Shor's algorithm is probabilistic one, i.e., it gives the correct answer and the probability of failure decreases as the algorithm is repeated. It is expected that Shor's algorithm, when executed on a quantum computer, will be able to factor large numbers exponentially faster than a conventional computer.

Therefore, it is believed that the implication of quantum computers for secure communications is twofold: (1) Shor's algorithm makes the widely used RSA encryption scheme vulnerable; (2) QKD can counter its effects by offering truly secure communication between two parties.

A quantum network provides a way of distributing private keys in conjunction with public or private networks employing internet protocol suites. To incorporate QKD into an Internet security protocol, one must implement an interface between existing Internet protocols and QKD channel protocols used for key distribution. For example, in implementations of point-to-point QKD, two communicating parties must be connected by a direct optical link. This may be realized by either an optical fiber or direct free-space communication link (see Chapter 14).

A cryptographic link consists of a sender and a receiver connected by a "quantum channel." The throughput of the system depends on the quality of single-photon sources and detectors. The quantum channel carries information encoded in the state of a single photon as opposed to classical optical communications, in which many photons carry the same information.

The first quantum cryptography protocol that has been proposed in 1984 by Bennett and Brassard is known as BB84. Since then, a number of different protocols have been developed for secure quantum key distribution [13.8].

In the BB84 protocol, for each photon sent the sender chooses randomly a bit value and a basis in which that photon is prepared. There may be, for example, two bases for polarized photons: H/V basis and diagonal ($\pm$ 45°) basis. Photons may be in two states in each of the basis, i.e., *0* (for horizontally polarized photon in H/V basis and +45° in the diagonal basis) and *1* (for vertically polarized photon in the H/V basis and − 45° in the diagonal basis).

At the receiver end a detector is activated to record incoming photons. The receiver chooses randomly the basis that is used to analyze the incoming photons, i.e., it may be chosen between the H/V (+) basis and the diagonal basis (×). This is schematically illustrated in Figure 13.33.

Sender   Receiver

Basis ⊕

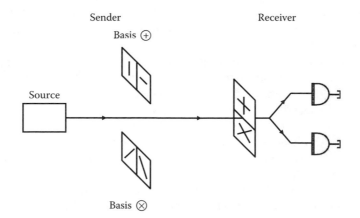

Basis ⊗

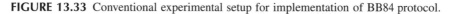

FIGURE 13.33 Conventional experimental setup for implementation of BB84 protocol.

The receiver records the result and the basis that was used in a successful detection. After sending a sufficiently large number of photons, the receiver publicly announces the cases in which photons were detected and the basis that were used for the measurements. However, the receiver does not reveal which results were recorded. The sender compares event by event to determine whether or not the receiver's analyzer was compatible to the sender's choice of bases. If they are incompatible or if the receiver failed to detect the photon, the bit is discarded. For the remaining bits, both the sender and the receiver know for sure that they have the same values. This set of bits forms the sifted key.

An example of such a protocol implementation is shown in Table 13.5. The horizontal (*H*) an vertical (*V*) basis is indicated by the sign (+); the diagonal +45 and −45 degree basis is marked by (×). Number (*1*) indicates the detection of the photon; (*0*) indicates the failure of the detection by a detector. The first row in

TABLE 13.5
Example of BB84 Protocol Application

Sender's encoder basis	×	+	+	×	+	×	×	+	×	+	+	×
Sender's photon polarization	+45	V	V	−45	H	+45	−45	H	+45	H	V	−45
Sender's bit sequence	1	1	1	0	0	1	0	0	1	0	1	0
Receiver's decoder basis	+	+	×	×	+	×	+	×	×	+	+	×
Receiver's photon polarization		V		−45	H	+45			+45	H	V	−45
Receiver's bit sequence		1		0	0	1			1	0	1	0
Sifted key		1		0	0	1			1	0	1	0

the table lists the sequence of basis that the sender uses. The second row shows the polarization orientation for each photon that was sent. The third row indicates the sender's bit sequence.

The bases that were used randomly by the receiver are listed in the fourth row. The polarization orientation that was recorded by the receiver is given in the fifth row. The sixth row shows the decoded bit sequence. This bit sequence allows for the identification of the sifted key.

The security of the key distribution is based on the fact that a measurement of an unknown quantum system will disturb the system: If the sender's and the receiver's sifted keys are perfectly correlated, it means that no eavesdropping has taken place and the key can be used for encoding a confidential message using the one-time pad. On the other hand, if the sifted keys are not 100% correlated, then, depending on the recorded quantum bit error rate (QBER), the sender and the receiver can either distill a secret key by error correction or the key is discarded and a new transmission is initiated.

The first experimental demonstration of quantum cryptography took place in 1989 at IBM where the so-called B92 protocol was realized. Since then, several experimental tests demonstrated that quantum cryptography is possible over distances of tens of kilometers. All experiments relied on strongly attenuated pulses that contained less than one photon on average to imitate single photons [13.9].

It is possible to use the quantum entanglers to generate keys at the sender's and the receiver's ends. Use of a quantum decoder and measuring the Bell states may identify the presence of an eavesdropper. A schematic of an experimental setup of a quantum cryptographer is shown in Figure 13.34. A quantum entangler is placed midway between the sender and the receiver. It generates pairs of photons with the same polarization. The sender and the receiver choose the basis of their polarization-sensitive detectors and record the results and measurement times. They exchange between themselves the information on the time and the

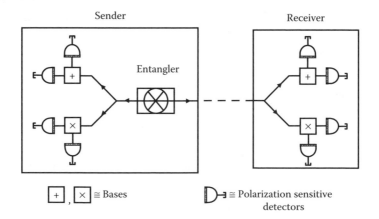

FIGURE 13.34 Photonic cryptographer.

basis that they used for each detected photon. The setup is similar to the one used for teleportation with the exception that the sender and the receiver each have to choose from two different bases.

Depending on the bases chosen for each specific photon pair, the measured data is either used to establish the sifted key or it is discarded [13.9].

Although classical communications, computing, and cryptography technologies will not vanish with the advent of quantum photonics, there is no doubt about the huge economic potential of this new technology. It is predicted that quantum cryptography will become commercially viable within a couple of years, whereas quantum communications will become a reality within a 10-year time frame.

REFERENCES

1. Aspelmeyer, M., Ursin, R., Zeilinger, A., De Riedmatten, H., Zbinden, H., and Gisin, N., Long-Distance Bell Inequality and Quantum Teleportation Field Trials, QuComm Consortium Report # IST-1999, 2003.
2. Hong C. K., Ou Z. Y., and Mandel, L., Measurement of subpicosecond time intervals between two photons by interference, *Phys. Rev. Lett.*, 59, 2044, 1987.
3. Knill, E., Laflamme, R., and Milburn, G.J., A scheme for efficient quantum computation with linear optics, *Nature* (London), 409, 46, 2001.
4. Pittman T.B., Jacobs B.C., and Franson J.D., Demonstration on nondeterministic quantum logic operations using linear optical elements, *Phys. Rev. Lett.*, 88, 257902, 2002.
5. Pittman, T.B., Jacobs, B.C., and Franson, J.D., Probabilistic quantum logic operations using polarizing beam splitters, *Phys. Rev.*, A 64, 062311, 2001.
6. Gasparoni, S., Pan, J. W., Walther, P., Rudolph, T., and Zeilinger, A., Realization of a photonic controlled-NOT gate sufficient for quantum computation, *Phys. Rev. Lett.*, 93, 020504, 2004.
7. Zhao, Z., Zhang, A.N., Chen, Y.A., Zhang, H., Du, J.F., Yang, T., and Pan, J.W., Experimental demonstration of a nondestructive controlled-NOT quantum gate for two independent photon qubits, *Phys. Rev. Lett.*, 94, 03501, 2005.
8. Curcic, T., Filipkowski, M.E., Chtchelkanova, A., D'Ambrosio, P.A., Wolf, S.A., Foster, M., and Cochran, D., Quantum Networks: From quantum cryptography to quantum architecture, *ACM SIGCOMM Computer Commun. Rev.*, 34, No. 5, 3, 2004.
9. Tittel, W. and Weihs, G., Photonic entanglement for fundamental tests and quantum communication, *Quantum Inf. Computation*, 1, 3–56, 2001.

14 Future Systems and Their Applications

14.1 MICROPHOTONICS IN SPACE

The main technology drivers for space systems are size, weight, and reliability. The costs of space systems are proportional to their size and mass.

As mentioned in the previous chapters, a new approach to microphotonics is based on a silicon-on-insulator (SOI) platform that utilizes novel concepts such as photonic-band-gap structures, thin-film smart materials, and microoptoelectro-mechanical systems (MOEMS). These technologies provide active and passive components that can form the basis of sophisticated microphotonic integrated circuits (micro-PICs) with higher functional densities than traditional integrated optics. Integration of a photonic system substantially minimizes the system mass and volume, reduces the number of external interconnects, and facilitates greater redundancy for higher system reliability, as well as providing robust optical alignment.

Since the early 1990s, a paradigm shift has occurred within space agencies regarding the primary measure by which spacecraft are judged, changing from purely performance to performance per unit cost. This paradigm shift, combined with the shrinking budget of various space agencies, has made smaller satellites, capable of accomplishing missions of slightly reduced scope at greatly reduced costs, very attractive.

This change in the overall approach to space programs has generated a strong interest in small satellites and satellite constellations. The trend in small satellites is towards generic multiuse platforms that can be reconfigured for various specific mission requirements. Today, all electronic devices and computer chips are housed in "black boxes" aboard the spacecraft and are connected to other boxes and actuators via cables. A *multifunctional structure* (MFS) is defined as a structure that integrates electronics and thermal and structural functions. In the MFS-integrated structure and electronics concept, all the electronic subsystems, including data transmission, command and data management, and power distribution and control, are directly integrated with the load-carrying thermal structure.

Distributed satellite systems are defined as constellations or clusters of satellites working together to achieve a mission. Distributed satellite systems appear suitable for several different types of missions: (1) global positioning and navigation, (2) global communication and information transfer, (3) global reconnaissance, and (4) space science [14.1]. It is expected that by distributing functions

TABLE 14.1
Projected Mission Requirements

Mission	Mass (kg)	Power (W)	Downlink Rate (Mb/sec)	Onboard Memory (MB)
Astronomy	640	243	0.5–2.0	100–300
Earth science	225	215	0.5–2.0	10–100
Solar physics	120	115	0.2–10.0	100–500
Interplanetary	300	257	0.05–0.1	500–3000
Communications	Constellations	Constellations	Application dependent	Application dependent
Geostationary	2062	4200	Application dependent	Application dependent
Low Earth orbit	585	2200	Application dependent	Application dependent

among a large group of satellites, mission risk and cost can be reduced while system reliability and adaptability can be increased.

Table 14.1 shows the projected mass and power requirements for some future satellites missions [14.1]. Design of scientific satellites is driven by the desire to obtain as much scientific data as possible under tight budget constraints. Commercial satellite designs (i.e., mostly used for communication applications), on the other hand, are driven by a profit motive and to serve as many customers and generate as much revenue as possible. It has been observed that the design of scientific and commercial satellites is diverging, with commercial satellites being projected as requiring the most mass and power.

It is apparent that the small satellites will become an important segment in the overall satellite business in the near future. *Small satellites* are those of mass less than 500 kg. A commonly accepted classification of satellites is summarized in Table 14.2.

TABLE 14.2
**Mass Classification
of Satellites**

Class	Mass (kg)
Large	> 1000
Medium	500–1000
Small	> 500
Mini	100–500
Micro	10–100
Nano	1–10
Pico	0.1–1.0
Femto	> 0.1

To consider the advantages and applicability of microphotonics to spacecraft systems, one has to first consider the basic structure of a spacecraft. A spacecraft consists of the bus structure and one or more payloads attached to the bus. The bus structure performs the basic generic functions and housekeeping, while the payloads perform specialized tasks for specific commercial and/or scientific purposes.

The spacecraft bus entails the following subsystems:

1. Power generation and distribution (deployable solar panels, storage battery, and power control)
2. Attitude and position sensors (gyroscopes, accelerometers, star trackers, sun sensors, etc.)
3. Propulsion or thruster system to control the orbit of the spacecraft
4. Thermal-control system (passive radiators, thermal blankets, and active microlouvers or smart thermal radiator (STR) devices)
5. Data transmitter (downlink) and control signal receiver
6. Data storage memory
7. Electrical harness for communications and data and power interface to the payloads
8. Antenna system

For constellations of satellites, the spacecraft will also require a satellite-to-satellite link.

The use of microphotonics to miniaturize bus and payload subsystems can provide a tremendous benefit to space systems. It maximizes functional densities and facilitates a higher frequency of missions, greater redundancy in mission-critical components, and greater scientific or commercial benefit per mission.

Photonic systems, such as optical spectrometers and interferometers, have been traditionally employed in space as scientific payloads to perform various measurements in Earth observation and planetary exploration. These tend to be fairly bulky discrete optical systems. Discrete implementations of photonic circuits tend to be sensitive to vibrations and temperature variations. Therefore, additional environmental controls are required for their deployment in space. However, there are concrete benefits for the use of photonic systems in space, including:

- Reduced susceptibility of systems to electromagnetic interference (EMI), electrical discharges, and charged particles
- Substantial reduction in the weight of signal harnesses
- Substantially higher information storage capacity
- Substantially higher information transmission (download) capacity
- Optoisolation of critical spacecraft subsystems
- High-speed optical processing of RF and microwave signals
- Distributed optical sensor systems and microsystems for satellite control, such as laser gyroscopes, microelectromechanical system (MEMS) magnetometers and accelerometers, and fiber-optic sensors (stress and temperature)

A nanosatellite is essentially a spacecraft made up of a stack of Si wafers, containing all of the functions of a typical satellite subsystem. However, they could be semiautomatically fabricated much as computer chips are at reasonable unit costs, and would drastically decrease launch costs by allowing many satellites to be placed into orbit at once. Such architecture would be ideal for deploying large satellite constellations.

Thus, miniaturization technologies such as microphotonics and nanotechnology reduce satellite size and cost by making all components smaller and lighter. A generic modular approach to the structure of nanosatellites will greatly reduce overall mission costs. Microphotonic systems provide the added benefits of having low loss (< 1 dB/km), being lightweight, being wavelength-division-multiplexing (WDM) scalable, having high data rates (gigahertz), and having freedom from EMI with reduced shielding requirements.

It is expected that microphotonic systems will be used to:

- Enhance the performance of existing electronic systems
- Offer greater functionality, miniaturization, and reliability for future space systems
- Enable new advanced functionalities and sensors for next-generation spacecrafts

Specific areas in which microphotonics can be used successfully for space applications include hyperspectral Earth and planetary observation, intersatellite communications, downlink communications, radar surveillance, laser gyroscopes, and smart microsystems. In addition to the devices that have been reviewed in Chapter 12, systems such as MEMS microthrusters, MEMS docking ports, waveguide gyroscopes, and a series of microphotonic sensors have been specifically developed for space applications.

Microphotonic sensing devices include various low-mass sensors for temperature, vibration, displacement, spectral, and chemical measurement. Distributed fiber-optic sensors for temperature and pressure are currently enjoying considerable interest for space deployment.

Table 14.3 summarizes the major avenues for microphotonics in space applications, ranked in order of priority (taking into account the scope of applicability, the degree of innovation, and the improvements that could be drawn from the use of microphotonics).

The following sections present examples of the use of microphotonics to improve current space systems:

1. Optical interconnects
2. Satellite optical communication links
3. Quantum communication links in space
4. Optical beamformers for synthetic aperture radar (SAR) antennas
5. Photonic sensors
6. Navigation systems
7. Smart thermal radiators
8. Sun shields

TABLE 14.3
Summary of Microphotonic Applications in Space

Application	Devices	Expected Benefits	Critical Issues	Field of Interest

Routing and switching	Time-delay units
	Multiplexers/
	  demultiplexers
	Wavelength-
	  recognizing switches
	Photonic DCUs
	Wavelength converters
		Smaller mass
		Smaller size
			Reliability
			Reduced
			  insertion loss
				Telecommunication

Optical-beam-forming network	Time-delay units
	Multiplexers/
	  demultiplexers
	Wavelength-
	  recognizing switches
	Photonic DCUs
	Wavelength converters
		Smaller mass
		Smaller size
		Lower power
		  consumption
			Number of
			  channels
				Telecommunication

Integrated sensors	Gyroscopes	Compactness	Feasibility	Generic

Integrated spectrometers	Superprisms
	Filters
	Echelle gratings
		Improved
		  mechanical
		  stability
			Optical
			  bandwidth
			  resolution
				Emerging

Image processing (filtering and correlation)	Photonic VLSIs	Reduced size Added flexibility	Integration	Onboard processing for Earth observation missions

Processing for imaging interferometry	Couplers
	Asymmetric splitters
	Modulators
	Filters
	Photodetectors
		Improved
		  mechanical
		  stability
		Added modularity
		Added flexibility
			Insertion losses
			Optical
			  bandwidth
			  resolution
				New technique for
				Earth observation
				missions

Optical computing	Optical processors	Increased
		  processing speed
		Reduced power
		  consumption
			New concept	Generic

14.2 OPTICAL INTERCONNECTS FOR SPACECRAFT

One area that is receiving considerable attention is the spacecraft harness that enables box-to-box communications between systems within a spacecraft. Spacecraft contain various interconnected electronic modules in close proximity to each other, many of which can emit substantial EMI, which add to the natural radiation levels encountered in space. Moreover, spacecraft are subject to various types of external radiation (VUV, atomic oxygen, and cosmic rays) that can cause ionization and erode protective thermal blankets and coatings. Exposure to electrons, ions, and protons can cause substantial buildup of surface electronic charge on the spacecraft.

It is clear that electrical integration of the spacecraft subsystems constitutes a major part of the overall mission costs. A time-consuming procedure is followed to ensure that an electrical connection will not damage the electrical components within any other flight hardware. Moreover, subsystems on a spacecraft can have different grounds, resulting in potential differences between boxes and possible ground loops in the harness or shield, which can cause a subsystem to either cease operation or function improperly. Electrical harnesses also require shielded wires and careful bundling and routing to minimize cross talk and signal degradation.

On the other hand, optical interconnects within a spacecraft significantly reduce the possibility of damaging subsystems during integration and the overall integration time and costs. Optical links also offer substantial performance advantages because fiber-optic cables do not emit EMI and are not susceptible to interference from other cables or wires. Routing of signals is simplified with significantly improved signal integrity without the use of heavy shielding. Using optical links between electronic modules also effectively isolates the sensitive signal lines of the various subsystems to prevent a discharge that can damage the entire spacecraft. Table 14.4 summarizes the advantages of a photonic box-to-box harness.

In summary, for space systems, optical interconnects have significant advantages over electrical signal routing in terms of weight, ease of system integration, and fewer EMI concerns. Electrical integration of spacecraft subsystems is costly because of the long and tedious procedures that must be followed for safe electrical connections of flight hardware. Integration of optical harnesses is much

TABLE 14.4
Comparison of Electrical and Optical Interconnects

Parameter	Electrical Harness	Optical Interconnects
Typical weight	10 kg	< 0.2 kg
EMI	Requires special routing and shielding	Not applicable
Ground loop problems	Exist	Do not exist
System integration	Time consuming and costly	Simplified signal routing and integration
Data rates	MHz	> 1 GHz
Scalability	Requires change in system architecture	Extra capacity can be added
Signal integrity	Requires significant shielding and data coding	Enables high data transfers

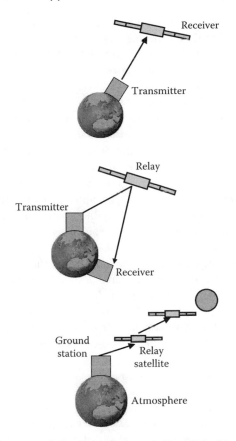

FIGURE 14.1 Satellite communication schemes: (a) direct link, (b) single relay, (c) interplanetary relay link.

simpler; it eliminates conductive paths and ground loop problems between hardware modules. The possibility of damage to hardware modules because of misconnections is also significantly reduced.

14.3 SATELLITE OPTICAL COMMUNICATION LINKS

Communication operations are a major part of the activities of the space vehicle bus. This can consist of both up- and downlinks to ground stations and links to adjacent satellites. Figure 14.1 shows schematically the three most commonly used satellite communication schemes, i.e., direct link, single relay link, and interplanetary relay link.

The idea for a global system of communication satellites was developed in the 1970s and resulted in the launch of the Tracking and Data Relay Satellite System in 1983. That system consisted of seven tracking and data relay satellites. The constellation of satellites provided global communication and data relay services for space shuttles, the International Space Station, the Hubble Telescope, and a multitude of low-Earth-orbiting satellites.

Most commercial satellites are used to relay data between terrestrial points. A satellite communication system can be attractive because there are no cable-laying costs and a single satellite can cover a very large area. Maximizing downlink time is usually accomplished using a constellation of satellites in a geostationary orbit (GEO). Compared to terrestrial systems, GEO satellite networks have much simpler delivery paths. A satellite in a geosynchronous orbit follows a circular orbit over the equator at an altitude of 35,000 km, completing one orbit every 24 h, i.e., the time that it takes the Earth to rotate once. Moving in the same direction as the Earth's rotation, the satellite remains in a fixed position over a point on the equator, thereby providing uninterrupted contact between ground stations in its line of sight. Satellite communications currently make exclusive use of active systems, in which each satellite carries it own equipment for reception and transmission.

To keep up with the ever-increasing demand for satellite phones and TV and data services, commercial companies have designed satellites that can support a continually growing number of users at the expense of greater spacecraft size and power. This is especially true for satellites with GEOs, where larger transmission powers and high-gain-antenna diameters are required to overcome the tremendous path loss that takes place as the signal propagates from Earth to the satellite and back.

Many satellites use a band of frequencies of about 6 GHz for uplink and 4 GHz for downlink transmission. Other bands at 14 GHz for uplink and 12 GHz for downlink are also in use, mostly with fixed ground stations. A band at about 1.5 GHz is used for both uplink and downlink with small, mobile ground stations such as ships, land vehicles, and aircrafts. Communication satellite systems have entered a period of transition from point-to-point high-capacity trunk communications between large and costly ground terminals to multipoint-to-multipoint communications between small and less expensive stations. The development of the time-division multiple access (TDMA) method has allowed this transition. With TDMA, each ground station is assigned a time slot on the same channel for use in transmitting its communications. All other stations monitor these slots and select the communications directed to them. By amplifying a single carrier frequency in each satellite repeater, TDMA provides the most efficient use of the satellite power supply.

Scientific payloads such as remote sensing and hyperspectral imaging entail the downloading of measured data, usually to dedicated ground stations. Typical downlink rates are on the order of 1 to 6 Mb/sec using RF links. The amount of data that can be transferred depends on the line-of-sight time between the satellite and the receiving ground station. In many instances, the science that can be performed is compromised by the capacity of the satellite to downlink the resulting data. High-resolution remote sensing and spectral imaging can generate large amounts of data between downlink data transfers. This requires either

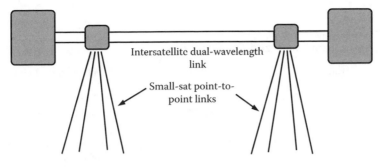

FIGURE 14.2 RF/optical communication link.

compressing or compromising the data set or employing multiple downlink stations to increase the time available for data transfer.

The future markets for both exploratory and telecommunication applications depend on the rapid growth of two-way interactive links. A current example is specific downloads from an Internet site with the user requesting certain data sets and the site responding with the requested material. This usually entails some latency between the two sites. Each user has different preferences for the specific data to be transferred.

Such requirements may be realized by supplementing RF multipoint communications with scalable WDM point-to-point optical links (see Figure 14.2).

Unlike microwave systems, data rate growth is unconstrained by the carrier frequency of optical links. Extremely-high-gain antennas are achievable at optical frequencies, facilitating long-range communication with reduced antenna size and weight and/or reduced transmitter power requirements compared to microwave systems. The optical beam divergence of the laser source is a critical parameter for space communications.

Optical wavelengths for communication links are typically 1 to 1.5 μm, whereas a typical microwave wavelength is 30 mm at 10 GHz. The angular divergence of the optical signal is almost 10^4 times smaller than that of the microwave.

Table 14.5 presents a trade-off comparison of RF and optical links. RF links are optimal in applications in which the same data is sent to many users such as passive direct TV. Optical links are better suited to point-to-point communications. The advantages of optical links, as summarized in Table 14.5, include significantly smaller aperture sizes, lower power requirements, lower typical mass, significantly higher data-transfer capabilities, and scalability for future requirements.

As it is indicated in Table 14.5, optical laser links provide secure point-to-point communications. The apertures for receiving and sending optical signals, typically less than 4-in. O.D., are much smaller than those employed for RF signals. RF transmitters require an increasingly large antenna aperture to provide a finer beam. The power efficiency can also be much greater because optical power is concentrated in a relatively collimated laser beam of small cross section

TABLE 14.5
Comparison of RF and Optical Links

Parameter	RF Link	WDM Optical Link
Data-transfer capacity	150 Mb/sec	2.5–10 Gb/sec
Beam divergence	$3 \times 10^4\,\mu m/dB$	$1.5\,\mu m/dB$
Power loss	1	10^{-8}
Scalability for future needs	Poor	Proportional to the number N of wavelengths
Power requirements	kW	< 20 W
Weight	> 100 kg	< 5 kg
Aperture size	Several square meters	Typically 4-in. O.D.
Single to many point	Good	Poor
Point to point	Poor; needs large aperture	Good; most power efficient
Redundancy for high quality of service	Poor; single-event upset can stop communications	Good; can reroute signals to different points and have several optical links on one satellite bus

rather than a diffuse RF beam. The power requirements could be reduced to watts from kilowatts. Optical links can facilitate much higher information transfer rates (Gb/sec) than traditional microwave or RF links (Mb/sec) between ground stations and a satellite system and between clusters of satellites in a constellation. A single optical wavelength can carry encoded information at rates exceeding 10 GHz.

WDM can further extend information capacity more than tenfold. A single optical link can facilitate over 80 wavelength channels from 2.5 to 10 GHz each. Owing to the small aperture size (< 4 in. O.D.) and power requirements (< 20 W) associated with optical links, they could be provided using lower-cost small-satellite platforms (see Figure 14.3). A costly RF communication satellite could be upgraded by integrating it with small-satellite optical links.

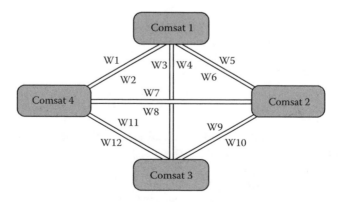

FIGURE 14.3 Multiwavelength optical links in a satellite constellation.

WDM seems to be the most feasible route for high-capacity broadband satellite systems operating above 10 Gb/sec. This facilitates using optical-wavelength-routing principles of networking across intersatellite links (ISL) by means of onboard optical-switching processors. Wavelength routing enables continuous optical paths to be established between any two satellites to create an optical transport network (OTN). The OTN enables single-hop connections with no intermediate down-conversion to electrical signals. WDM interlinks between satellites enable signal rerouting to favorably positioned satellites for downlinks to ground stations. This can enable a high-quality service that is comparable to and compatible with terrestrial fiber-optic networks. The main benefits identified for WDM routing of satellite links include the following:

- Simplified routing with minimal overhead for routing control
- Reduced queuing delays
- Redundant bandwidth for higher quality of service
- Higher traffic load capacity
- Signal rerouting for high-quality service and maximum downlinks

The internal optical bus can be considered as a local area or metro network (see Figure 14.4). Source-to-receiver distances will tend to be short (< 50 m); however, there can be considerable data switching and rerouting between the various subsystems. This is especially true for space systems because redundancy and rerouting is required to counter catastrophic failure of components. The ring network will therefore consist of several redundant fiber-optic lines interconnecting the subsystem nodes. Ideally, the system will interconnect to the satellite transmitter or receiver subsystem and mass-storage subsystems. Signal distribution to components within the subsystem may involve TDM and WDM and passive $1 \times N$ signal distribution.

Reception may entail converting the data to an electrical signal or directly using the optical signal for processing. For example, optical data could be directly interfaced to a DVD or CD laser writer for mass storage. It should be noted that a single terrestrial WDM node typically occupies an entire electronics rack. Obviously, this is not acceptable for space systems.

Figure 14.5 shows a conceptual design of a WDM transceiver that illustrates the tremendous advantage of microphotonics for space applications. The WDM OADM transceiver consists of the following blocks:

- Er-doped optical waveguide amplifiers (EDWAs)
- VO_2 or MEMS-based variable optical attenuators (VOAs)
- Bidirectional wavelength-division multiplexers or demultiplexers (superprism, echelle grating, and waveguide Bragg grating [WBG])
- 2×2 MEMS crossbar switches for dynamic signal add/drop
- Hybrid mounted detectors for signal detection
- Hybrid mounted low-power laser diodes for signal sourcing

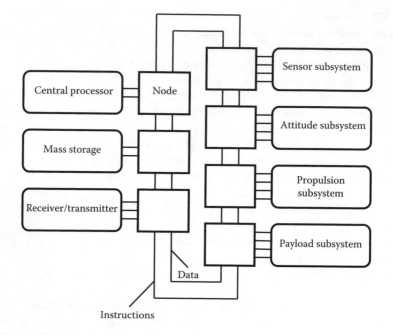

FIGURE 14.4 Schematic of satellite intranet optical bus.

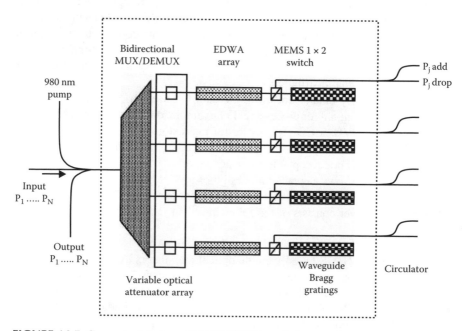

FIGURE 14.5 Conceptual design of OADM WDM optical transceiver.

Waveguide optical amplifiers (i.e., EDWAs) and waveguide Bragg gratings (WBGs) are fabricated using active glass channel waveguides (Er, Yt, and Ge doping or ion exchange). These can be fabricated in multichannel arrays on Si. In the design shown in Figure 14.5, each EDWA may operate on a single-wavelength channel, eliminating the problems of cross talk that are encountered with EDFAs that amplify the entire band simultaneously. A number of the EDWAs can share a single pump source because the pump power required per wavelength is significantly less than what a broadband EDFA would require.

The WDM OADM transceiver integrates an EDWA with a planar WBG for signal filtering and pulse compression. These WBGs are easier and cheaper to pattern than fiber Bragg gratings (FBGs) because the substrate is much more rigid than optical fiber. The WBG improves the signal and also reflects it back through the EDWA. Therefore, signals that pass through the node are amplified in two passes through the EDWA, reducing the length required for the EDWA by 50%. The single EDWA acts as both a signal preamplifier for the dropped channels and as an amplifier for the add channels. The MOEMS x-bar switch allows the input signal to be dropped at selected wavelengths after undergoing one pass of amplification through the EDWA. An external circulator can be employed to add a new signal using the same line. The added signal then undergoes one pass of amplification before being routed into the multiplexer. This innovative EDWA OADM architecture minimizes the number of optical circulators required over typical nodes by at least 50%. Costs are further reduced with regard to discrete components by sharing the wavelength-division multiplexer and EDWA for input and output. Potential size reduction is over an order of magnitude relative to commercial off-the-shelf (COTS) discrete components.

The main challenges for high-capacity optical links include the following:

1. Signal tracking and pointing
2. Atmospheric turbulence
3. Atmospheric absorption
4. Scattering by water droplets, ice, dust, and aerosols
5. Microvibrations

Efficient free-space optical coupling of the optical signal from a light source or optical device into a single-mode (SM) optical fiber or waveguide is a challenging task under the conditions that can be encountered in the various potential space applications (see Table 14.6). Figure 14.6 presents the link elements to be considered in the design, from uplink transmitter to the end user.

The most challenging aspect is optical downlinks through the atmosphere because these encompass additional effects due to the atmosphere: absorption, turbulence, and scattering. Figure 14.7 shows the typical spectral characteristics for transmission through the atmosphere. Atmospheric absorption due to water vapor, CO, CO_2, etc., can decrease the signal intensity There are transmission spectral windows just below 1 μm, near 1.3 μm, and just above 1.5 μm.

TABLE 14.6
Sources of Perturbations

Link (0.8- to 1.6-μm wavelength range)	Perturbations
Ground–space	Optical wavefront distortion due to atmospheric turbulence
	Atmospheric absorption and scattering
	Spacecraft microvibrations
	Ground station vibrations
Space–space	Spacecraft microvibrations

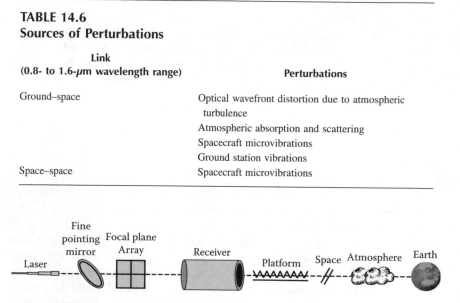

FIGURE 14.6 Parameters of end-to-end connectivity in a deep space link.

Atmospheric turbulence and scattering by dust and aerosols also degrades the wavefront quality. This causes the resulting image at the focal point of the collection optics to be much larger than the theoretical aberration limit. However, these are challenges that are faced daily by astronomers and have been resolved by using active adaptive optics.

Adaptive optics such as pairs of MEMS micromirror arrays or electro-optical arrays can be employed to correct wavefront errors in the optical beam due to atmospheric turbulence.

A schematic of an advanced fiber-coupling device (AFCD) is shown in Figure 14.8. The preliminary AFCD subsystem consists of the following:

1. Active MEMS or smart material adaptive optical system for correction of beam divergence and waveform distortion
2. Optional rotatable λ/2 plate for optical coupling to PM or SM optical fiber (for interferometric applications)
3. Hybrid MEMS-based vibration-damping system
4. Lightweight monolithic support for optics and micromechanical stages
5. Integrated fiber collimator consisting of aspheric or achromatic lens system integrated with PM or SM fibers
6. MEMS flexure x-y tilt stage to support and align fiber collimator
7. Four-quadrant donut-shaped photodetector for coarse beam alignment

An additional pointing or tracking mechanism (x-y, rotation, and tilt) is required to align the source input beam with respect to the adaptive correction

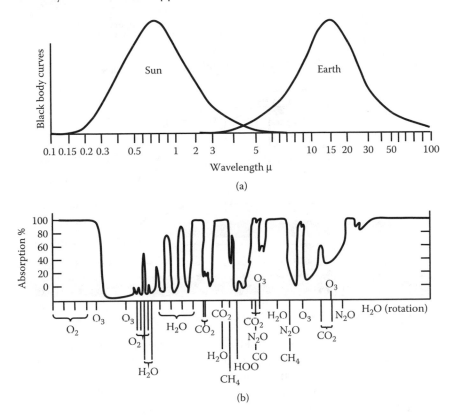

FIGURE 14.7 Atmospheric absorption: (a) solar and thermal radiation, (b) integrated.

optics. Additional subsystems include autonomous algorithms for vibration damping and for beam fiber alignment. The feasibility of the AFCD depends on the pointing accuracy and the beam quality of the optical source. Using a collimated source reduces the accuracy required for x-y positioning relative to the optical planes of the AFCD. Active feedback between the source and receiver could be employed to establish and track the optical alignment with time.

The adaptive optics module includes controllable beam shaping to adjust beam divergence to provide a minimal focal spot at the receiving fiber. It can also include an N × M programmable array of variable delay cells to correct for wavefront variations due to atmospheric turbulence and optics nonidealities. This can employ electrorefractive cells such as liquid-crystal arrays, electro-optical cells, or a combination of MEMS micromirror arrays. Active wavefront correction entails measurement and subsequent active correction of wavefront error. For space-to-space and short-distance coupling, this may not be required. For coupling to PM fibers, a rotatable $\lambda/2$ plate facilitates rotation of the beam polarization to match that of the perpendicular slow and fast axes of the receiving fiber. The adaptive optics module also provides the desired beam diameter to the fiber collimator.

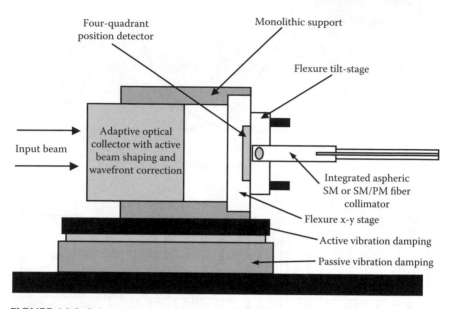

FIGURE 14.8 Schematic of a single-mode advanced fiber-coupling device (AFCD).

Coarse pointing alignment could be provided using a retroreflector at the receiver and a tracking beam at the sender. The sender forwards a tracking beam that is retroreflected back to the sender by the receiving satellite or station. This sets up the initial pointing alignment between the sender and the receiver. Fine alignment between the adaptive optics and the receiving SM or PM fiber collimator can be provided using a compact flexure positioning stage with x-y and planar tilt degrees of freedom. A four-quadrant photodetector with a central opening for the beam can be used to assist initial coarse alignment. This could employ piezoelectric or other MEMS technologies that have been developed for stable precision fiber alignment, such as miniature piezomotors and amplified MEMS membranes.

A schematic of the basic components comprising an optical link terminal (satellite or ground station) is shown in Figure 14.9.

For satellite-to-satellite links, effects such as atmospheric absorption and turbulence are not important. The resulting collection optics can be simpler, and the required optical power can be significantly lower.

14.3.1 DOWNLINK COMMUNICATIONS

Because optical systems can employ a parabolic collector similar to the antennas employed for RF signals, RF and optical communication systems can share some of the front-end collection optics, facilitating the ergonomic coexistence of both the RF and optical systems. The command data for the uplink generally does not require high transmission rates and can be based on the well-established RF links currently being used. In most space missions, the achievable science is limited

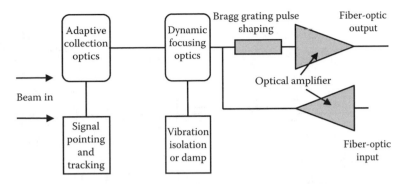

FIGURE 14.9 Schematic of a basic optical link terminal for space systems.

by the downlink capacity. An optical downlink can significantly reduce the size and weight of the required satellite antenna while increasing the potential downlink transmission rate.

For the large distances involved in planetary exploration and the small optical beam divergence, typically about 7 μrad, antenna pointing and tracking systems and algorithms become important.

Mistracking and satellite vibrations can contribute several decibels of signal loss. One possibility is to employ a second optical wavelength to provide a tracking beacon (see Table 14.7).

A schematic of the optical transmitter is shown in Figure 14.10. It is based on recent developments in high-performance fiber lasers (FLs). The FL employs special Bragg fiber gratings to stabilize the output wavelength. Linewidths below 0.005 nm have been achieved (see Table 14.8 for the system requirements). Optical pumping of the FL master oscillator is provided by fusion-spliced SM 980-nm laser diodes. Multiple diodes can be employed to provide redundancy.

TABLE 14.7
Uplink and Downlink Scenarios

	Scenario 1	
Uplink	RF X-band or Ka-band command data	
Downlink	RF-modulated optical carrier for telemetry and data	$\lambda 1$
	Optical beacon for tracking and clock synchronization	$\lambda 2$
	Scenario 2	
Uplink	RF-modulated optical carrier for command data	$\lambda 1$
	Optical beacon for tracking and clock synchronization	$\lambda 2$
Downlink	RF-modulated optical carrier for telemetry and data	$\lambda 3$
	Optical beacon for tracking and clock synchronization	$\lambda 4$

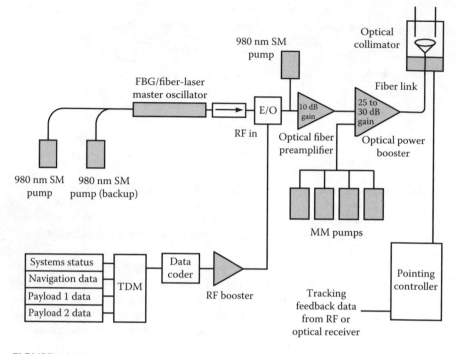

FIGURE 14.10 Schematic of fiber laser transmitter.

TABLE 14.8
Summary of Typical Space Requirements and the Characteristics of a Fiber Laser System

Parameter	Space Functional Requirements
Output power	1–5 W
Coupling efficiency of laser diodes into optical fiber	> 60% (1060-nm system)
Wavelength	1060 nm
Spectral width	≤ 0.05 nm (FWHM)
Spectral stability	< 0.1 nm
Modulation	Intensity, non-return-to-zero (NRZ) line code
Modulation data rate	300 Mb/sec (up to 2.5 Gb/sec)
Output irradiance profile	Gaussian
Output wave state	Plane

The FL can be configured to operate either near 1060 nm by using a SM Yb-doped active fiber or near 1550 nm by using a Er-Yb-doped active fiber. Power conversion at 1060 nm is more efficient, and the beam divergence is also much lower than at 1550 nm. Thus, the lower wavelength is advantageous for long-distance communications.

For data rates below 2.5 GHz, the laser diodes can be modulated directly. At higher rates, an electro-optical modulator can be employed to provide rates in the gigahertz range. However, this typically results in a 3-dB loss. Data from satellite memory buffers, representing stored scientific data, sensor, navigation, and system status information, is time-division-multiplexed and coded. This RF signal is amplified and modulated onto the optical carrier. The optical carrier is preamplified to 100 mW using a SM fiber amplifier and is then boosted into 1-W range using a multimode booster. Fusion splicing of all the laser pumps provides high-reliability and low-loss pumping. Using multiple pumps provides redundancy to facilitate long-term reliability. The output of the FL is Gaussian and circular in cross section, ensuring good collimation.

The output beam is expanded and collimated by the output optics and parabolic antenna to provide acceptable beam divergence. Pointing the antenna makes it possible to use the received RF uplink data or a separate optical beacon for feedback to verify the forward-looking angle for data downlink transmission. An MEMS mirror array, located between the collimator and the output antenna, can be used to provide fine pointing and to compensate for satellite jitter. The collimated output is fed to a Cassegrain antenna to provide the required beam diameter.

Figure 14.11 shows the detailed schematic of the master oscillator for optical links as realized using fiber-optic components. It requires a number of discrete components that must be temperature-compensated and spliced together. The master oscillator uses waveguide optical amplifiers and gratings. This can provide an order-of-magnitude reduction in size and mass.

A microphotonic alternative design of the discrete-component system shown in Figure 14.12 may include hybrid integration of the components, such as pumps, drivers, and EDWA, on a single SOI chip. Such a microphotonic design provides much higher mechanical reliability than the discrete-component system.

14.4 QUANTUM COMMUNICATION LINKS IN SPACE

Up to now, all fundamental demonstrations of long-distance quantum communication protocols (such as quantum state teleportation, quantum key distribution, or quantum dense coding) have utilized optical fibers to distribute quantum entanglement. However, the distance over which single photons can travel through such fibers is quite limited to some hundred kilometers (see Chapter 13). A possible solution to overcome this limitation is the use of optical free-space links. Together with the utilization of space infrastructure such as satellites it will be possible to implement quantum communication links in space.

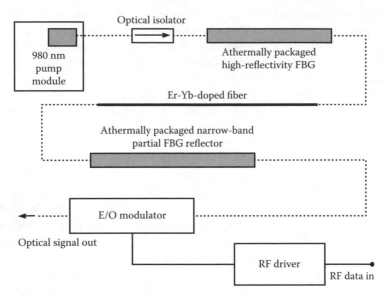

FIGURE 14.11 Schematic of a master oscillator for optical space links.

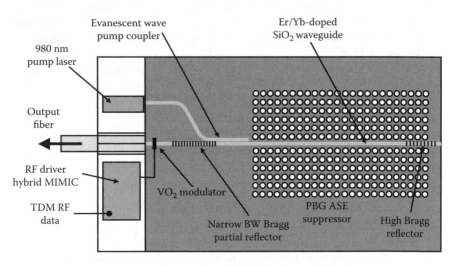

FIGURE 14.12 Schematic of a fully integrated EDWA master oscillator.

Quantum communications in space has become a new technological challenge in the evolving field of quantum photonics. Its main goal is to achieve the distribution of single photons or entangled photon pairs from satellites to implement quantum technologies such as quantum cryptography. The main benefit of a space infrastructure is that it allows for much larger photon propagation distances than are achievable with Earth-bound quantum links, thus eventually enabling the buildup of a global quantum communication network.

Several implementation scenarios for quantum photonic communication systems in space have been proposed. These scenarios are adapted from the existing ground–satellite, intersatellite, and interplanetary schemes, which have been briefly summarized in Figure 14.1.

For example, an Earth-based transmitter terminal shares quantum entanglement between (1) ground and satellite, (2) two ground stations, or (3) two satellites. In these scenarios the transmitter terminal distributes entangled photon pairs to the receivers that can perform an entanglement-based quantum communication protocol. As mentioned in Chapter 13, the principle of a repeater operation is that it redirects qubit states without detecting them.

According to a report prepared by a team from Vienna University of Technology [14.2], a straight uplink to one satellite-based receiver can be used to implement secure quantum key distribution between the transmitter station and the receiver. One of the photons of the entangled pair is detected at the transmitter, and thus the entangled-photon source is used as a triggered source for single photons. A similar protocol can be established for communication between two distant Earth-based communication stations by using the satellites as relay stations (see Figure 14.13).

In another scenario a transmitter with an entangled-photon source may be placed on a space-based platform. This makes it possible to cover longer distances because of the reduced influence of atmospheric turbulence.

In principle, a quantum key exchange can be performed between arbitrarily located ground stations. The use of entangled states sent to two separate ground stations allows instantaneous key exchange between these two communicating Earth-bound parties.

The required shared entanglement can be established either by two direct downlinks or by using additional satellite relay stations. Quantum entanglement can also be distributed between a ground station and a satellite or between two

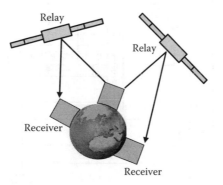

FIGURE 14.13 Quantum communication link based on satellite relays.

satellites. A more elaborate scheme that includes a third party might also be implemented, which allows performing a Bell-state analysis on two independent photons. This would enable quantum state teleportation and entanglement swapping for a truly global quantum communication network [14.2].

The design of the transmitter is similar to that described in Chapter 13 (Figure 13.17), comprising the entangler and modules for polarization manipulation and detection of single photons. A schematic diagram of a transmitter for space quantum photonic link is shown in Figure 14.14. A set of conventional satellite instrumentation would have to be added, such as a pointing acquisition and tracking system (PAT) and adaptive collection optics with vibration isolation and dynamic focusing optics (DFO). A reference laser (REF) may be used to compensate for any orientation misalignment. A central control module (CONTROL) is interfaced with a data acquisition and storage (DAS) module.

The receiver of a space quantum photonic link consists of a single-photon analysis and detection module (POL and DET) combined with all the conventional satellite subsystems that have also been used in the transmitter. A schematic diagram of the receiver is shown in Figure 14.15.

A reference laser is also included to compensate for any misalignment between transmitter and receiver. The reference laser of the receiver operates at a wavelength differing from that of the transmitter in order to clearly separate the two distinct signals for a more accurate polarization analysis of the transmitter reference laser [14.2].

14.5   OPTICAL BEAMFORMERS FOR SAR ANTENNAS

Environmental monitoring, Earth resource mapping, and military systems require broad-area imaging at high resolutions. It is also desirable to have the capability

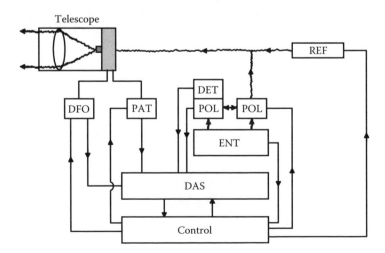

FIGURE 14.14 Transmitter of a space quantum photonic link.

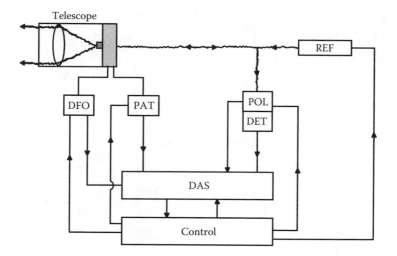

FIGURE 14.15 Receiver of a space quantum photonic link.

of acquiring an image in inclement weather and during night as well as day. Synthetic aperture radar (SAR) provides such a capability. SAR systems take advantage of the long-range propagation characteristics of radar signals and the complex information processing capability of modern digital electronics to provide high-resolution images. SAR complements photographic and other optical imaging capabilities because of the minimum constraints on time-of-day and atmospheric conditions and because of the unique responses of terrain to radar frequencies. SAR has provided terrain structural information for mineral exploration to geologists, oil-spill boundaries on water to environmentalists, sea state and ice hazard maps to navigators, and reconnaissance and targeting information to military operations. Some applications, particularly civilian, have not yet been adequately explored because lower-cost electronics are just beginning to make SAR technology economical for smaller-scale use.

SARs produce a two-dimensional image. One dimension is called *range* (or *cross track*) and is a measure of the "line-of-sight" distance from the radar to the target (see Figure 14.16). Range measurement and resolution are achieved in SAR by precisely measuring the time delay from the transmission of a pulse to the receiving of the echo from a target. The simplest SAR range resolution is determined by the transmitted pulse width, i.e., narrow pulses yield fine range resolution. The other dimension is called *azimuth* (or *along track*) and is perpendicular to the range. It is the ability of SARs to produce relatively fine azimuth resolution that differentiates it from other radars. To obtain fine azimuth resolution, a physically large antenna is needed to focus the transmitted and received energy into a sharp beam. The sharpness of the beam defines the azimuth resolution. Similarly, optical systems, such as telescopes, require large apertures (mirrors or lenses, which are analogous to the radar antenna) to obtain fine imaging resolution. Because SARs are much lower in frequency than optical

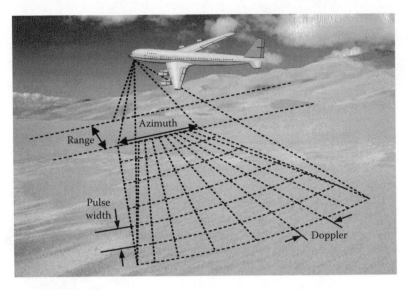

FIGURE 14.16 Schematic of SAR.

systems, even moderate SAR resolutions require an antenna physically larger than can be practically carried by an airborne platform, i.e., antenna lengths several hundred meters long are often required. However, airborne radar systems can collect data while flying and then process the data as if it came from a physically long antenna. The distance the aircraft flies in synthesizing the antenna is known as *synthetic aperture*. A narrow synthetic beam width results from the relatively long synthetic aperture, which yields finer resolution than is possible from a smaller physical antenna. A target's position along the flight path determines the Doppler frequency of its echoes; targets ahead of the aircraft produce a positive Doppler offset and those behind produce a negative offset. As the spacecraft flies a distance (the synthetic aperture), echoes are resolved into a number of Doppler frequencies. The target's Doppler frequency determines its azimuth position.

The scientific application drivers for SAR measurements are listed as follows:

- *Geology:* measurement of topography and topographic changes and hazards assessment such as flood potential, volcanoes, earthquakes, etc.
- *Oceanography:* study of ocean currents, winds, ocean surface features, sea ice thickness and coastal processes, tracking of ships entering a country's coastal waters, and supporting sea search and rescue initiatives
- *Hydrology:* measurement of soil moisture and snow water equivalence
- *Ecology:* land cover classification, inundation mapping, and biomass measurements
- *Glaciology:* study of snow faces, seasonal melt, icebergs, surface morphology, ice velocity, and surface topography

A key technological challenge in SAR systems is the antenna system. The physical size requirements change with frequency, orbit altitude, swath width, and antenna gain. However, owing to the desire to limit the level of azimuth and range ambiguities in the SAR signal, the physical size of the antenna cannot be smaller than certain prescribed limits. A typical size for a space-borne L-band SAR antenna is about 10×2 m. Examples of space-borne SAR include Radarsat and SEASAT, ERS-1, JERS-1, and SIR-C.

The L- and C-band phased arrays employ multiple transmit/receive (T/R) modules that are distributed across the physical apertures of the antennas. These distributed T/R modules also provide electronic beam-steering capability.

It has been suggested that the key technological challenges for future antennas are the following:

- Reduction in antenna weight
- Accommodation of a large antenna within the envelope of the launch vehicle shroud
- Reliability of the deployment mechanism
- Reduction in loss through antenna feed network

One of the possible solutions is the use of inflatable antenna technology. Development of this technology can provide significant advantages in reducing the volume of the stowed antenna at launch; this will allow SAR systems to fly in smaller launch vehicles. However, several key technological challenges need to be addressed before inflatable antenna technology can be realized as a workable option for SAR:

- An appropriate approach to antenna feed
- A design allowing electronic beam steering
- Lifetime of the material used
- Mechanical control of the antenna
- Overall system reliability

In particular, the incorporation of electronic beam-steering capability with the required fast beam-switching time presents a formidable challenge.

The development of so-called phased-array antenna systems provided a very efficient method for improvement in space-borne SAR and communication satellite systems. These systems consist of a number of independent small antennas that can be electronically scanned to program the way in which signals of the individual elements are processed. Each of these antenna elements are individually excited by an RF signal of specific phase and amplitude, which are calculated using the relative position of the element in the antenna and the desired beam direction. The relative phases and amplitudes of the respective signals feeding the antennas are varied in such a way that the effective radiation pattern is reinforced in a desired direction and suppressed in undesired directions. The constructive and destructive interference effects among the signals radiated by

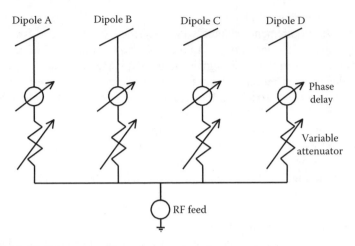

FIGURE 14.17 Schematic of four-element phased-array antenna.

the individual antennas determine the effective radiation pattern of the array. This allows a phased array to be used to point a fixed radiation pattern or to scan rapidly in azimuth or elevation (two-dimensional array). Rapid scanning at microsecond rates is desirable for mobile phone communication systems and for tracking a moving target using a high-resolution narrow beam. A phased array has the ability to focus beams in specified spatial coordinates while simultaneously placing nulls along other spatial coordinates, making it immune to jammers, and providing a significant reduction in multipath interference.

In the conventional microwave approach, the feed distribution network is implemented using microwave waveguides and phase shifters, as shown schematically in Figure 14.17.

The practical use of phased arrays has been severely limited because phase-shifting electronics are intrinsically narrowband. Microwave circuit components and waveguides tend to be cumbersome, especially at lower microwave frequencies. Distribution of the microwave signal to the various elements in the array requires the use of relatively lossy RF waveguides. This has limited the widespread use of large phased arrays especially in airborne applications, in which size and weight are primary concerns. Another significant limitation is that microwave circuits are inherently narrowband, limiting the bandwidth of the beam-forming system to a few percent of the RF carrier frequency. The use of narrowband components results in a beam-pointing error known as *squint*, where each frequency component points in a different spatial direction. Modern radar tracking and surveillance systems require wideband performance (in the range of 2 to 18 GHz bandwidth) making phase-steered systems almost useless.

One way to eliminate the squint error is to employ true-time-delay beam-forming techniques. An array antenna is "steered" by applying an appropriate amount of time delay to each signal used to drive the elements of the array. In order to achieve time delay of a microwave signal, without the necessity of digitizing it and storing it in a computer memory for later reconstruction, one must provide an appropriate path length for the signal to propagate.

Photonic systems may provide the solution to the preceding problem. By replacing phase shifts with true time optical delay lines, the signals from different antenna elements can be correlated independent of frequency. The broadband nature of optical delay lines enable beam steering independent of RF frequency. However, the available optically controlled phased-array antennas (OCPAA) consist of very large, bulky, and costly systems, which are yet to be demonstrated with a scale and capability comparable to similar, fully electronic microwave systems.

Microphotonic beamformers hold great promise in providing the ultimate solution that will allow reducing both weight and volume of phased-array networks. A beamformer will in general control the timing and amplitude information for each antenna element. There are many advantages when using optically-phased-array antennas (PAA):

- The losses of optical channels are about 0.2 dB/km, which is much lower than the losses of RF coaxial cable.
- The weight of an optical channel is much less than the weight of a typical coaxial cable or a RF waveguide.
- The bandwidth of an optical channel is over 1 THz, which is much higher than that of most conventional cables
- Communication over optical channels is immune to EMI.

There are three types of photonic solutions being proposed for beamforming:

- Bulk-optics photonic systems
- Fiber-optic networks
- Microphotonic processors

14.5.1 OPTICAL BEAMFORMERS BASED ON BULK OPTICS

Bulk optics can be used to synthesize a two-dimensional phase and amplitude distribution that is sampled and transferred to a microwave signal using heterodyning. In order to process the two-dimensional set of light beams for the different array elements, microstructures that interact with optical beams perpendicular to the surface of the device can be used. Liquid-crystal technology can be used to create an array of light shutters or an array of optical phase shifters. MEMS-based arrays of mirrors can be used to redirect light beams. Also, acousto-optical Bragg cells can redirect optical beams. In a Bragg cell, an acoustic standing wave induces a refractive index grating. This grating deflects a light beam that passes through the Bragg cell. Using the acoustic frequency, one can control the angle of deflection. With this method, a reconfigurable optical delay line can be created.

14.5.2 Optical Beamforming Networks Based on Fiber-Optic Components

RF phase control for the elements of a PAA has been accomplished by stretching the fiber. Fixed networks of optical fibers, serving as time delays, have been used to create a number of fixed beams at a PAA. Point-to-point optical microwave lines, consisting of a laser followed by a certain length of fiber and terminated with a photodetector, have been used as delay elements in electrically switched delay lines. Dispersive fibers have been applied as a wavelength-dependent delay lines. Also, beamformers have been made, in which a FBG and a circulator have replaced the dispersive fiber. Also, multiple tunable lasers have been combined with one Bragg grating to create a beamformer.

As shown in Figure 14.18, the fiber Bragg grating (FBG) provides variable time delays. Each wavelength experiences different time delays. This can be used in combination with a tunable FL or laser diode and an RF electro-optical modulator to provide a variable RF true time delay. The net insertion loss for the FBG is less than 1 dB; so, it is very power efficient. The peak-to-peak ripple, representing the deviation of the FBG characteristic from a linear variation of time delay, with the optical wavelength in long gratings can be under ±10 psec. This is about the limit attainable with chirped FBG gratings. The desired ripple in the time variation is under ±2 psec.

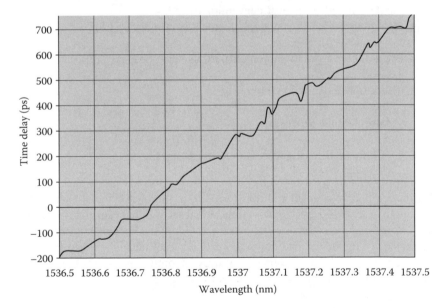

FIGURE 14.18 Time-delay variation vs. wavelength for a fiber Bragg grating.

14.5.3 Microphotonic Beamformers

Microphotonics can facilitate a very compact integrated optical processor for providing programmable RF time delays and various RF filtering functions. This approach to programmable optical time delay minimizes the number of control junctions required to achieve a given time interval resolution (t_{step}). In this concept a chain of binary-coded time-delay elements is employed. For a system consisting of N_s binary-coded time-delay elements:

$$t_{step} = T_{max}/2^{Ns}$$

where N_s is the total number of binary-coded delay lines and T_{max} is the time delay of the longest delay line. The time-delay contribution of the nth delay line is

$$\Delta t(n) = T_{max}/2^n$$

The maximum time delay is given by

$$\Delta t_{max} = 2\, T_{max} - T_{max}/2^{Ns}$$

In this architecture, fairly high time-delay resolutions, on the order of 8 to 12 bits and higher, are feasible using a minimum number of switching gates. The ripple can be below a picosecond.

The binary-coded programmable optical delay line can be realized using a channel waveguide structure, as shown in the Figure 14.19. The integrated programmable optical delay line consists of two basic elements, a high index channel waveguide that facilitates sharp bends and a 2 × 2 active optical cross-connect switch. Using the preferred Si waveguides in a SOI waveguide structure, the channel waveguides can be made by high-aspect reactive ion etching. This facilitates right-angle bends using a mirror structure. The high index of the Si results in a very low optical numerical aperture (NA) within the Si waveguide core (NA = 0.11/3.45).

Multiple programmable time-delay lines can be fabricated simultaneously on a single SOI chip for distribution to several antenna elements. The accuracy using an E-beam printed photomask is about 0.5 μm.

Using high-index channel waveguides such as Si can increase the time delay attainable per unit length of fiber because the velocity of light within the medium varies inversely with the refractive index n. This reduces the size of the Si delay line by a factor of about 2.5 relative to a silica delay line, with a commensurate reduction in propagation losses. The time delay is given by

$$t_d = n\, \Delta L/c$$

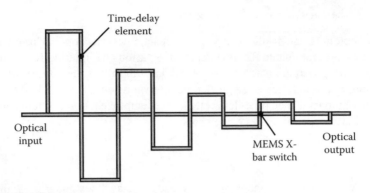

FIGURE 14.19 Integrated MEMS–waveguide optical delay line.

TABLE 14.9
Time Delay per Length of SiON and Si Waveguides

Waveguide Structure	Time Delay (psec/cm)
Si on SiO$_2$/Si	115
SiON on SiO$_2$/Si	48.3

where n is the refractive index of the medium, ΔL is the effective length of the delay line, and c is the velocity of light in vacuum. Time delay per length values for two waveguide materials are given in Table 14.9. A schematic of a conceptual design of an optically controlled phase-array antenna is shown in Figure 14.20.

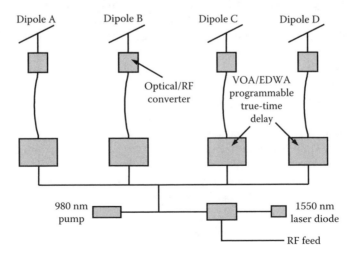

FIGURE 14.20 Optically controlled phased-array antenna.

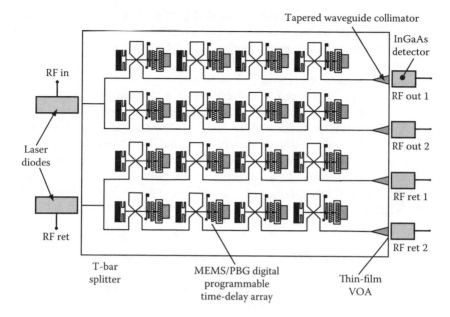

FIGURE 14.21 Conceptual design of micro-PIC T/R module.

With the advance of MOEMS, photonic-band-gap structures, and integrated waveguide technologies, there is now the possibility of a practical implementation of miniature integrated optical circuit to provide RF true-time-delay phase shifting and programmable amplitude control for optical beamforming. The future T/R modules and phased-array systems will be based on a unique combination of microphotonic devices integrated on low-cost SOI wafers.

Figure 14.21 shows a conceptual layout of a micro-PIC for SAR T/R. The illustrated micro-PIC integrates the following functions:

- VOA for amplitude control
- Multibit programmable true time delay (6 bit or higher)
- Passive signal splitting and distribution
- Multiple parallel RF signal processing lines

The presented design consists of the following microphotonic components:

1. Total internal reflection (TIR) Si-channel waveguides as delay elements
2. T-bar signal splitter
3. 2×2 PBG MEMS voltage-controlled optical switch
4. MEMS (or VO_n voltage-controlled) optical attenuator
5. Tapered waveguide beam collimators for coupling to output detectors or output optical fibers

The basic microchip can provide several parallel lines of 6 bit (or more) programmable true time delay and integral programmable amplitude control. Moreover, it is possible to have separate transmitting and receiving lines, as shown in Figure 14.21, eliminating the need for the T/R switches and possibly leading to a better SAR signal processing and architecture.

The inputs to the micro-PIC can consist of either the desired RF signals (transmitted and received) or the RF-modulated optical carrier contained within an optical fiber. For frequencies of 2.5 GHz and below, a 1550-nm laser diode can be directly modulated with the desired RF signal. For higher RF frequencies, an electro-optical modulator or shutter is required. It may be possible to also integrate the optical modulators directly on the SOI chip. Work is currently underway at several laboratories to develop high-speed optical modulators for Si. There are three potential approaches for the microphotonic modulators that are being considered:

1. Si waveguide Mach–Zehnder interferometer using low-power MOS current injection
2. Si waveguide Fabry–Perot voltage-controlled filter using PBG reflectors as waveguide mirrors
3. VO_n electrochromic film

The Si modulators use the variation of the index of refraction of Si with free-carrier density. As described in Chapter 5, the refractive index of Si increases with the injected-carrier concentration. A MOS structure can be used to modulate the free-carrier concentration of the underlying Si in either the cavity of a Fabry–Perot filter or in one arm of a Mach–Zehnder interferometer. This causes a change in the resulting optical interference, resulting in a change in the output intensity.

The delayed optical output can be coupled to optical fiber for remote conversion to RF. Alternatively, InGaAs detectors can be coupled to the waveguide output to provide the optical-to-RF conversion near the chip. The latter approach will provide the best signal throughput.

The signal loss introduced by the micro-PIC chip can be compensated using a short strand or Er-doped fiber pumped at 980 nm to provide optical amplification of the RF-modulated optical carrier.

Figure 14.22 shows the architecture for implementing a microwave microphotonic subsystem and integrating it with the SAR RF subsystems (one channel of the multichannel system).

The system employs a master optical carrier source. A highly coherent FL source at 1550 nm is used. The optical carrier can be used for both transmitting and receiving functions; the 1550-nm FL ensures high overall phase coherence for the SAR system. A fiber-optic fused directional coupler is used to add a 980-nm high-power pump signal to the fiber laser. Several 980-nm diode pumps can be used to provide redundancy. With a typical power conversion efficiency (electrical to optical) exceeding 25%, the 980-nm laser diodes are much more efficient

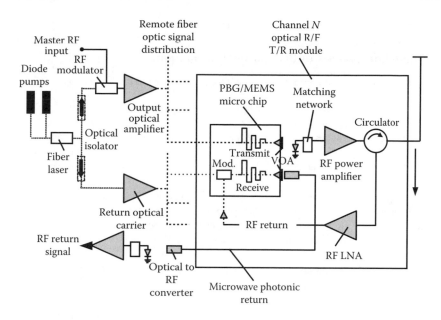

FIGURE 14.22 Schematic of microwave microphotonic T/R module.

than laser diodes operating at 1550 nm (< 3% conversion efficiency) and have much better reliabilities for high-power applications (> 100,000-h lifetime). The purpose of the 980-nm bias optical signal is to drive Er-doped fiber-optic amplifiers and FL. These can be used to compensate for the optical losses sustained by the programmable time-delay subsection of the micro-PIC chip.

Signal distribution to the antenna T/R modules employs fiber-optic links that are significantly lighter than RF waveguides, have very low loss (<1 dB/km), and are insensitive to EMI. RF interference on complex antennas between adjacent elements requires substantial shielding of RF components. The use of optical fiber links can substantially reduce the weight of the antenna structure by means of the reduced shielding requirements. Nominal SAR structures employ an array of heavily-machined Al enclosures to house RF-matching networks, tuning stubs, etc., which add substantially to the cost and weight of the antennas. Moreover, fiber-optic links are flexible and are therefore more suitable for inflatable membrane antennas. Optical fibers can be also used to strengthen the membrane antenna because of the high strength of telecommunication optical fiber. For space applications, optical fibers can be clad with polyimide, Al, or Cu.

Table 14.10 summarizes the characteristic of microwave and photonics-controlled phased-array structures. The photonic technique can provide significant advantages in terms of weight reduction, improved signal quality due to EMI minimization, minimal squint error, and resulting improved pointing accuracy. Power requirements should also be lower. Weight reduction is a result of the reduced weight of fiber-optic links relative to microwave waveguides and the reduction in EMI shielding and RF tuning requirements. Beam-steering speeds

TABLE 14.10
Trade-Off Analysis of Beamforming Techniques

Parameter	Microwave[a]	Bulk Optics	Fiber Optics	Microphotonics
Weight	3 kg/m^2	2 kg/m^2	<1 kg/m^2	< 0.2 kg/m^2
Bandwidth	20 to 100 MHz	GHz	GHz	GHz
Tunable delay	600 psec (8 bit)	Size dependent	1100 psec (per grating)	115 psec/cm (12-bit resolution)
Method	Phase delay	True time delay	True time delay	True time delay
EMI sensitivity	High	Low	Low	Low
RF frequency sensitivity	High squint error	Frequency independent	Frequency independent	Frequency independent
RF matching	Impedance-matching stubs	None (direct conversion to RF)	None (direct conversion to RF)	None (direct conversion to RF)
Shielding	Machined Al enclosures for all critical RF components	Minimal	Minimal	Minimal
Switching speed	2 μsec	nsec using electro-optical switching	nsec using electro-optical switching	msec to μsec (MEMS); psec (VO$_2$/MEMS)

[a] Based on typical size and weight of SAR structures, i.e., size of 20×1.5 m and weight of 80 kg.

depend on the application. For TDM systems such as mobile phone communications with distribution to many points, high beam-switching speeds are required. Fiber- and integrated optics solutions are ideal for the next generation of inflatable membrane antennas because of the flexibility of fiber-optic links and reduced shielding requirements.

14.6 PHOTONIC SENSING SYSTEMS

Photonic sensors have substantial merit for various space and terrestrial applications. For example, fiber-optic sensors can be applied to a number of measurement tasks including temperature, pressure, gas flow, gas volume, and structural strain. The numerous advantageous characteristics of fiber-optic sensors include the following:

- Immunity to EMI
- Resistance to aggressive environments and chemicals
- Electrical and fire safety
- Small, lightweight sensing elements

Photonic sensor systems can also benefit from the recent advances in optical sources, signal multiplexing and passive distribution, optical switching, and signal processing that have been developed for terrestrial optical telecommunication applications. This progress facilitates relatively low-cost multisensor systems that can be imbedded in mechanical systems and structures to provide continuous monitoring of system parameters and performance. Photonic sensors can form the backbone of advanced distributed sensor systems that are gaining significant utilization in varied terrestrial applications and space, including monitoring of structural health, spacecraft propulsion systems, and combustion engines.

There are two diverging applications for photonic sensor systems:

- Distributed optical sensor systems for monitoring large space structures (rocket propulsion systems and health monitoring of membranes and spacecraft skins)
- Miniature integrated sensors (i.e., gyroscopes and accelerometers) for attitude control of small satellites and nanosatellites

Photonic sensors can be all optical, such as FBGs for temperature measurement, or a hybrid combination of MEMS and fiber optics such as MEMS accelerometers with fiber-optic readouts. In addition to the sensors themselves, the systems require integrated processors to distribute the signals and to process the return optical data.

14.6.1 DISTRIBUTED FIBER-OPTIC SENSOR SYSTEMS

An optical multisensor system for rocket propulsion subsystems may be used as an example of distributed sensors. This sensor system is used for temperature, pressure, and volume monitoring (see Figure 14.23). The main purpose is to replace the current electronic sensors with optical sensors to improve safety margins in the propulsion system and to improve measurement accuracy in the propellant and pressurant tanks. Electronic sensors require additional shielding and encapsulation for operation within the propulsion system because of the risk of an electronic spark or discharge. Sparking can result in catastrophic failure of the spacecraft. Shielding of various electronic sensors may reduce the measurement accuracy of critical parameters, such as the remaining volume of rocket fuel.

Photonic multisensor systems require central processors that can distribute or multiplex optical signals to the various sensors, modulate the optical signals to improve measurement accuracies, and process the return optical signals from the various sensors, while interfacing with the spacecraft electronic central processing system.

Basic subsystems of the distributed sensor with integral optical bus and optical processor include the following:

- Channel waveguide input coupling to signal source
- Signal modulator

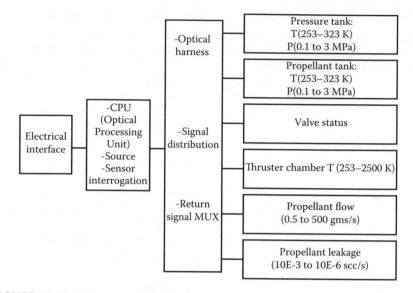

FIGURE 14.23 Photonic sensing system for rocket propulsion.

- Three-port optical circulator to separate output and return signals
- Signal distribution to sensors (active, passive, and WDM)
- Output channel waveguide fan-out
- Return signal analyzer (WDM and OTDR)
- Linear detector array

A schematic of an optimized $1 \times N$ active (or passive) signal distribution and demultiplexing system using a tunable, narrowband optical source is shown in Figure 14.24. Only an array of N photodetectors is required at the sensor output side.

MEMS techniques can be employed to control the sensitivity of fiber gratings to temperature and pressure variations. Athermal packaging techniques can be employed to reduce FBG sensitivity to temperature fluctuations, allowing sensitive pressure measurements. Thermomechanical amplification techniques are used to increase FBG sensitivity to temperature from the nominal value of 0.014 nm/K to more than 0.15 nm/K.

14.6.2 GYROSCOPES

Gyroscopes are used to determine changes in angular direction, usually by applying a rapidly spinning heavy mass. Conventional spinning mass gyroscopes, originally the gyroscopes of choice for space applications, require lubrication and eventually wear out.

The gyroscope is one of the most successful commercializations of miniaturized devices. The miniaturized gyroscopes have been extensively used for inertial navigation and guidance systems. At present there are increasing numbers

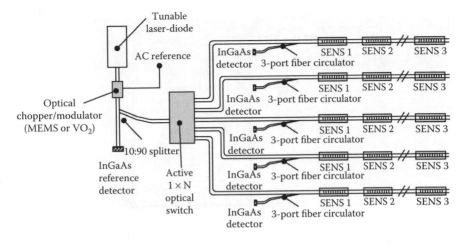

FIGURE 14.24 Schematic of optimized $1 \times N$ signal distribution and demultiplexing.

of applications for smaller and less expensive gyroscopes or angular rate sensors because of the emergence of new consumer and automotive products that demand angular velocity information.

Recently, traditional gyroscopes have been challenged by MEMS-based gyroscopes. MEMS-based gyroscopes do not have any specific life-limiting features. The resulting long life of more than 15 years is a significant plus, particularly for space applications. Although uniquely designed for continuous space operation, these new gyroscopes are lighter, cheaper, better performing, and have fewer components than their conventional counterparts.

Almost all of the MEMS based gyroscopes have been based on the principle of vibratory operation. The basic structure of a vibratory gyroscope can be modeled as a spring mass structure. The mass, which is suspended by springs, can vibrate in both the x and y directions. For operation, the mass is excited into vibration in the x direction (the vertical vibration). The vibration in the y direction (the lateral vibration) is induced by the Coriolis force. Therefore, the angular rate can be obtained by measuring the vibration amplitude in the y direction. The gyroscope can also be driven into vibration in both the x and y directions, and the angular rate is found by detecting the phase variation in the lateral vibration signal. These mechanical and MEMS-based gyroscopes can exhibit some sensitivity to vibration and shock.

Optical gyroscopes are envisioned as the next-generation gyroscopes for various aerospace and commercial applications. Optical gyroscopes combined with processors provide the best dynamic range and sensitivity and offer the prospect of low manufacturing cost.

Optical gyroscopes are based on interferometers. One arm of the interferometer can have sensors that are sensitive to an external stimulus. External stimuli, such as pressure, temperature, and specific chemicals, can cause a relative delay in the signal between the two arms of the interferometer by affecting the refractive

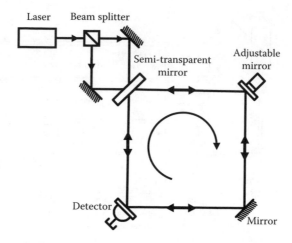

FIGURE 14.25 Schematic of an optical gyroscope.

index at one arm of the interferometer. This net change can be detected at the output of the device and provide a measure of the external stimulus. The Sagnac effect may be used as the measuring principle of external stimuli in optical gyroscopes.

The Sagnac effect states that in a rotating frame an optical path length difference is experienced by light beams propagating in opposite directions (see Chapter 3). The exact theoretical explanation for the Sagnac effect is still under debate and is providing some controversy for the Special Theory of Relativity. However, the effect has been experimentally proven and is unique to a rotational frame of reference. In a rotational frame of reference, there can be relative motion between photons and matter. This path difference, ΔL, is directly proportional to the absolute rotation around the axis perpendicular to the plane of the optical loop. Measurement of this path difference, or correspondingly the phase difference, is the basis of the new generation of optical gyroscopes (see Figure 14.25).

Recently, new types of optical gyroscopes have been implemented:

- Fiber-optic gyroscopes (FOGs)
- Ring laser gyroscopes
- Waveguide gyroscopes
- Microphotonic (PBG/MOEMS) gyroscopes

FOGs use many loops of optical fiber to convert the rotation rate to a phase shift induced by the Sagnac effect.

Ring laser gyroscopes convert the Doppler wavelength shift of two counter-propagating laser signals to the rotation rate by using a multimirror cavity. In the ring laser gyroscope, the two counter-rotating optical beams travel around a closed circuit or ring, which is usually rectangular or triangular (see Figure 14.26).

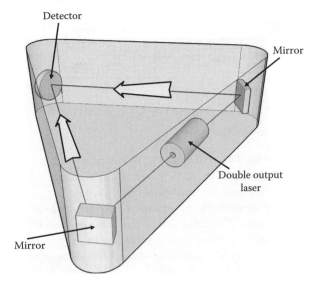

FIGURE 14.26 Ring laser gyroscope.

Mirrors are located at each corner to reflect the beams. At one corner, there is a detector or an output sensor. In an actual application such as an aircraft autopilot, three laser gyroscopes may be used to sense changes in pitch, roll, and yaw. In addition, there could be three accelerometers to measure longitudinal, lateral, and vertical motion (see Figure 14.27). A set of three gyroscopes and three accelerometers provides a complete navigation and guidance system (see the following section)

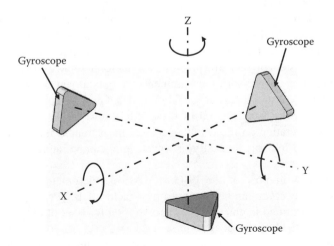

FIGURE 14.27 Three-dimensional laser gyroscope.

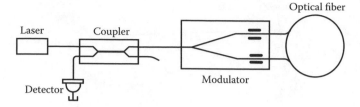

FIGURE 14.28 Waveguide gyroscope.

The laser gyroscope is a viable option for almost all military and commercial applications, including aircrafts, tactical and strategic missiles, naval and marine vehicles, land vehicles, weapon platforms, and spacecraft.

Waveguide ring resonators are an alternative structure to interferometers based on ring resonators. The optical signal is split into two beams. Each beam couples from the respective channel waveguide into the ring resonator, one portion of the split beam traveling in a clockwise direction and the other traveling in the counterclockwise direction.

The waveguide gyroscopes can be formed, for example, in LiNbO$_3$. They may be multifunctional because they provide beam splitting, polarization, and modulation. These devices are adequately robust for the aerospace environment. Figure 14.28 shows a schematic of a waveguide-based gyroscope that is based on a LiNbO$_3$ optical modulator.

Microphotonic gyroscopes are envisioned as the next-generation gyroscopes for aerospace, military, and commercial applications. These microphotonic-chip-like devices provide the required accuracy and are suitable for very high-volume production. They may be interfaced with GPS data to provide a map display in an automobile that indicates its location and progress.

A tunable FL is used for interrogation of microphotonic sensing systems. Measurands such as the axial acceleration and chip temperature are converted to wavelength shifts and are thus independent of the signal intensity. The laser interrogation system offers some potential for microphotonic gyroscopes by facilitating the Sagnac phase-shift measurements vs. optical wavelength.

PBG/MOEMS structures may be fabricated that provide optically interrogated accelerometers. By using several of these PBG/MOEMS devices in series, the effects of acceleration on the resulting optical measurands, such as a change in the relative transmittance intensity or wavelength, can be intensified to increase the device accuracy.

By using N number of the PBG/MOEMS accelerometers in parallel, the measurement accuracy can be improved by a factor of $1/SQRT(N)$. Figure 14.29 shows the transmitted intensity versus acceleration for a set of (1) one, (2) three, and (3) six PBG/MOEMS devices in series.

A further benefit is that this micro-PIC structure could contain accelerometers tuned to different acceleration ranges through the design of the MOEMS springs, such as micro-g, nominal g, and high g. Therefore, the overall accuracy could be

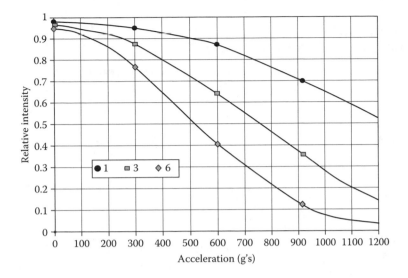

FIGURE 14.29 Transmitted intensity ratio vs. acceleration for PBG/MOEMS accelerometer.

much better than current systems, despite the small size of the optical integrated circuit.

Microphotonic gyroscopes can be as small as the MEMS-based devices, but with the added benefits of higher fabrication yields, leading to lower unit costs, greater long-term reliability (due to the absence of mechanical components), and greater immunity to vibration, EMI, and shock can be achieved. The use of the so-called reentrant resonant rings could facilitate even higher accuracy. In this architecture, the optical pulse is injected into the ring resonator and allowed to travel many times before being extracted from the ring. This provides a high value of reentrants, such as that provided by using many loops of optical fiber.

Table 14.11 compares the PBG/MOEMS accelerometer to electrical MEMS and to strain sensors. The PBG/MOEMS optical accelerometer can be made much smaller and more robust than the traditional MEMS electrical accelerometer. Traditional MEMS accelerometers rely on many interdigitated fingers to convert acceleration to a change in the capacitance of the device. This requires considerable substrate area and complex electronics, and results in a complex electrode structure. In contrast, the optically read PBG/MOEMS accelerometer employs a much simpler and more reliable actuator structure.

14.7 SATELLITE NAVIGATION SYSTEMS

The navigation system is a generic and critical component of a spacecraft. The exact details of the navigation system design may vary based on the pointing requirements of a satellite for its payload and/or guidance requirements. The main

TABLE 14.11
Comparison of Various Gyroscopes

Sensor Type	Size (mm)	Mass (g)	Power (mW)	Range (g)	Resolution (% F.S.)
MEMS (Analog devices)	10×15	1	10	± 40	0.40
Strain	$25 \times 25 \times 25$	35	60	± 40	0.25
Optical PBG/MOEMS	1×2	10^{-4}	0.1	0–1200	0.40
Multiple set of N optical PBG/MOEMS	1×2 (per device)	10^{-4} (per device)	0.1 (per device)	0–1200	$0.40/\mathrm{SQRT}(N)$

task of the navigation system is to determine (1) spacecraft position and (2) spacecraft velocity, but the basic parameters to be determined are as follows:

1. Spacecraft altitude
2. Geographical position
3. Orbit trajectory and current velocity
4. Orientation and pointing towards Sun and Earth
5. Acceleration and rotation status
6. Relative position for formation flying

Spacecraft navigation is accomplished by integrating the output of a set of sensors to compute position, velocity, and attitude. The sensors used are gyroscopes and accelerometers. Gyroscopes measure angular rotational rates with respect to inertial space, whereas accelerometers measure linear acceleration with respect to an inertial frame. Typically, an inertial reference unit will contain three gyroscopes and three accelerometers. For space system applications, some additional redundancy may be required.

The following example illustrates the potential benefits that may be provided by implementing microphotonic technologies in the navigation systems. The satellite navigation micro-PIC is comprised of a number of basic components or functional blocks that are interlinked using the SOI channel waveguides. As shown in Figure 14.30, the spacecraft navigation micro-PIC is a multifunction optical integrated circuit that consists of a number of critical devices on a single SOI chip:

- Input fiber–waveguide coupler
- SOI ridge channel waveguide
- T-bar splitter
- T-bar Mach–Zehnder interferometric radiation sensor
- PBG temperature sensor

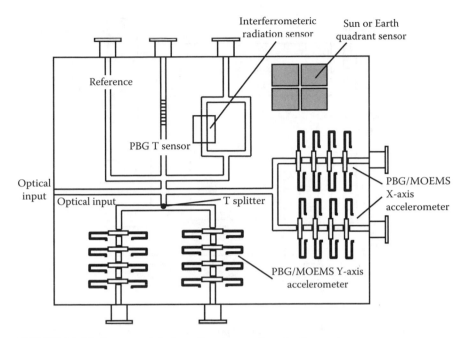

FIGURE 14.30 Conceptual design of a spacecraft navigation system.

- PBG/MEMS optical accelerometers
- Waveguide optical gyroscope
- Si PIN detector array for the Sun sensor
- VO_n bolometer array for the Earth sensor
- Output coupling or hybrid integration with InGaAs or Ge detectors

Because the output coupling is directly to photodetectors, fiber–waveguide output coupling is not an issue, and the net signal throughput and SNR can be very high.

The end result is an order-of-magnitude saving in the system mass and volume. This facilitates greater device redundancy to reduce the possibility of a system failure. Moreover, redundancy can also be used to improve the net performance and accuracy of the navigation measurement system.

A hybrid system is envisioned that combines inertial guidance as provided by the integrated optical accelerometers and gyroscopes with GPS positioning data. A block diagram of a microphotonic three-axis navigation system is shown in Figure 14.31. The design presents a compact navigation system, including a laser source, micro-PIC inertial and pointing sensors, the microprocessor, and signal-interfacing and signal acquisition subsystems. The two micro-PICS provide inertial navigation data in the two orthogonal planes (X-Y and Y-Z). This provides data about the instantaneous state and trajectory of the spacecraft. This can be supplemented using miniature GPS receiver chips to provide

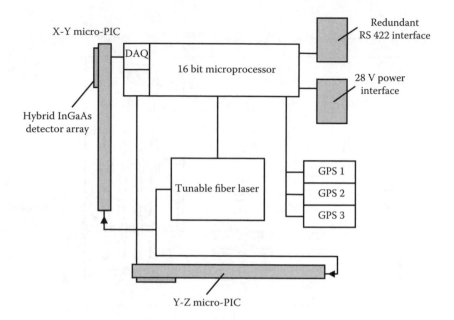

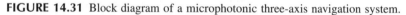

FIGURE 14.31 Block diagram of a microphotonic three-axis navigation system.

geographically linked positional coordinates. Using differential GPS, position accuracies to about ± 2 m are feasible, with typical accuracies of about 10 m.

The main subsystems include:

1. A 16-bit microprocessor for system control with data acquisition and processing electronics
2. X-Y and Y-Z inertial sensing micro-PICS with redundant sensors
3. Tunable fiber laser using 980-nm redundant pumps
4. Triple redundant GPS receivers for spacecraft position determination and validation

The tunable laser provides the signal source and the interrogation system for the micro-PIC sensors. Typically, micro-PIC sensors convert the desired physical parameter, such as an axial acceleration, to a shift in the spectral wavelength. This significantly reduces the effects of variations in signal levels on measurement accuracy.

Additional components that could be considered for integration on the navigation micro-PICS include:

1. InGaAs detector array through hybrid integration
2. Schottky or PIN array for Sun sensing and tracking
3. VO_n/MEMS bolometer Earth sensor array

Micro-PICs increase the integration of optical components on a single Si substrate to provide multifunctional optical processing and switching similar to electronic integrated circuits. This reduces the number of external optical interconnections required and minimizes sensitivity to external vibrations and overall system size and weight while maximizing the system information capacity, optical throughput, and reliability.

System integration entails the selection of an architecture that minimizes size within the constraints of adequate optical transmission. Technological development is currently converging on the SOI platform. This benefits from compatibility with high-speed CMOS integrated electronics as well as a broad foundation of established design and fabrication tools. The recent demonstration of superlattice strained interfaces that enable the deposition of single-crystal LiNbO$_3$ and GaAs on SOI has further enhanced the appeal of SOI.

14.8 THERMAL RADIATOR DEVICES

The efficient thermal control of spacecraft is a critical issue that impacts both the performance and longevity of internal subsystems and payloads. Although a spacecraft can be subjected to external temperature swings from about $-150°C$ to $+150°C$, the corresponding internal temperature must be regulated over a nominal range, typically from $-10°C$ to $+30°C$ (see Figure 14.32). The tighter the temperature control of the spacecraft, the better the performance ratings and greater the lifetime of the spacecraft subsystems.

All spacecrafts rely on radiative surfaces to dissipate excess heat through thermal radiation into dark space. Passive thermal management systems currently

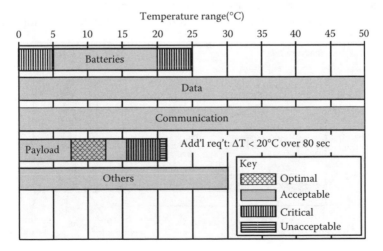

FIGURE 14.32 Typical spacecraft component temperature limits.

employ thermal radiators with a fixed thermal emissivity to maintain the space-craft internal temperature within acceptable boundaries. These are typically sized for the spacecraft's maximum power dissipation requirements. This requires a compromise between efficient heat dissipation at acceptable upper-temperature extremes and maintaining the spacecraft's temperature at the low-temperature extremes.

The spacecraft thermal load consists of the following:

- Heat generated by internal components
- Absorbed solar radiation (1420 W/m² of solar radiation primarily in the UV and visible)
- Absorbed reflected solar radiation due to finite planetary Albedo radiation (i.e., planetary 35% reflection of solar radiation)
- Absorbed IR planetary radiation (340 W/m² of planetary infrared radiation, primarily in the 6- to 18-μm range)

The typical thermal environment of a spacecraft is schematically illustrated in Figure 14.33. Thermal control is especially important for satellites in a low Earth orbit (LEO) in which, with orbital periods ranging from 60 to 100 min, half the time the spacecraft is in direct sunlight and the other half in Earth's shadow. Nominally, in the Earth's shadow, nonessential systems will be turned off to conserve power. The resulting wide fluctuation in the spacecraft heat load over the period of an orbit can result in large temperature fluctuations of the spacecraft components. As the external temperatures fluctuate widely, the satellite must prevent itself from being alternately overheated and frozen.

For a spacecraft with a fixed IR radiator emittance, the spacecraft temperature decreases rapidly as the heat load decreases from the nominal value of 100 W, as shown in Figure 14.34. When the spacecraft is subjected to high temperatures,

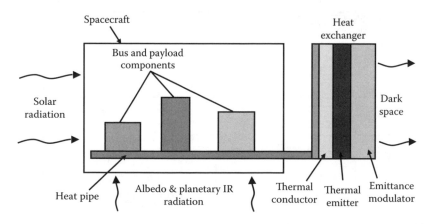

FIGURE 14.33 Schematic of typical thermal environment of a spacecraft.

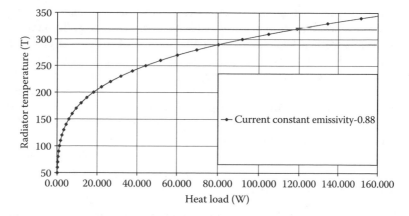

FIGURE 14.34 Variation in spacecraft temperature for a fixed-emittance radiator.

it can stress the internal electronics and reduce performance and lifetimes. Therefore, the design must consider robust and costly spacecraft components.

In the example shown in Figure 14.34, a fixed radiator of $\varepsilon = 0.88$ is sized to provide an operating temperature of 310 K. However, as the spacecraft operating conditions change, e.g., if the spacecraft goes into an eclipse, then the net power can decrease substantially. For a decrease in the net power to 40 W during a cold swing, the spacecraft would cool to about 240 K. To compensate for the heat-load change in this example, additional supplementary heaters of 60 W are required. Relative to the spacecraft power budget of 100 W, the required heater power of about 60 W is a significant fraction of the overall spacecraft power resources. Typically, the heater power is provided using batteries; these batteries use a substantial fraction of the available satellite mass budget. This reduces the mass and power available for the payloads. A summary of typical power requirements for small spacecraft is given in Table 14.12.

TABLE 14.12
Summary of Typical Characteristics for Small Spacecraft

Parameter	Small Satellite	Microsatellite	Nanosatellite
Mass	150–250 kg	50–100 kg	10–25 kg
Payload mass	100 kg	20–30 kg	3–8 kg
Typical power	100 W	50 W	15 W
Corresponding radiator area	0.5 m²	0.2 m²	0.1 m²
Payload power	40–50 W	20–25 W peak	10 W peak
Lifetime	5 yr	3 yr	1–2 yr
Total typical hot/cold cycles in low Earth orbit	29,200	17,520	5,840

Small spacecraft require a cost-effective dynamic thermal management system that maximizes the resources available for the payloads while minimizing the spacecraft thermal fluctuations to enable the use of lower-cost commercial components. The thermal-control system needs to provide reliable dynamic operation for over 30,000 thermal vacuum cycles in a LEO environment.

There are several innovative technologies that have been developed in the recent years with an aim to improve the overall management of the spacecraft power systems. These include the use of (1) mechanical louvers, (2) MEMS-based microlouvers, and (3) smart-coating-based radiators.

A brief summary of the radiator devices is given in the following sections.

14.8.1 MECHANICAL LOUVERS

Traditional active-thermal-control systems have employed various forms of mechanical louvers to regulate heat dissipation from radiators into deep space. The louvers are opened mechanically to increase thermal emission into deep space and closed to reduce heat loss. These louvers allow for net emittance tunability from about 0.10 to about 0.65 (see Figure 14.35). The louvers have a useful lifetime of about 10,000 cycles. Active control requires access to the spacecraft power. Traditional mechanical louver systems are 6.5 cm × area of louver; their weights are in the range of 3.3 kg/m².

14.8.2 MEMS-BASED LOUVERS

Microlouver systems based on MEMS technologies are also being considered to facilitate miniature micromechanical thermal radiators. These minilouvers and their open/close mechanism resemble micromirrors. Typically, a hinge and a stop pillar, above a silicon surface, support them (see Figure 14.36). Their lids are coated with aluminum to reflect heat and light. The silicon is coated with a high-emissivity material [14.3, 14.4]. When heat needs to be dumped, the louvers facing away from the sun are opened by electrostatically twisting their hinges to expose the high-emissivity coating. As silicon is relatively transparent to IR radiation, a heat source below the coating radiates IR through the silicon into space, cooling the spacecraft.

Typical MEMS-based device sizes are 200×200 μm^2. Therefore, a relatively large number of microlouvers would be required to replace a 0.5- to 1-m² conventional louver-based radiator.

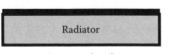

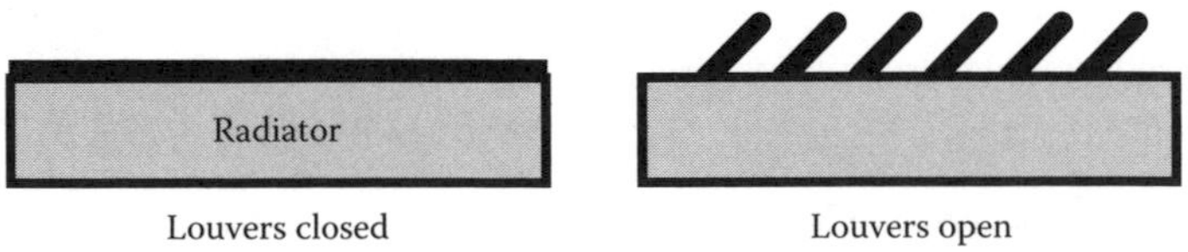

Louvers closed   Louvers open

FIGURE 14.35 Mechanical louvers.

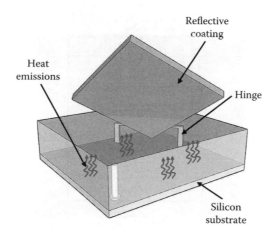

FIGURE 14.36 MEMS-based microlouver.

14.8.3 Smart Thermal Radiators

There are several "smart materials" that have been considered for thermal-control applications:

- Ceramic-tiles-based $La_{1-x}M_xMnO_n$ perovskite materials
- Multilayer electrochromic active devices based on WO_3
- VO_2-based thermal radiators

14.8.3.1 $La_{1-x}M_xMnO_n$

Some experimental work on the use of $LaSrMnO_3$ and $LaCaMnO_3$ for thick-film devices has been performed. A variation in the thermal emissivity from 0.20 (near 150 K) to 0.6 (above 300 K) was obtained. The material was formed as a 0.2-mm tile that was subsequently attached to the space structure [14.5, 14.6]. The weight of the thick-film radiator is in the range of 2 kg/m².

Optical absorption of $LaSrMnO_3$ is high in the visible (VIS) and NIR range. The resulting material has a high solar absorption of 0.84. This requires an additional cold mirror reflector coating that is IR transparent to reduce the solar absorptance. Some attempts have been made to develop the selective VIS reflector. However, this is technically challenging because of the number of layers required.

14.8.3.2 WO_3

Electrochromic structures based on WO_3 exhibit relatively low VIS and NIR optical absorption in the insulating state and a high reflectance in the metallic state [14.7]. The switching mechanism involves the motion of ions, such as H^+ or Li^+, under the influence of an electric field to form HWO_3 (or $LiWO_3$) complexes that are metallic and highly reflective in the VIS and NIR. The transition from an insulating to a metallic state is controlled by the application of a voltage.

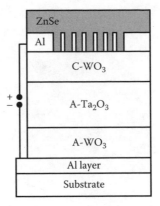

FIGURE 14.37 All-solid-state electrochromic device.

There is a current flow during switching, typically about 4 mA/cm², that must be provided by the satellite battery.

The electrochromic technique employs a multilayer structure involving a Li⁺-containing polymer to convert WO_3, which is IR transparent, to $LiWO_3$, which is IR reflective. The conversion is realized through the voltage-induced transport of Li⁺ between two indium tin oxide (ITO) electrodes separated by an ion conductive layer. Electrochromic devices require the use of an electrolyte to facilitate the migration of the colorant ions (H⁺, Li⁺, and Na⁺) from the reservoir to the WO_3. Figure 14.37 shows the layer structure of the all-solid-state electrochromic device for thermal emissivity modulation with a protective surface layer (shaded layer within the Al top electrode grid) [14.8].

14.8.3.3 VO₂-Based STR

An innovative approach to the thermal radiators has been developed through the development of a nanoengineered smart thin-film coating based on crystalline VO_2 [14.9]. The VO_2-based STR (V-STR) passively regulates heat transfer between the spacecraft and dark space in response to the spacecraft temperature. A high net dynamic variation in heat transfer between the spacecraft and dark space is achieved through control of the metal–insulator transition in the VO_2 structure by using innovative nanoengineering techniques to exploit the high concentration of electrons in the metallic state. The V-STR significantly reduces the heat loss at lower temperatures to enable spacecraft designers to eliminate or reduce internal heating requirements.

This nanoengineered smart thin-film coating can be applied to existing spacecraft thermal radiators to facilitate dynamic thermal control. This approach has significant advantages in terms of (1) reliability, (2) thermal-control performance, (3) minimal added mass (<100 gm/m²), (4) passive operation requiring no additional power, (5) direct integration with the space structure, and (6) structural simplicity.

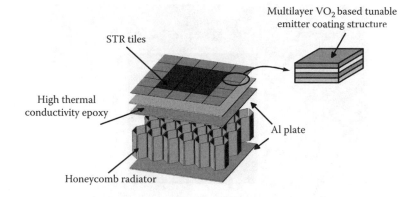

FIGURE 14.38 Passive V-STR structure.

The V-STR can be deposited directly on radiator tiles (30 × 30 mm); these tiles can be fixed to the honeycomb panel or other radiator panels to provide smart radiators of various sizes. The assembled V-STR structure, as shown in Figure 14.38, controls the heat exchange to dark space.

Figure 14.39 shows a schematic of the operation of the V-STR for the low-temperature case and high-temperature case.

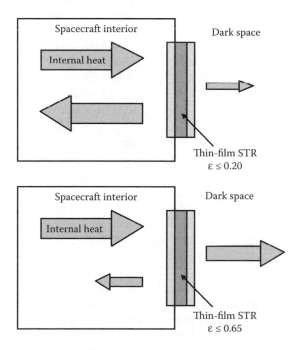

FIGURE 14.39 Schematic of operation of V-STR.

FIGURE 14.40 Scanning electron microscope picture of VO_2 crystallites.

Structural nanoengineering is accomplished through the selection of the substrate or alloy, the preparation and texturing of the substrate, and the subsequent selection of the VO_2 deposition conditions to provide a two-dimensional or three-dimensional nanocrystalline structure, as shown in Figure 14.40. Through nanoengineering, the VO_2 film is prevented from undergoing a full transition into the metallic state. Instead, an intermediate semimetallic state is induced at the critical electron density. The high density of free electrons that form at the onset of the transition to the metallic state, within the semimetallic regime, provides very high infrared optical absorption and corresponding IR emittance.

Figure 14.41 shows the emittance tunability in a single layer of a nanoengineered undoped VO_2 coating on Al. As shown in Figure 14.41, the emittance may be tuned from below 0.20 in the insulating state at lower temperatures to above 0.65 above the transition temperature in the semimetallic state.

The effect of this innovative radiator on the overall spacecraft temperature variation is illustrated in Figure 14.42. ANSYS simulations of the variation in spacecraft temperature with changing net heat load have been performed for (1) a traditional fixed radiator ($\varepsilon = 0.88$) sized for 120 W (about 0.5-m^2 radiator area) and (2) a simple single-layer V-STR, as shown in Figure 14.42.

The experimentally measured emittance tunability of 0.45 has been used in the simulations. The results shown in Figure 14.42 indicate that with the V-STR approach it is possible to compensate for a decrease in heat load from 80 to 40 W. This corresponds to a significant reduction of the resulting temperature fluctuations of the spacecraft subsystems and payloads from ±50°C to the required range of ±20°C. With a fixed radiator, as has been shown previously in Figure 14.34, an additional heater power of 40 W would be required to maintain the same thermal stability. This is a significant satellite resource cost in terms of

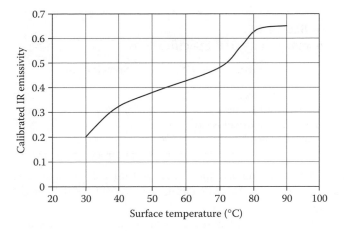

FIGURE 14.41 Emittance tunability in a single layer of undoped V-STR.

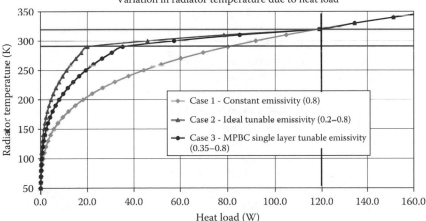

FIGURE 14.42 Variation in spacecraft temperature for (1) fixed radiator ($\varepsilon = 0.88$) and (2) single-layer V-STR (with $\Delta\varepsilon = 0.45$).

added batteries and decrease of power availability for the satellite payloads. With the V-STR solution, additional payloads or subsystems can be accommodated on the spacecraft to increase the functional density of the spacecraft and provide higher value to the users.

An additional advantage of the V-STR technology is the optoelectronic switching characteristics of the VO_2 that can be tailored by doping with acceptors and donors. This is a feature similar to that observed in the case of crystalline Si in microelectronics. The VO_2 metal–insulator transition temperature can be tailored from below –20°C to above 70°C through doping. This provides

TABLE 14.13
Summary of V-STR Specifications

Parameter	Specifications
Solar absorbance	< 0.25
Transition temperature	0°C to 65°C (for VO_2 doped with W)
High temperature emissivity	> 0.75
Emittance tunability	0.45

considerable leeway to optimize the V-STR performance. Structured single layers on VO_2 have yielded broadband IR emittance tunability of 0.45 (see Table 14.13 for specifications).

In summary, the V-STR has several significant benefits for spacecraft:

- Low intrinsic solar absorption, with $\alpha < 0.3$ in thinner VO_2 layers on Al
- Flexible film structure due to metallic bonding to minimize cracking under device thermal shock or vibration
- High-reliability emissivity cycling (> 10^8 theoretically estimated, > 10,500 vacuum thermal cycles experimentally tested) that can withstand severe thermal shock
- Robust mechanical integration with the spacecraft radiator to minimize effects of shock and vibration
- High dynamic variation in heat transfer, with $\Delta \varepsilon > 0.45$ observed for single-layer crystalline nanostructures on Al
- VO_2-based STR tiles on Al as a direct substitute for current OSR fixed-emittance thermal radiator tiles

The V-STR approach is a generic one that can be applied to improve the performance of various future spacecraft as well as individual payloads that require improved thermal stability while minimizing resources such as heater power.

Table 14.14 summarizes the performance of the mechanical louvers and V-STR in terms of reliability, added mass, volume, and structural simplicity.

With the trend toward greater functional densities for space systems, as well as the use of smaller, high-performance satellites, there is a need for lightweight thermal-control systems that can dynamically tune their characteristics in response to changes in their operating environment while providing reliable operation. The aforementioned microphotonic radiator devices provide low-cost solutions to the heat management of various types of spacecraft, focusing on small and microsatellites. The efficient thermal control of a spacecraft is a critical issue that impacts both the performance and longevity of internal subsystems and payloads.

TABLE 14.14
Comparison of Mechanical Louvers and V-STRs

Parameter	Mechanical Louver	V-STR
Added mass	4.500 kg/m^2	< 0.005 kg/m^2
Reliability	10^4 cycles	> 10^8 cycles
Transition region width	Emittance change over 10 to 15°C in pulsed-like spacecraft temperature control	30°C to 50°C, gradual emittance variation and proportional-like spacecraft temperature control
Mechanical integration with spacecraft	Mechanically attached to radiator using screws.	Direct replacement for OSR radiator tiles
Added volume	7 cm × radiator area	No added volume required

14.9 SUN SHIELDS

The aim of the sun shield is to protect satellite antennas against the absorption of solar radiation while maintaining RF performance and electrostatic discharge (ESD) protection. When the spacecraft is subjected to cycling between large temperature extremes, thermal stresses develop in the antenna, reducing its performance and lifetime.

The antennas are exposed to external temperature swings from about −150°C to +150°C. On the high-temperature side, the solar radiation increases the antenna temperature and reduces the RF performance while accelerating antenna surface aging. On the low-temperature extremes, the increasing sheet resistivity causes a charge accumulation and increases the risk of an ESD that causes cracking and fissuring of the antenna sheet. A breakdown of antenna functionality can result in a catastrophic failure of the satellite mission. By improving the reliability and increasing the antenna's lifetime, the sun shield reduces the mission costs.

An example of the sun-shield application on an antenna feed horn, is illustrated in Figure 14.43 and Figure 14.44. Figure 14.43 presents a photograph of a sun-shield covering an antenna feed horn [14.10]. Figure 14.44 illustrates a typical quasi-optical telescope, including the telescopic optics, the feed horn, and the cavity, with a sun shield protecting the entrance of the horn.

Typically a small spacecraft will undergo over 30,000 thermal vacuum cycles between temperature extremes. A reliable sun shield must withstand these variations without degradation of its performance. In addition to the temperature cycles, vacuum (outgassing risk) and space plasma (gamma radiation, protons, electrons, and atomic oxygen) cause performance degradation in time, with the additional risk of film erosion. LEO and GEO environments are summarized in Table 14.15.

The sun shield can be applied as a small sheet (4- to 20-cm diameter) covering the entrance circle of an antenna feed horn. The sun shield covers the circular input area to protect the antennas from ESD while rejecting heat from solar

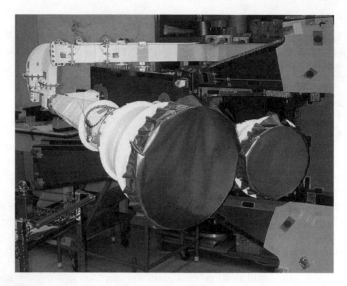

FIGURE 14.43 Photograph of a sun shield applied to antenna feed horns. (From Samson, S., Buissières, F., Markland, P., and Emond, A., EMS Technologies Canada Ltd., unpublished document 2005.)

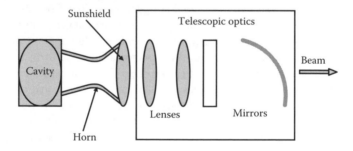

FIGURE 14.44 Schematic of a quasi-optical telescope.

radiation and minimizing RF losses. The values listed in Table 14.15 may change with variation of solar activity.

Feed horns are used to provide low side-lobe levels as well as a highly symmetric beam pattern. Often front-end optics placed between the horns and mirrors are used to precisely shape and control the beam. In some cases, there is no room for such optics, and so the horns have a particularly important role to play in determining the telescope beam pattern. However, solar radiation causes the horn to heat up, inducing its degradation.

The requirements for a sun shield can be summarized as follows:

- Low RF loss (good RF transmittance sheet resistivity > 10^5 ohm/sq, at room temperature)

TABLE 14.15
Summary of LEO and GEO Space Environments

Parameter	LEO	GEO
Vacuum	10^{-5} Torr	10^{-6} Torr
Temperature range	$-150°C$ to $+150°C$	$-175°C$ to $+175°C$
Atomic oxygen	Fluence: 9×10^{17} particles/cm^2	Negligible
	Flux: 10^{12} to 10^{14} atom/cm^2/min	
X-rays	10^{-3} W/cm^2	
VUV radiation	0.75 μW/cm^2 (for λ between 100 and 150 nm)	
	11.00 μW/cm^2 (for λ between 200 and 300 nm)	
Protons and ions	1.1×10^{10} proton/cm^2/yr (15 kilo rads/yr)	
Electrons	1.5×10^{13} electron/cm^2/yr	

- Good heat rejection (high emittance > 0.8)
- Low solar radiation absorptance (< 0.55)
- Protection against ESD (Low sheet resistivity < 10^9 ohm/sq at $-150°C$)

The specifications of the performance characteristics for the sun shield are summarized in Table 14.16.

14.9.1 CURRENTLY USED SUN SHIELDS

To address these problems, the industry uses such traditional approaches as (1) white antistatic paints, (2) using no sun shield at all and exposing the antenna feed and/or reflector to the harsh space environment, and (3) Ge-coated Kapton.

White paint sunshields are relatively heavy, increase the antenna distortion, and are costly. When no sun shield is used at all, then the temperature excursions impair the antenna's RF performance and cause problems such as the entrapment of solar radiation in the feed horns. ESD risk is a major concern for large area antennas because the crack created by the ESD can increase, causing total antenna

TABLE 14.16
Summary of Performance Characteristics for Sun Shields

Parameter	Specifications
Solar absorbance	< 0.5
Sheet resistivity for ESD	< 10^9 (ohms/sq at $-150°C$)
Sheet resistivity for RF transmittance	<10^5 (ohm/sq at 20°C)
IR emissivity	≥ 0.79
RF insertion loss	<0.05 dB per pass at 30 GHz
Temperature range performance	$-175°C$ to $+175°C$

failure. A new solution developed recently uses Ge-coated Kapton. Although providing a solution to solar radiation control, the Ge-film has a very high sheet resistivity at very low temperature ($\leq 100°C$), increasing the risk of ESD. Also, Kapton and Ge films introduce some RF losses.

These approaches, although they avoid the basic problems, have their own drawbacks:

1. White paints
   a. Costly to implement
   b. Relatively heavy
   c. Increases thermal distortion of the reflectors because of their large coefficient of thermal expansion, thus affecting the RF performance
   d. Difficult to implement in a feed horn
2. No sun shield
   a. Larger temperature excursions, thus affecting the RF performance
   b. Risk of solar entrapment in the feed horn
   c. Risk of solar collimation on the feed, caused by solar specular reflection of the reflector
   d. Requires extensive thermal modeling to properly simulate solar reflections and gradients on the exposed hardware
3. Ge-coated Kapton
   a. Significant losses at higher frequencies (i.e., > 30 GHz)
   b. Increased surface resistivity at low temperatures
   c. Surface charging when exposed to space radiation

A new passive approach has been developed that can be applied directly to existing antennas, resulting in negligible added weight to the space structure (< 20 g/m^2). This approach uses VO_2-based thin films to control the rejection of heat from solar radiation while maintaining RF performance and preventing ESD risk (by a shallow doping of the thin film). This innovative approach has significant advantages over competitive technologies in terms of performance in the space environment, mass, cost, structural simplicity, reliability, and compatibility with current designs.

14.9.2 MICROPHOTONIC SUN SHIELDS

This technology is based on a very successful innovative approach to sun shields through the development of nanoengineered thin-film coatings based on crystalline VO_2. It has been demonstrated that this nanoengineered thin-film coating can be applied as a sun shield to existing antenna membranes. The VO_2 coating is used to reject — with a minimal mechanical intrusion and a very low RF loss — the heat produced by the solar radiation and at the same time protect against ESD. This approach has significant advantages in terms of (1) ESD protection reliability, (2) high solar heat rejection, (3) minimal additional weight (<20

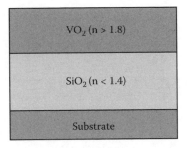

FIGURE 14.45 Microphotonic sun-shield structure.

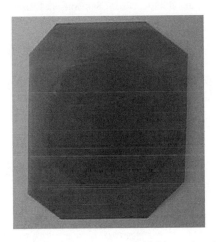

FIGURE 14.46 Photograph of a VO_2-based sun-shield coating deposited on a Kapton membrane.

gm/m²), (4) very low RF loss, (5) enabling of direct integration with the antenna feed horns, and (6) structural simplicity.

A schematic of the multilayer nanoengineered sun-shield coating is illustrated in Figure 14.45. The VO_2 layer is used for high emittance and ESD protection, whereas the SiO_2 layer increases the solar reflectance.

Nanoengineered VO_2 thin films were deposited on 8-cm-diameter Kapton membranes (see Figure 14.46).

Figure 14.47 illustrates the resistivity performance data of the nanoengineered sun shield together with that for the Ge-coated sun shield. More than a three orders of magnitude in sheet resistivity improvement at low temperatures suggests that this microphotonic sun shield (MSS) is an attractive alternative. At the same time, MSS complies with the other performance requirements.

MSS has successfully passed space lifetime and space environment tests. It is generic and can be applied to all kinds of missions and satellites to improve the RF performance of the antennas.

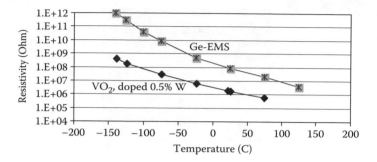

FIGURE 14.47 Sheet resistivity variation with temperature for (1) microphotonic sun shields and (2) Ge-coated sun shields. (From Samson, S., Buissières, F., Markland, P., and Emond, A., EMS Technologies Canada Ltd., unpublished document, 2005.)

Specifically, the MSS has significant advantages in terms of the following:

1. ESD protection (sheet resistivity $< 1 \times 10^9$ ohm/sq)
2. Minimal added mass (less than 50 gm/m^2)
3. Good RF performance
4. High IR emittance
5. Low intrinsic solar absorptance
6. Good resistance to thermal shock and mechanical bending
7. Atomic oxygen resistance

A major advantage of the MSS is that it can be relatively easily interfaced with small feed horns that are used in the K or L frequency bands.

REFERENCES

1. Jilla, C.D. and Miller, D.W., Satellite design: past, present and future, *Int. J. Small Satellite Eng.*, Vol. 1, 1997.
2. Aspelmeyer, M. et al., Quantum Communications in Space, Executive summary report prepared for the European Space Agency under ESTEC/Contract No. 6358/02/NL/SFe, 2003.
3. Osiander, R. et al., Variable emissivity through microelectromechanical system (MEMS) technology, *Research and Development Symposium*, John Hopkins University, Applied Physics Laboratory, November 1999.
4. Birur G.C. et al., Micro/nano spacecraft thermal control using a MEMS-based pumped liquid cooling system, *SPIE Conference on Micromachining and Microfabrication*, San Francisco, CA, October 21–24, 2001.
5. Shimazaki, S. et al., Design and preliminary test results of variable emittance device, *Proceedings of the 8th International Symposium on Materials in a Space Environment*, Arachon, France, 2000.
6. Shimakawa, Y. et al., A variable-emittance radiator based on a metal-insulator transition of (La, Sr)MnO$_3$ thin films, *Appl. Phys. Lett.*, 80, 4864–4866, 2002.

7. Franke, E. et al., Low orbit-protective coating for all-solid-state electrochromic surface heat radiation control device, *Surf. Coat. Technol.*, 285–288, 2002.

8. Granquist, C.G. et. al., Recent advances in electrochromics for smart windows applications, *Solar Energy*, 63, 199–216, 1998.

9. Kruzelecky, R.V., Haddad, E., Jamroz, W., Soltani, M., Chaker, M., and Colengelo, G., Thin-film smart radiator tiles with dynamically tunable thermal emittance, *Proceedings of the 35th International Conference on Environmental System 33-ICES*, Paper 03ICES-242, Rome, Italy, July 2005.

10. Samson, S., Buissières, F., Markland, P., and Emond, A., *Sunshield RF Performance*, EMS Technologies Canada Ltd., unpublished document, 2005.

15 Conclusion

The future for microphotonic devices and systems looks certain. It is expected that within a few years, a number of basic applications will start making an appearance in the market. Among these will be the devices and systems based on technologies that have been briefly summarized in this book, i.e., PBG structures, Si microphotonics, optical computing, and quantum photonics.

PBG-based lasers will be among the first devices that will find their way to the market. They will be accompanied by PBG-based waveguides and PBG-based passive components, such as filters, benders, and splitters. Industry observers expect that the first PBG-based commercial system will be an optical spectrometer and a set of chemical and biological microsensors. At the same time, the telecom industry will start to use PBG-based add-drop filters and Mach–Zehnder interferometers.

In parallel, Si microphotonics will emerge as the prime technology of the future and perhaps the platform upon which the next technological revolution will be built. One may expect that Si will be essential in a number of significant developments. The concept of "all-Si optical circuits" will soon become a reality.

In the near future, optical computers will most likely be hybrid optical/electronic systems that use electronic circuits to preprocess input data for computation and to postprocess output data for error correction before outputting the results. The promise of all-optical computing is highly attractive, and the goal of developing optical computers continues to attract a lot of attention.

It is expected that within the next 20 years there will be a demonstration of the first quantum photonic logic circuit and a prototype of a quantum computer. Further development will continue to make optical qubits an attractive system and to bring these quantum circuits to the level of operation needed for linear optics quantum computing (LOQC). The realization of simple quantum gates and adequate quantum key distribution protocols has already been demonstrated.

Many problems in developing appropriate materials and devices must be overcome before the aforementioned systems and devices will be in widespread commercial use. Nevertheless, the future for microphotonics looks bright.

Index

Related Titles

Other related titles of interest include:

Biomedical Photonics Handbook
Tuan Vo-Dinh
ISBN: 0-8493-1116-0

Microlithography, Second Edition
Kazuaki Suzuki and Bruce W. Smith
ISBN: 0-8247-9024-3

Micro-Optomechatronics
Hiroshi Hosaka, Yoshitada Katagiri, Terunao Hirota, and Kiyoshi Itao
ISBN: 0-8247-5983-4